THE COLLECTED PAPERS OF

Albert Einstein

VOLUME 17

THE BERLIN YEARS:

WRITINGS & CORRESPONDENCE,

JUNE 1929–NOVEMBER 1930

DIANA KORMOS BUCHWALD

GENERAL EDITOR

THE COLLECTED PAPERS OF ALBERT EINSTEIN

English Translations Published to Date

VOLUME 1
The Early Years, 1879–1902
Anna Beck, translator; Peter Havas, consultant (1987)

VOLUME 2
The Swiss Years: Writings, 1900–1909
Anna Beck, translator; Peter Havas, consultant (1989)

VOLUME 3
The Swiss Years: Writings, 1909–1911
Anna Beck, translator; Don Howard, consultant (1993)

VOLUME 4
The Swiss Years: Writings, 1912–1914
Anna Beck, translator; Don Howard, consultant (1996)

VOLUME 5
The Swiss Years: Correspondence, 1902–1914
Anna Beck, translator; Don Howard, consultant (1995)

VOLUME 6
The Berlin Years: Writings, 1914–1917
Alfred Engel, translator; Engelbert Schucking, consultant (1997)

VOLUME 7
The Berlin Years: Writings, 1918–1921
Alfred Engel, translator; Engelbert Schucking, consultant (2002)

VOLUME 8
The Berlin Years: Correspondence, 1914–1918
Ann M. Hentschel, translator; Klaus Hentschel, consultant (1998)

VOLUME 9

The Berlin Years: Correspondence, January 1919–April 1920

Ann M. Hentschel, translator; Klaus Hentschel, consultant (2004)

VOLUME 10

The Berlin Years: Correspondence, May–December 1920

and

Supplementary Correspondence, 1909–1920

Ann M. Hentschel, translator; Klaus Hentschel, consultant (2006)

VOLUME 12

The Berlin Years: Correspondence, January–December 1921

Ann M. Hentschel, translator; Klaus Hentschel, consultant (2009)

VOLUME 13

The Berlin Years: Writings & Correspondence, January 1922–March 1923

Ann M. Hentschel and Osik Moses, translators; Klaus Hentschel, consultant (2012)

VOLUME 14

The Berlin Years: Writings & Correspondence, April 1923–May 1925

Ann M. Hentschel and Jennifer Nollar James, translators; Klaus Hentschel, consultant (2015)

VOLUME 15

The Berlin Years: Writings & Correspondence, June 1925–May 1927

Jennifer Nollar James, Ann M. Hentschel, and Mary Jane Teague, translators; Andreas Aebi and Klaus Hentschel, consultants (2018)

VOLUME 16

The Berlin Years: Writings & Correspondence, June 1927–May 1929

Jennifer Nollar James, William D. Brewer, and Steven Rendall, translators (2021)

VOLUME 17

The Berlin Years: Writings & Correspondence, June 1929–November 1930

Jacquelyn Bussone, William D. Brewer, Jennifer Nollar James, Steven Rendall, and Jennifer Wunn, translators (2024)

THE COLLECTED PAPERS OF

Albert Einstein

VOLUME 17

THE BERLIN YEARS:
WRITINGS & CORRESPONDENCE,
JUNE 1929–NOVEMBER 1930

ENGLISH TRANSLATION
OF SELECTED TEXTS

Diana Kormos Buchwald, Ze'ev Rosenkranz,
Joshua Eisenthal, József Illy, Daniel J. Kennefick, A. J. Kox, Jennifer L. Rodgers, Tilman Sauer, and Barbara Wolff
EDITORS

Jacquelyn Bussone, William D. Brewer, Jennifer Nollar James, Steven Rendall,
and Jennifer Wunn
TRANSLATORS

Princeton University Press
Princeton and Oxford

Published by Princeton University Press, 41 William Street, Princeton, New Jersey 08540
In the United Kingdom: Princeton University Press, 99 Banbury Road, Oxford OX2 6JX
press.princeton.edu

ISBN-13: 978 0 691 24616 1

Library of Congress catalog card number: 87160800

This book has been composed in Times.

The publisher would like to acknowledge the editors of this volume for providing the camera-ready copy from which this book was printed.

Princeton University Press books are printed on acid-free paper and meet the guidelines for permanence and durability of the Committee on Production Guidelines for Book Longevity of the Council on Library Resources.

Printed in the United States of America

1 3 5 7 9 10 8 6 4 2

CONTENTS

PUBLISHER'S FOREWORD

We are pleased to be publishing this translation of selected documents of Volume 17 of *The Collected Papers of Albert Einstein*, the companion volume to the annotated, original-language documentary edition. As we have stated in all earlier volumes, these translations are not intended for use without the documentary edition, which provides the extensive editorial commentary necessary for a full historical and scientific understanding of the source documents. The translations strive first for accuracy, then literariness, though we hope that both have been achieved. The documents were selected for translation by the editors of *The Collected Papers of Albert Einstein.*

We thank Jacquelyn Bussone, William D. Brewer, Jennifer Nollar James, Steven Rendall, and Jennifer Wunn for their dedication in making Einstein's writings and correspondence available to the English-speaking world; Diana Kormos Buchwald for her revisions and the production of camera-ready copy; A. J. Kox, Ze'ev Rosenkranz, Sini Elvington, and Jacquelyn Bussone for conforming the editions and creating the index. Many thanks to Beth Gianfagna and Trevor Perri for their careful copyediting of the manuscript, and to Terri O'Prey for seeing it all to completion.

Finally, we are most grateful to the National Historical Publications and Records Commission, Washington, D.C., Award No. PE-103459-22, for their support of our project.

Princeton University Press
November 2023

LIST OF TEXTS

In this list, writings are indicated by a **bold** running number.

Vol. 5, 256a. To Heike Kamerlingh Onnes, 26 February 1911

Vol. 5, 272a. To Heike Kamerlingh Onnes, 16 July 1911

Vol. 5, 285a. To Heike Kamerlingh Onnes, 20 September 1911

Vol. 9, 63a. To Georg Lockemann, 19 June 1919

Vol. 12, 113a. To Wander J. de Haas, before 1 April 1921

Not selected for translation

Vol. 16, **90a.** "General Theory of Relativity and Equations of Motion." Fragment of early version of *Einstein 1928b* (Vol. 16, Doc. 91), before 24 November 1927

Not selected for translation

Vol. 16, **234a.** Poem on Scharbeutz, between July and late September 1928

Not selected for translation

Vol. 16, **234b.** Calculations, between July and late September 1928

Not selected for translation

Vol. 16, **314a**. Notebook, after ca. 15 November 1928

Not selected for translation

Vol. 16, **354a.** Aphorism on the Allure of Theoretical Research, 1929

Vol. 16, **367a.** On *Einstein 1929n* (Vol. 16, Doc. 365), after 10 January 1929

Vol. 16, **473a.** To the Children at MOPR, 27 March 1929

Vol. 16, 500a. To Elsa Einstein, 17 April 1929

Not selected for translation

Vol. 16, **521a.** "Unified Field Theory" [Fragment I], ca. 10 May 1929

Not selected for translation

Vol. 16, **521b.** "Unified Field Theory" [Fragment II], 10 May 1929

SELECTED TEXTS

Vol. 5, 256a. To Heike Kamerlingh Onnes[1]

Zurich, 26 February 1911

Dear Colleague,

The days I was allowed to spend with you in Leiden will always be counted among the most beautiful in my life.[2] I thank you and your wife belatedly for the beautiful hours you have offered me in your home, and you in particular for having given me in your laboratory an insight in your wonderful sphere of activity.

Mr. Keesom[3] was so kind to share with me at length his thoughts on opalescence. I am very much enjoying his derivation, whose rigorousness I believe to be sufficient. It would be very desirable if he were to present his derivation in the Annalen der Physik, while referring to his priority. That would certainly be more useful than a remark by me in this regard without any factual content.

As soon as I find the time, I will try to calculate, from the Boltzmann principle, the deviation from the equation of state in the proximity of the critical point.[4] With what you say with regard to Eötvös's law,[5] I am completely in agreement. You conclude that the diameter of the molecular sphere of action is proportional to the molecule's diameter. More precisely, one could say *that the number of molecules on which a molecule acts through molecular attraction is independent of the nature of the substance*.[6] This is what the law of corresponding states requires. This requirement can, however, be met in two different ways, namely:

(1) if the (sphere of force) sphere of action is truly proportional to the molecular diameter;

(2) if the molecule's sphere of force is so small that it is only able anyhow to act on one immediately adjacent to it.

Assumption (1) is a priori improbable. After all, whatever applies to the entire molecule's sphere of action also applies to those of the individual atoms comprising it. So a single atom's sphere of action should increase with the size of the molecule to which the atom belongs, which one should consider as a priori truly improbable. Therefore, if the second assumption, which is actually equally compatible with the law of corresponding states, is already a priori more probable, then its probability increases greatly because it provides a good order of magnitude for the value of the Eötvös constant, actually without any further assumption about the nature of the attractive forces.[7]

Once again, with my heartfelt thanks, sincerely yours,

A. Einstein

Vol. 5, 272a. To Heike Kamerlingh Onnes

Prague, 16 July 1911

Dear Prof. Kamerlingh-Onnes,

I would like to allow myself today to share with you a few reflections that tie directly into your beautiful investigations on the electrical conductivity of metals which, through you, could possibly lead to results.

Your most amazing result is, in my opinion, that the specific resistance of impure metals assumes a fixed threshold value at low temperatures.[1] This result is surely related to the weak temperature dependence of the specific resistance of alloys at normal temperatures. How can your result be explained by means of the electron theory?

In Drude's formula,[2] $\sigma = \frac{1}{2}\frac{\varepsilon^2 n\lambda}{m\ u}$, u is prop. $\sqrt{T}$. Should n (number of free electrons), λ (mean free path), or both change in such a way that the factor $\frac{1}{\sqrt{T}}$ will be exactly compensated for alloys and at low temperatures? For n, it is improbable that it will change as $\sqrt{T}$ does; with λ, it is improbable, in my opinion, that it will change at all. It seems much more likely that each added molecule means an inhomogeneity of specific extension; these inhomogeneities *alone* determine the mean free path at low temperatures, specifically, in the simplest view, independently of the temperature. The behavior of weakly alloyed substances at low temperatures is, then, the simplest thinkable, but in contradiction with the electron theory.

Now, it should be remembered, though, that Sommerfeld correctly derived the energy of g-rays and X-rays under the assumption that any elementary collision needs a time t, with $E\tau = h$[3]

E = kinetic energy of the projectile

h = Constant of Planck's radiation formula = $6.55 \cdot 10^{-27}$.

According to this hypothesis, every collision of a metal electron needs the time $\frac{2h}{mu^2}$ so that each electron only flies freely during the time

$$\frac{\frac{\lambda}{u}}{\frac{\lambda}{u} + \frac{2h}{mu^2}} = f$$

per unit of time. According to Sommerfeld's hypothesis, Drude's expression for the conductivity must be multiplied by f. If one does this, and if one remembers that, at low enough temperatures, the first part of the denominator f vanishes against the second one in alloys (const. λ), then one obtains for sufficiently low temperatures

$$\frac{1}{\sigma_{(\mathrm{limit})}} = \frac{4h}{\varepsilon^2 n} \cdot \frac{1}{\lambda^2}$$

so, indeed independent of T, if λ and n are.

Furthermore, if the free mean path in alloys at low temperatures is solely determined by the addition, then $\frac{1}{\lambda}$ must be proportional to the amount of the addition, given a specific choice of the addition; $\frac{1}{\sigma}$ *is thus proportional to the square of the amount of the addition*. But this is such a strict criterion that testing it would be of great interest.

The same additional factor f naturally occurs with a certain choice of the addition, such that the Wiedemann-Franz law[4] for pure metals and alloys must retain its validity down to absolute zero—at least, as long as the metals' insulating thermal conductivity is not taken into account, which actually can be calculated with some probability.

It would also be of great interest, moreover, to study the thermo-electrical properties of weak alloys in the region of temperature-independence of their electrical conductivity. Thermo-elements made of silver alloys, for example, could be investigated with equal types of addition, but different quantities added, so that, for example,

silver with 1% lead,

compared with " " 1% " .

When only one type of electron is considered, the electromotive force E of a closed chain with solder joints of temperature T' and T between the alloys a and b is, according to Drude,[5] (Ann. d. Phys. 4. 1. 1900, p. 595),[6]

$$E = 2\underbrace{\frac{R}{\varepsilon N}}_{\text{el. equivalent}}\left\{T'\lg\frac{N_a'}{N_b'} - T\lg\frac{N_a}{N_b} - \int_T^{T'} T\frac{\partial\log\frac{N_a}{N_b}}{\partial T}dT\right\},$$

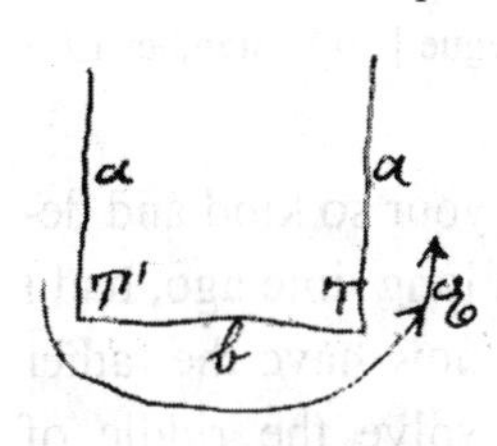

where N denotes the electron densities in question. We have to understand as free electron density the electron density multiplied by Sommerfeld's "freedom factor," so, in our region of temperature-independence of the alloys

$$N_a = N_a^0 \frac{mu\lambda_a}{2h},$$

$$N_b = N_b^0 \frac{mu\lambda_b}{2h}.$$

Because, in the case that interests us, the electron densities apart from Sommerfeld's factors would be approximately the same in both alloys ($N_a^0 = N_b^0$),

$$\frac{N_a}{N_b} = \frac{\lambda_a}{\lambda_b} = \frac{c_b}{c_a},$$

when one denotes, with c_a resp. c_b, the concentrations of the additions in both alloys. So, the formula above simply yields

$$\boxed{E = \frac{2R}{\varepsilon N} \lg\left(\frac{c_b}{c_a}\right)[T' - T]},$$

in which the numerical factor is not exact, however, because of the nonexact nature of the theory. This law is of such a simplicity as to be ideal for a beautiful proof of the theory.

I will not burden you with further conclusions, but I would really like to ask you to accept, in a fatherly way, what was presented here so far, though it must be admitted that the theory stands on somewhat weak empirical legs. If the matter is confirmed, we will achieve a very valuable enrichment of our knowledge about elementary motions. If, however, it is not confirmed, it seems that it will make the inadmissibility of Sommerfeld's hypothesis likely.

With heartfelt greetings, yours,

A. Einstein

My respectful greetings to Prof. H. A. Lorentz[7] and kind regards to Messrs. Perier and Keesom.[8] I have read your and Mr. Perier's latest paper, as well as Mr. Keesom's published work in the Annalen,[9] with enjoyment. I am working with Mr. Hopf[10] on calculating a second approximation for the opalescence formula.

Vol. 5, 285a. To Heike Kamerlingh Onnes

[Prague,] 20 September 1911

Dear Prof. Kamerlingh-Onnes,

Please accept my apologies for waiting so long to reply to your so kind and detailed note. I did have much to do, but I would have written a long time ago, had I something good to write. But that's what is missing. Also, I now have the rather firm impression that Sommerfeld's hypothesis[1] will not solve the riddle of conductivity.

Your formula, in any case, is the simplest and most natural one that can be drawn up without fundamentally calling Drude's theory into question. Acording to Drude,[2] when

$$\text{conductivity} = \frac{n\lambda}{u} \cdot \text{const.}$$

at high temperatures $n\lambda$ must then be proportional to $\frac{1}{u}$ or $\frac{1}{\sqrt{T}}$. Taking into account the behavior at low temperatures, you naturally replace this by $\frac{1}{\sqrt{\frac{h\nu}{e^{\frac{h\nu}{kT}}-1}}}$. What seems questionable to me about this is the difficulty to explain the behavior you found for weak alloys at low temperatures.[3] There, one would have to put $n \cdot \lambda$ proportional to u. But I cannot wrap my head around this. One is very tempted there to take λ constant; should one possibly set n proportional to u, but, for pure metals set $n\lambda$ proportional to $\frac{1}{u}$ (at high temperatures)? If one puts λ = const. and n prop. to u for weak alloys and low temperatures, then for pure metals one must also posit n prop. to u; to get from there to your formula, one must posit for pure metals

$$\lambda \text{ prop. } \frac{1}{\sqrt{T}} \cdot \frac{1}{\sqrt{\frac{h\nu}{e^{\frac{h\nu}{kT}}-1}}},$$

which is very unnatural. The end is, therefore, that we know nothing.

There are actually several quantities available that, unfortunately, we cannot individually follow up with complete certainty. Recently, it occurred to me that perhaps one could find out something about the mean free path in a similar way as in the theory of gases. At sufficiently small gas densities, collisions of viscous flow with the walls of narrow tubes play a role. Something similar is to be expected with electrical conduction in thin wires. In such wires, the resistance at absolute zero, for example, cannot become zero. The greater the mean free path, the greater its influence at given dimensions of the conductor. However, when the mean free path changes with the temperature, the influence of the surface of the conductor will change differently with temperature than the "inner" resistance, which could be determined very precisely. Because one has wires with a diameter of $\frac{1}{1000}$ mm and the mean free path can barely be below 10^{-6} mm, it can be presumed that the increase of the resistance resulting from the surface effect is not really all that small. I hope that, through systematic investigations of this type, the length of the mean free path and its dependence on the temperature can be determined. I will arrange for a local gentleman who wants to work with me to carry out a few relevant measurements at "civilized" temperatures (room & boiling temperature). The lowest temperatures, however, are obviously inaccessible to us. Perhaps this interesting matter could also be taken up at some point at your place. The only theoretically questionable thing about the whole business is that one cannot, in fact, know a priori whether in collisions at the surface of the wire the axial component of the velocity of the electrons (i.e., the component in the direction of the wire's axis) is being influenced. If not, then the effect would indeed not be present.

In any case, I believe that we are far removed from having a valid theory of metal conductivity. It is very puzzling that the specific heat of electrons practically disappears without the apparent presence of a very great dependence of the electron density on the temperature. One should be very eager to know whether the Franz-Wiedemann law is valid down to the lowest temperatures, since, in the theory for this law, the hypothesis is essential that the mean kinetic energy of the electron is $\frac{3}{2}\frac{RT}{N}$.

With best regards, I remain sincerely yours,

A. Einstein

Please do convey my greetings to your wife, also on behalf of my wife.[4] Also, best regards to Mr. Perier and Prof. Lorentz[5] and his family.

Vol. 9, 63a. To Georg Lockemann[1]

[Berlin,] 19 June 1919

Dear Professor,

I believe that one can best understand the current state of the ether question by considering it from the historical viewpoint. Before Maxwell, the ether was thought to be an all-pervasive material with the essential properties of an elastic solid, and whose transversal waves were supposed to be light waves. Maxwell himself and those who elaborated his theory made prolonged but unsuccessful attempts to interpret the Maxwell equations mechanically. Under the pressure of this failure, one became accustomed to considering electromagnetic fields as states of the ether *while foregoing a mechanical analysis of those states.* Then came the question of the state of motion of the ether relative to matter; the special theory of relativity is based on the realization that one cannot speak of such a state of motion. In other words: If one wishes to continue to speak of an ether as the "carrier" of the electromagnetic phenomena, then that carrier must be something quite different from what we otherwise think of as a material body, since one cannot speak of its "motion." Under those circumstances, it would appear preferable to drop the whole concept and speak only of electromagnetic and gravitational fields, not to regard them as states of something else; for only these fields occur in the natural laws, while the notion of a carrier for them (at least in the present state of science) would appear quite unnecessary.

Yours respectfully,

A. Einstein.

Vol. 12, 113a. To Wander J. de Haas

[Berlin, between fall 1917[1] and 21 January 1928[2]]

[Not selected for translation.]

Vol. 16, 90a. "General Theory of Relativity and Equations of Motion." Fragment of early version of *Einstein 1928b* (Vol. 16, Doc. 91)

[before 24 November 1927][1]

[Not selected for translation.]

Vol. 16, 234a. Poem on Scharbeutz

[Scharbeutz, between early July and late September 1928]

[Not selected for translation.]

Vol. 16, 234b. Calculations

[Scharbeutz, between early July and late September 1928][1]

[Not selected for translation.]

Vol. 16, 314a. Notebook

[Gatow and Berlin?, after ca. 15 November 1928][1]

[Not selected for translation.]

Vol. 16, 354a. Aphorism on the Allure of Theoretical Research

[Berlin, 1929]

The enormous excitement of theoretical research is based on the fact that seeking the intellectually simple can lead us to understand reality.

A. Einstein 1929

Vol. 16, 367a. On *Einstein 1929n* (Vol. 16, Doc. 365)

[Berlin, after 10 January 1929][1]

I owe [the solution] to this dilemma to a remark by Mr. Müntz[2] that, according to Cauchy's general theorem, starting from *n* algebraically independent differential equations for *n* field variables is generally permissible, including in the present case of generally covariant equations. Mr. Müntz will explain this more precisely himself. The derivation of the system of equations is, then, simply as follows:

We posit

$$\overline{\mathfrak{B}}^{\alpha}_{\mu\nu} = h(\Lambda^{\alpha}_{\mu\nu} + \varphi_\nu\delta^{\alpha}_{\mu} - \varphi_\mu\delta^{\alpha}_{\nu}) - \varepsilon h(\varphi_\nu\delta^{\alpha}_{\mu} - \varphi_\mu\delta^{\alpha}_{\nu}) \qquad \ldots (1)$$

or, more briefly,

$$\overline{\mathfrak{B}}^{\alpha}_{\mu\nu} = \mathfrak{B}^{\alpha}_{\mu\nu} - \varepsilon\mathfrak{W}^{\alpha}_{\mu\nu} \qquad \ldots (1)$$

Vol. 16, 473a. To the Children at MOPR[1]

[Berlin,] 27 March 1929

Dear children,

You have made me very happy with your congratulations and splendid drawings. I would like to praise some in particular, but that would be neither kind to the rest nor conducive to the brotherly spirit, which is the most important thing. Allow yourselves to be led by the best. Read the letters of Rosa Luxemburg,[2] and never lose sight of the fact that people differ from one another more through their external fate than through feelings and actions!

Vol. 16, 500a. To Elsa Einstein[1]

[Düsseldorf, early Wednesday, 17 April 1929]

[Not selected for translation.]

Vol. 16, 521a. “Unified Field Theory” [Fragment I][1]

[Berlin, ca. 10 May 1929][2]

[Not selected for translation.]

Vol. 16, 521b. "Unified Field Theory" [Fragment II][1]

[Berlin, 10 May 1929][2]

§3. The Hamiltonian Principle[3]

Let $\mathfrak{H}$ be a scalar density which depends only upon the $h_{s\mu}$ and the $\Lambda^{\alpha}_{\mu\nu}$ (but not upon derivatives of those quantities). Then it is easy to see that it contains the $h_{s\mu}$ only in the combination $g_{\mu\nu}$, because $\mathfrak{H}$ must not contain any free Roman indices,[4] so that it is also invariant with respect to rotations of the local systems. For every variation of the fundamental tensor that vanishes at the boundaries of a chosen region, we have

$$\delta\left\{\int \mathfrak{H} d\tau\right\} = 0. \tag{32}$$

We define the tensor densities $\mathfrak{H}^{\mu\nu}$ and $\mathfrak{H}^{\mu\nu}_{\alpha}$ through the equation

$$\delta\mathfrak{H} = \mathfrak{H}^{\mu\nu}\delta g_{\mu\nu} + \mathfrak{H}^{\mu\nu}_{\alpha}\,\delta\Lambda^{\alpha}_{\mu\nu}. \tag{33}$$

If we insert the expression for $\Lambda^{\alpha}_{\mu\nu}$ into (33) and the resulting expression into (32), then, after some calculations that we need not reproduce here, we obtain the following expression for the variation of the integral:

$$2\int \mathfrak{G}^{\mu}{}_{\alpha} h^{\alpha}_{s}\,\delta h_{s\mu} d\tau, \tag{34}$$

where we have set

$$\mathfrak{G}^{\mu}{}_{\alpha} = \mathfrak{H}^{\mu}{}_{\alpha} - \mathfrak{H}^{\mu\nu}_{\alpha/\nu}. \tag{35}$$

From this, we can then derive the field equations

$$\mathfrak{G}^{\mu}{}_{\alpha} = 0. \tag{36}$$

Furthermore, as in the current theory of relativity, we obtain four identities by expressing the fact that the variation (34) must vanish *identically* for those variations of the h that can be generated by simple variations of the coordinate system (which vanish at the boundaries[--)]. In this way, one obtains the identities

$$D_{\mu}(\mathfrak{G}^{\mu}{}_{\beta}) = \mathfrak{G}^{\mu}{}_{\beta/\mu} + \mathfrak{G}^{\mu}{}_{\sigma}\Lambda^{\sigma}_{\beta\mu} \equiv 0, \tag{37}$$

where D_{μ} ⟨the symbol⟩ is an abbreviation for the operator given after the equals sign. The calculation that leads to (37) will be shown as an example of the application of the calculus developed above.

From what we have said, the identity (cf. (34))

$$\int \mathfrak{G}^{\mu}{}_{s} \delta h_{s\mu} d\tau \equiv 0$$

should hold when $\delta h_{s\mu}$ is a "transformation variation," i.e., when

$$\delta h_{s\mu} = h_{s\mu,\beta} \xi^{\beta} + h_{s\beta} \xi^{\beta}{}_{,\mu},$$

where (ξ^{β}) is a vector that vanishes at the limits of integration. This yields the ⟨equation⟩ identity

$$(\mathfrak{G}^{\mu}{}_{s} h_{s\beta})_{,\mu} - \mathfrak{G}^{\mu}{}_{s} h_{s\mu,\beta} \equiv 0 .$$

⟨Due to (30) and (31), this yields⟩

The quantity in parentheses can be written as $\mathfrak{G}^{\mu}{}_{\beta}$. The first term then takes on the form

$$\mathfrak{G}^{\mu}{}_{\beta/\mu} + \mathfrak{G}^{\mu}{}_{\sigma} \Delta^{\sigma}_{\beta\mu}$$

after the introduction of the covariant divergence. In the second term, if we replace $\mathfrak{G}^{\mu}{}_{s}$ with $\mathfrak{G}^{\mu}{}_{\sigma} h_{s}{}^{\sigma}$, then it takes on the form

$$-\mathfrak{G}^{\mu}{}_{s} \Delta^{\sigma}_{\mu\beta},$$

from which, taking account of (23), the formula (37) results.

Simplest Hamilton function. In contrast to the Riemannian theory, in the present theory, there is a tensor $\Lambda^{\alpha}_{\mu[-]}$ that contains only the *first* derivatives of the fundamental tensors. The simplest invariants will thus also contain only such first derivatives and are functions of the h and Λ. Among these, those that are quadratic in the Λ's are the simplest. In choosing the Hamilton function, we limit ourselves to these. We then find that the $\mathfrak{H}$ must be a linear combination of the three scalar densities

$$\left.\begin{aligned} \mathfrak{J}_{(1)} &= h \Lambda^{\alpha}_{\mu\beta} \Lambda^{\beta}_{\underline{\mu}\underline{\alpha}} \\ \mathfrak{J}_{(2)} &= h \Lambda^{\alpha}_{\mu\beta} \Lambda^{\underline{\alpha}}_{\underline{\mu}\underline{\beta}} \\ \mathfrak{J}_{(3)} &= h \Lambda^{\alpha}_{\mu\alpha} \Lambda^{\beta}_{\underline{\mu}\beta} . \end{aligned}\right\} \quad \ldots\ldots \qquad (38)$$

Here, in order to retain a clear and simple notation, I have indicated the "raising" or the "lowering" of indices by underlining the corresponding index.[5] This notation will be used throughout the following.

Before I insert this general linear combination into Hamilton's principle, a certain formal preparation is necessary. We first consider the linear combination[6]

$$\mathfrak{H}_{(1)} = \frac{1}{2}\mathfrak{J}_{(1)} + \frac{1}{4}\mathfrak{J}_{(2)} - \mathfrak{J}_{(3)} . \qquad (39)$$

We assert that the Hamilton tensor belonging to $\mathfrak{H}_{(\beta)}$ (in its purely contravariant form) is a symmetric tensor.[7] Keeping in mind that the fundamental tensor, as a factor in the second term on the right-hand side of (35), can be put within the differentiation, we find the general result

$$\mathfrak{G}^{\mu\alpha} = \mathfrak{H}^{\mu\alpha} - \mathfrak{H}^{\mu\nu}_{\underline{\alpha}/\nu}. \tag{35a}$$

$\mathfrak{H}^{\mu\alpha}$ is, according to its definition, always a symmetric tensor, but this is not true of $\mathfrak{H}^{\mu\nu}_{\underline{\alpha}/\nu}$. For $\mathfrak{H}_{(1)}$ as the Hamilton function, however, as we shall prove shortly, this tensor is also symmetric. By direct calculation, we find

$$\cancel{\mathfrak{H}^{\mu\nu}_{(1)\underline{\alpha}} = h\left(\frac{1}{2}\Lambda^{\nu}_{\underline{\mu}\underline{\alpha}} - \frac{1}{2}\Lambda^{\mu}_{\underline{\nu}\underline{\alpha}} + \frac{1}{2}\Lambda^{\alpha}_{\underline{\mu}\underline{\nu}} - \varphi_{\underline{\nu}}\delta_{\underline{\alpha}}{}^{\mu}\right)},$$

or

$$\mathfrak{H}^{\mu\nu}_{(1)\underline{\alpha}} = \frac{1}{2}h[(\Lambda^{\mu}_{\underline{\alpha}\underline{\nu}} + \Lambda^{\alpha}_{\underline{\mu}\underline{\nu}} - \varphi_{\underline{\alpha}}\delta^{\nu}_{\mu} - \varphi_{\underline{\mu}}\delta^{\nu}_{\underline{\alpha}} - 2\varphi_{\nu}\delta^{\mu}_{\underline{\alpha}}) + (\Lambda^{\nu}_{\underline{\mu}\underline{\alpha}} + \varphi_{\underline{\alpha}}\delta^{\nu}_{\underline{\mu}} - \Phi_{\underline{\mu}}\delta^{\nu}_{\underline{\alpha}})],$$

or, taking (22) into account, $\mathfrak{H}^{\mu\nu}_{(1)\underline{\alpha}} = \frac{1}{2}h(\Lambda^{\mu}_{\underline{\alpha}\underline{\nu}} + \Lambda^{\alpha}_{\underline{\mu}\underline{\nu}} - \varphi_{\underline{\alpha}}\delta^{\nu}_{\underline{\mu}} - \varphi_{\underline{\mu}}\delta^{\mu}_{\underline{\alpha}} + 2\varphi_{\underline{\nu}}\delta^{\mu}_{\underline{\alpha}})$

$$+ \frac{1}{2}\mathfrak{B}^{\nu}_{\underline{\mu}\underline{\alpha}}.$$

On the right-hand side of this equation, however, the first term is symmetric in the indices α and μ; the second has, according to (28), an identically vanishing divergence with respect to ν. Thereby, the symmetry of $\mathfrak{G}^{\mu\alpha}_{(1)}$ is now proven.

Second, we consider the linear combination

$$\mathfrak{H}_{(2)} = \frac{1}{2}\mathfrak{J}_{(1)} - \frac{1}{4}\mathfrak{J}_{(2)}.^{[8]} \tag{40}$$

From (38) and (4), we obtain by calculation

$$\mathfrak{H}_{(2)} = h\Lambda^{\underline{\alpha}}_{\mu\beta}\left(\frac{1}{2}\Lambda^{\beta}_{\underline{\mu}\underline{\alpha}} - \frac{1}{4}\Lambda^{\alpha}_{\underline{\mu}\underline{\beta}}\right)$$

$$= \frac{1}{4}h\Lambda^{\underline{\alpha}}_{\mu\beta}(-\Lambda^{\mu}_{\beta\alpha} - \Lambda^{\beta}_{\alpha\mu} - \Lambda^{\alpha}_{\mu\beta}).$$

Setting

$$\Lambda^{\mu}_{\underline{\alpha}\underline{\beta}} + \Lambda^{\beta}_{\underline{\mu}\underline{\alpha}} + \Lambda^{\alpha}_{\underline{\beta}\underline{\mu}} = S^{\mu}_{\underline{\alpha}\underline{\beta}}, \tag{41}$$

whose tensor is symmetric in all three indices, we find

$$\mathfrak{H}_{(2)} = -\frac{h}{4}\Lambda^{\underline{\alpha}}_{\beta\mu}S^{\mu}_{\underline{\alpha}\underline{\beta}},$$

or, finally,

$$\mathfrak{H}_{(2)} = -\frac{h}{12} S^{\underline{\mu}}_{\alpha\beta} S^{\mu}_{\alpha\beta} \;\ldots. \tag{42}$$

To conclude, in the interest of a uniform notation, we set

$$\mathfrak{H}_{(3)} = \mathfrak{J}_3 = h\varphi_\alpha \varphi^\alpha \;\ldots. \tag{43}$$

The most general Hamilton function that we can consider can now be written in the form:

$$\mathfrak{H} = \mathfrak{H}_{(1)} + \varepsilon_2 \mathfrak{H}_{(2)} + \varepsilon_3 \mathfrak{H}_{(3)}, \tag{44}$$

where ε_2 and ε_3 are provisionally undetermined constants.

§4. Specification of the Continuum.

Thus far, our investigation has been independent of the dimensionality considered. In order to *carry out* the transition to the spacetime continuum, we must now choose 4 as the dimensionality. Furthermore, we must take into account that the metric with respect to the local systems is pseudo-Euclidean. We do that, following Minkowski, by adopting the following convention: the three first local coordinates are real; the fourth is purely imaginary.

This then determines the real properties of all the tensors. We thus conclude: Tensor components with an even number of fourth indices (Roman or Greek) are real, while those with an odd number of fourth indices are purely imaginary. This holds particularly for the fundamental tensor.

With this, our specialization of the scheme developed above, adapted to the particular properties ⟨of our empirical⟩ spacetime continuum, is completed.

§5. The Field Equations.

The field equations that belong to the general Hamilton function (44) can now be thought of, according to equation (35a), as having the form

$$\mathfrak{G}^{\mu\alpha} = 0 \;\ldots. \tag{45}$$

and can be decomposed into their symmetric and antisymmetric parts. We first consider the antisymmetric part of this tensor equation, to which $\mathfrak{H}_{[-]}$, according to the results from §3, makes no contribution. We then find

$$\varepsilon_2(\mathfrak{G}^{\mu\alpha}_{(2)} - \mathfrak{G}^{\alpha\mu}_{(2)}) + \varepsilon_3(\mathfrak{G}^{\mu\alpha}_{(3)} - \mathfrak{G}^{\alpha\mu}_{(3)}) = 0. \tag{46}$$

Due to the symmetry of the $\mathfrak{H}^{\mu\alpha}$, this equation can also be written in the form

$$\varepsilon_2(\mathfrak{H}^{\mu\nu}_{(2)\underline{\alpha}} - \mathfrak{H}^{\alpha\nu}_{(2)\underline{\mu}})_{/\nu} + \varepsilon_3(\mathfrak{H}^{\mu\nu}_{(3)\underline{\alpha}} - \mathfrak{H}^{\alpha\nu}_{(3)\underline{\mu}})_{/\nu} = 0. \tag{46a}$$

It follows from (33) and (42) that:

$$\mathfrak{H}_{(2)\underline{\alpha}}^{\mu\nu} = -\frac{1}{2}hS_{\underline{\mu\nu}}^{\alpha} = -\frac{1}{2}\mathfrak{S}_{\underline{\mu\nu}}^{\alpha}. \tag{47}$$

From (33) and (43), it follows that

$$\mathfrak{H}_{(3)\underline{\alpha}}^{\mu\nu} = h(\varphi^{\mu}\delta_{\underline{\alpha}}^{\nu} - \varphi^{\nu}\delta_{\underline{\alpha}}^{\mu}), \ldots \tag{48}$$

$$\mathfrak{H}_{(3)\underline{\alpha}}^{\mu\nu} - \mathfrak{H}_{(3)\underline{\mu}}^{\alpha\nu} = 2h(\varphi^{\mu}\delta_{\underline{\alpha}}^{\nu} - \varphi^{\alpha}\delta_{\underline{\mu}}^{\nu}). \tag{49}$$

For the further calculation of the second term of (46a), we must take into account that

$$(\varphi^{\mu}hg^{\alpha\nu})_{/\nu} \equiv \varphi^{\mu}{}_{;\nu}hg^{\alpha\nu} + \varphi^{\mu}(hg^{\alpha\nu})_{/\nu},$$

$$(hg^{\alpha\nu})_{/\nu} = h(g^{\alpha\nu}{}_{;\nu} + \varphi_{\sigma}g^{\alpha\sigma}),$$

where the first term in the parentheses on the right vanishes. We then have

$$\varphi^{\mu}(hg^{\alpha\nu})_{/\nu} = h(\varphi_{\underline{\mu};\underline{\alpha}} + \varphi^{\mu}\varphi^{\alpha}),$$

so that we obtain

$$\mathfrak{H}_{(3)\underline{\alpha}}^{\mu\nu} - \mathfrak{H}_{(3)\underline{\mu}}^{\alpha\nu} = 2h(\varphi_{\underline{\mu};\underline{\alpha}} - \varphi_{\underline{\alpha};\underline{\mu}}) = 2f^{\mu\alpha}. \tag{49a}$$

(46) thus takes on the form

$$\varepsilon_2\mathfrak{S}_{\underline{\mu\nu}/\nu}^{\alpha} - 2\varepsilon_3 f^{\mu\alpha} = 0. \tag{46a}$$

We interpret $f^{\mu\alpha}$ as the tensor density of the electromagnetic field; ⟨it⟩ the electromagnetic field cleaves in the first approximation together to the potentials just as in the earlier theory, which is equivalent to the system of Maxwell's equations ("law of induction"). That also the other system of Maxwell's equations holds in the first approximation is easily discernible[.] In the first approximation, $\mathfrak{S}_{\underline{\mu\nu}/\nu}^{\alpha}$ can namely be replaced by $\mathfrak{S}_{\underline{\mu\nu},\nu}^{\alpha}$. If we then differentiate (46a) with respect to $x^{\bar{\alpha}}$, the first term vanishes identically, owing to the complete antisymmetry of $\mathfrak{S}$, so that we find, in first approximation, the field equation

$$f^{\mu\alpha}{}_{,\alpha} = 0.$$

Nevertheless, equation (46a) is not permissible from the physical point of view. It states, namely, that a four-potential exists (for $\mathfrak{S}_{\mu\nu}^{\alpha}$ also a four-vector) in the first approximation not only for $\varphi_{\mu\alpha}$, but also for the *dual* six-vector. It then follows from this that, according to (46a), *electric* masses are not possible, just as—due to the usual vector potential's existence—*magnetic* masses cannot exist.

This difficulty can, however, be overcome by taking the field equations to the limit $\varepsilon_3 = 0$. This is not quite the same as if one simply were to leave off the term that is multiplied by ε_3 in the Hamilton function, as we shall soon see.

If, in the equations (46a), we allow ε_3 to go to the limit 0, then, by this limiting transition, we obtain the equation

$$\mathfrak{S}^{\alpha}_{\underline{\mu}\underline{\nu}/\nu} = 0, \tag{50}$$

and one might first think that this equation must replace the antisymmetric system (46a). That is, however, not allowable, since these 6 equations can be fulfilled by ensuring that the 4 equations

$$\mathfrak{S}^{\alpha}_{\underline{\mu}\underline{\nu}} = 0 \tag{51}$$

are fulfilled;[9] they, together with the symmetric system of equations (10 equations), certainly yield too few equations for the determination of the fields. One can readily convince oneself of this, e.g., by a strict or approximate treatment of the spherically symmetric case.

The correct consideration of approaching the limit seems instead to be the following: With ε_3, not only does $\mathfrak{S}^{\alpha}_{\underline{\mu}\underline{\nu}/\nu}$, but also $\mathfrak{S}^{\alpha}_{\underline{\mu}\underline{\nu}}$, goes to zero in the limiting case as (proportion[al to] ε_3. We now employ the identity

$$D_{\mu}(\mathfrak{G}^{\mu\alpha}_{(2)}) \equiv 0(\equiv \mathfrak{G}^{\mu\alpha}_{(2)/\mu} + \mathfrak{G}^{\mu\sigma}_{(2)}\Lambda^{\sigma}_{\underline{\alpha}\underline{\mu}}) \ldots \tag{52}$$

which can be obtained from (37) after raising the index α. One can, due to (47), replace this with

$$D_{\mu}\left(\mathfrak{H}^{\mu\alpha}_{(2)} + \frac{1}{2}\mathfrak{S}_{\underline{\mu}\underline{\nu}/\nu}\right) \equiv 0 \ldots. \tag{52a}$$

If we apply the operation D_{μ} to (46a), and take (52a) into account,

$$\varepsilon_2 D_{\mu}(\mathfrak{H}^{\mu\alpha}_{(2)}) + \varepsilon_3 D_{\mu}(f^{\mu\alpha}) = 0 \ldots\text{[11]} \tag{53}$$

Now, however, owing to (42), $\mathfrak{H}^{\mu\alpha}_{(2)}$ is a quadratic function of $\mathfrak{S}^{\alpha}_{\mu\nu}$, and thus it also goes to zero, as does $\varepsilon_3^{\,2}$ in our limiting transition. Thus, if we divide (53) by ε_3 and then take the limit $\varepsilon_3 = 0$, then the first term vanishes in that limit, and the remaining limiting relation is

$$D_{\mu}(f^{\mu\alpha}) = 0 \ldots. \tag{54}$$

This equation thus forms, together with

$$\mathfrak{S}^{\alpha}_{\underline{\mu}\underline{\nu}} = 0, \tag{55}$$

the result of the transition to the limit (54) in the first approximation, identical to the [one] Maxwellian system of equations. Alongside these two systems of equations, we still have the symmetric system of equations,

$$\mathfrak{G}^{\mu\alpha} + \mathfrak{G}^{\alpha\mu} = 0, \qquad \langle(56)\rangle$$

which, on taking our limit, evidently goes to

$$\mathfrak{G}^{\mu\alpha}_{(1)} + \mathfrak{G}^{\alpha\mu}_{(1)} = 0$$

or (due to the symmetry of $\mathfrak{G}^{\mu\alpha}_{(1)}$) to

$$\mathfrak{G}^{\mu\alpha}_{(1)} = 0. \qquad (56)$$

Equations (54), (55), and (56) together form the complete system of field equations. Their compatibility results from their derivation through an extremal principle.

It is easy to show that the equations (56) *in the first approximation* agree with the gravitational equations

$$R_{ik} = 0$$

of the earlier theory. I have also determined alongside Mr. Münz[12] that Schwarzschild's solution for the centrally symmetric case with a vanishing electromagnetic field also satisfies these new field equations.

(Here, page (20a) is to be inserted!)[13]

[p. 20a] It is clear that the 18 field equations (54), (55), and (56) for the 16 field quantities $h_s^{\ \nu}$ cannot all be independent of each other. Owing to the general covariance of the whole system, $18-(16-4)$, i.e., 6 identity relations must hold among the field equations. This is indeed also the case.

First of all, due to (37), the 4 identities among the equations (56)

$$D_\mu(\mathfrak{G}^{\mu\alpha}_{(1)}) \equiv 0 \qquad (57)$$

indeed hold.

Second, (55) fulfills the identity

$$(\mathfrak{S}^{\alpha}_{\underline{\mu\nu}}\delta^{\mu\nu\alpha\beta})_{/\beta} \equiv 0, \qquad (58)$$

where $(\delta^{\mu\nu\alpha\beta})$ denotes that tensor density which is antisymmetric with respect to all pairs of indices and whose components take on the values ± 1. The proof of this is readily carried out by replacing the symbol $_{/\beta}$ by ordinary differentiation and expressing the $\mathfrak{S}^{\alpha}_{\mu\nu}$ in terms of the $h_s{}^{\nu}$. Third, between the equations (54) and (55), the identity

$$[D_{\mu}(f^{\mu\nu}) - f^{\mu\alpha}\mathfrak{S}^{\nu}_{\mu\underline{\alpha}}]_{/\nu} \equiv 0 \ldots \tag{59}$$

holds. In order to prove this equation, one must replace the first term in the square brackets by the expression in parentheses in (54) and then ⟨thereupon⟩ apply the divergence commutation relation proved in a previous article[1)] to the $f^{\mu\nu}{}_{/\mu/\nu}$; this yields

$$f^{\mu\nu}{}_{/\mu/\nu} \equiv \frac{1}{2}(f^{\mu\nu}\Lambda^{\alpha}_{\mu\nu})_{/\alpha}.$$

(57), (58), and (59) are the required 6 identities.

§5. Concluding Remarks.[14]

That the theory given above represents, from the formal point of view, a thoroughly natural and logical self-consistent continuation of the fundamental ideas of the general theory of relativity is certainly indisputable. Everything follows from the additional idea of teleparallelism; only the transition to the limit to $\varepsilon_3 = 0$ ⟨means⟩ has, from the formal viewpoint, the character of an arbitrary convention. Since the well-known field laws can be deduced in this way in a sufficient approximation in such an unconstrained manner, this theory is worthy of being further elaborated and [----]

[1)]Sitz. Ber. d. Preuss. Akad. 1929. I. Equation (5).

1. The Commutation Rule for Differentiation in Unified Field Theory

[mid-1929][1]

The commutation rule for differentiation in the unified field theory[2]

The central problem of the unified field theory based on a Riemannian metric, and distant parallelism is characterized by the question: What are the most natural or simplest differential laws to which a manifold of the type under consideration can be subjected? The corresponding question here is considerably more difficult to answer than it is for a Riemannian manifold. It cannot be answered without some arbitrariness, since the [criteria] of "[simplicity]" cannot [...]

2. To Arbeitsausschuss Deutscher Verbände[1]

[Berlin,] 1 June 1929

The true cause of the difficult situation in which Germany and the rest of Europe find themselves is not the Versailles Treaty, but the disastrous mess that preceded it.[2] For that reason, our efforts should be directed to improving European organizations and camaraderie, not criticism of mistakes made in 1919.[3]

3. To Semen L. Frank[1]

Berlin, 1 June 1929

Dear Prof. Frank,

Thank you very much for the article, which does not seem very credible after all.[2] The matter would be better suited for the first of April. It certainly seems to be true that great nervousness and uncertainty have again gripped Russia.

Kind regards, your

A. Einstein

4. From Max Planck

Berlin, 4 June 1929

[Not selected for translation.]

5. To Eduard Einstein[1]

[Berlin,] 5 June 1929

Dearest Tetel,

Although I write to you so infrequently, I am very fond of you and often think that you must be working hard to prepare for the graduation examination.[2] But it will also pass, and you'll be as free a little man as causality allows, though that is a considerable limitation. When will I most likely get to see you? The splendid sailboat will be ready at the end of this week.[3] It will come first to Gatow, on Plesch's estate, as will I.[4] You'll be able to relax and recuperate. I am now quite well, much better than I was even a few months ago. But I still have my old ticker problem.[5] The house is now being built in Caputh (*nomen est omen*). It's terribly expensive, and the city made it so odious that I refused to accept their gift.[6] I'm very tempted to move entirely to Caputh, because living in two places will become too expensive. You'll be delighted by the setting. But it will still be awhile—around September—before it is finished.

Mama's letter regarding the money from New York has still not arrived;[7] remind her of it, and give her my warmest greetings.

Write me a card (no need to do more in your present misery) soon, kisses from your

Papa

I liked it at Adn's home, and I have the impression that the two of them are quite happy. So things have turned out much less badly than I feared.[8]

6. To Hans Albert Einstein and Frieda Einstein-Knecht

[Berlin,] Tuesday, 5 June 1929

Dear Albert,

I didn't stop to see you on my way back from Brussels because, given the heat, I didn't want to move around unnecessarily.[1] But in exchange, I have an even more pleasant memory of my earlier visit.[2] I found your place very agreeable, your life and your things, your whole mode of existence. The Wagerl wanted to take revenge on me for my earlier recommendation.[3]

Yesterday, I spoke with Counselor Seligsohn, one of the greatest authorities on patents. You should go to see him as soon as you happen to come here. I'm sending you his book to study.[4] It will be the right one for the theoretical introduction.

I'm working a lot on my problem, insofar as the many disturbances related to public and private occasions leave me time and strength to do so.[5] The subject is very difficult, and, moreover, I'm unsure whether I've taken the right path. But it is enormously interesting.

Meeting the Belgian queen was very enjoyable. However, I remained very red.[6] If I knew a good way to do so, I'd like to immediately do "Bolshevism." But people are a devilishly difficult material.

I close with a fine joke about fascism: Three things are never all found together:[7]

Fascist (1), intelligent (2), honest (3)

If someone is (1) and (3), then he's not (2)

If he's (1) and (2), then he's not (3)

If he's (2) and (3), then he's not (1)

In the end, I refused to accept the city gift. We are building a little house in Caputh that will probably be only finished in September. It cost a dreadful lot of money. But the city got us into this, and I have to deal with the consequences.[8] I hope they'll have much joy coping with it. Maybe I'll move out altogether.

Regards to you both, from your

Papa

Warmest greetings from my women.[9]

7. To Elizabeth, Queen of the Belgians[1]

Berlin, 5 June 1929

Your Majesty,

I thank you heartily for the kind letter and the two very successful little pictures. They will be a lovely souvenir of the harmonious afternoon I was allowed to spend with you and your amiable companion.[2] It showed once again that, for real human beings, there is no difference in spheres of existence.

In the joyful expectation of seeing you and your husband[3] again and resuming our music making together with a larger orchestra, I remain with all reverence, best wishes, and greetings, your

A. Einstein

8. Poem for the Launch of the Sailboat *Tümmler*

[Friedrichshagen, around 7 June 1929][1]

It's already hastening away
Your proud, fat son
Even heavier than it is fast
Soon it will do battle with the waves
Thanks to you who funded it
And opened your pocketbook[2]
No less to you who thought it up
And with strong arms, made it[3]
Enter now into your life
⟨Now plunge into the water⟩
Porpoise shall be your name
Don't capsize, and stay good
Enjoy the cool flood

9. To Semen L. Frank

Berlin, 12 June 1929

Dear Prof Frank,

I am very grateful to you for taking the trouble to pursue this matter to its source.[1] The outcome has made a great impression on me. Here, we are undoubtedly dealing with a cynical crime committed by the government, a crime that is well-suited to shatter any confidence in such a judicial system. I have also heard a devastating judgment of the present Russian rulers made by another man who is particularly competent because of his devotion to justice.

With renewed thanks and best wishes, your

A. Einstein.

Revoked in light of further information. A. Einstein.[2]

10. From Joseph Sauter[1]

Wabern, 15 June 1929

Dear Mr. Einstein,

One month ago, I finally concluded my arduous study of Weyl's book, "*Space, Time, and Matter*," full of enthusiasm but not without regrets for the great deal of time that I had wasted in hapless verifications of his calculations.[2] Leaving aside the quantum theory, I will now venture on to your more recent works; I mean your notes to the Berlin Academy on teleparallelism in Riemannian geometry and your unified field theory.[3]

I decided to write to you not to "stick out my tongue," but rather merely to let you know how my *homo mathematicus* head, brought to paroxysms by our Swiss patent office, reacted to the first of those notes. Here, neither the fox nor the snake are speaking, only an experimental guinea pig. Where I may appear to be malicious and ironic, I am so consciously, but only to express my thoughts more clearly. I agree with your results, and my criticisms are only of their form, particularly of its details.

The first sentence of your first note gives me the impression that two different authors are speaking.[4] For the first author, the ideal of physics would seem to be a pure description of the phenomena, with or without logic, with or without coordination, not even the ideal of Auguste Comte.[5] The other author, a strict Kantian, distinguishes between nature as we mysteriously find it (phenomena) and the artificially established construct (noumenon) which is to be attributed to nature and which is intended to render, in the most unified manner possible, everything that has thus far been observed of nature.[6]

The expression "*n*-Bein" seems inauspicious to me. Let us assume that I own a dog (which is, in fact, not true) and a cat (which is true), and, in order to distinguish between the two of them, I call the dog "vertebrate" and the cat "mammal." What would you say to that? And what do you think of the relation:

vertebrate: mammal = *n* Bein: axis system of the x^{ν} ?

Your equation (1) is painful to me because, in it, we are to sum over 2 indices α, both of which are lower indices.[7] Why not write $h^{\nu\alpha}A_{\alpha}$ consistently, as Weyl does?[8] I can understand that you do not set $h_{\alpha}^{\nu}A^{\alpha}$, as we then would have to write $h_{\mu}^{\alpha}A^{\mu}$ in (1a), which would obscure the difference between the h_{α}^{ν} and their normalized subdeterminants, h_{ν}^{α} .

The first sentence of §2 sounds quite dissonant to me.[9] By setting up an *n*-Bein-field, one has not yet expressed the existence of teleparallelism as such. The following sentence nevertheless assumes that it has proved the following: that the equality of *A* and *B* emerges from the preceding formula (2) is correct, but not that the parallelism of *A* and *B* is a necessary logical consequence of the premises established in §1. The existence of teleparallelism is shown only by your definition or, also, your decree.

Here, I would like to add something: If we go from three-dimensional Riemannian geometry to two-dimensional spherical geometry, we could call an arbitrarily-oriented worm in Calcutta "parallel" to an arbitrarily oriented worm in London: an arbitrary element cannot be avoided in this definition. But even in the case that the considered points are infinitesimally near to each other, that arbitrariness still persists. In my studies of Weyl's book, one of my main difficulties was the lack of a purely objective basis for the concept of the affine connection.

In the fifth edition, Weyl admits that he had taken the concept of parallelism at infinity from the works of Levi-Cività, where the latter, in our case, had gone outside the sphere and called two neighboring segments on the sphere "parallel" when one of them forms the tangential component of the other segment, displaced in parallel in the three-dimensional space; I find that to be arbitrary.[10] Consequence: I am not sticking my tongue out at you; your teleparallelism has just as much of a right to exist as does the affine connection that Weyl derives—or wants to derive—from the metric.

One final point: it is clear to me that, according to (7a), the Riemannian tensor $R^{i}{}_{k,\,lm}$ that is formed from the displacement coefficient $\Delta^{\nu}_{\mu\nu}$ must vanish identically,[11] but it is *not* clear that one can then verify this. For the past week, I have been trying one computational trick after another, but that verification still escapes me. Can you show me the way? I cannot rest… This pain is the real reason for the present letter.

Apart from the pain just mentioned, I am well.

I hope that my letter will find you in the best of health and full of happiness, and I send you warm wishes.

Yours sincerely,

J. Sauter.

Mr. Besso is on vacation; he would have kept me from sending you this letter!

11. To Norbert Wiener[1]

[Berlin?,] 19 June 1929

Dear Colleague,

Thank you for your kind transmissions.[2] I am entirely blameless for the great fuss that has been made over my works, especially over my new theoretical experiments. It is very good when they are contradicted by experts.

Now, regarding teleparallelism, I am fascinated by this obvious and, up to this point, unexamined theoretical possibility.[3] At the same time, I must say openly that I am fully aware my efforts in this area are not yet complete. However, I do not consider it impossible that something deep and true might result from this, even if genuine success has yet to justify my confidence. I think the problem of the connection between electricity and gravitation is already a meaningful problem at the macroscopic level.

Best regards, your

12. Acceptance Speech on Being Awarded the Planck Medal

"Ansprache von Prof. Einstein an Prof. Planck"
[*Einstein 1929nn*]

DATED [before 28 June 1929]
PUBLISHED 20 July 1929

IN: *Forschungen und Fortschritte*, V (1929): 248–249.

Speech by Prof. Einstein, addressed to Prof. Planck:1)

How can I put into words how moved I am, standing at this moment before my honored master and my friend, with whom I have been united by our common goals over so many years?[1] It was twenty-nine years ago that, as a young man, I was enraptured by how he brilliantly applied Boltzmann's statistical method in such a novel way to derive the radiation formula, making it possible to precisely determine the atoms' absolute size for the first time.[2] At that time, honored master, you recognized that one of the most important goals of the coming decades would necessarily be clarifying the fundamental structural property hidden behind the constant h that you had introduced. Carrying out that program constituted—as seen from a fundamental point of view—the substance of the most important branch of development of modern theoretical physics.

Who would be surprised that I, too, was compelled to enter the circle of your research efforts? Starting from the most general formal viewpoint, I struggled to

1)On the occasion of the fiftieth anniversary of the conferral of Prof. Planck's doctoral degree, on 28 June 1929, the German Physical Society of Berlin, together with the German society for Technical Physics, hosted a festive meeting. The Planck medal, which had been created to celebrate the seventieth birthday of Prof. Planck, was awarded by the Chairman of the German Physical Society, Prof. Heinrich Konen-Bonn, first to Plank himself on the occasion of that meeting. Afterward, Prof. Planck presented a second copy of the medal to Prof. Albert Einstein.

achieve a deeper understanding of the connections which, stimulated by your example, led physics to an ever-growing productivity. There were two particular thoughts around which my fervid efforts were grouped. The events of nature appear to be so largely determined that not only their temporal progression, but also their initial state is fixed by laws to a great extent.[3] I believed this concept would most necessarily be served by searching for overdetermined systems of differential equations. The postulate of general relativity and the hypothesis of a unified structure of physical space or fields were to be my guideposts in this search.[4] This goal has yet to be attained, and it would be hard to find a colleague who shares my hope of arriving at a deeper understanding of reality by that route. What I have discovered in the quantum domain are only insights or, in some measure, fragments that were byproducts of my fruitless attempts to find the solution to that great problem. It shames me to receive such a significant honor for those efforts.

Nevertheless, I strongly believe that we will not be brought to a halt before a subcausality; instead, we will eventually arrive at a supercausality in the sense I have indicated—admiring highly, as I do, the achievements of the younger generation of physicists which are subsumed under the heading of quantum mechanics, and I believe in the deeper truth of that theory. I just think its limitation to *statistical* laws will only be a passing one.

One more word to you, dear and honored Prof. Planck: I know—and I am glad to be able to express it—that, as an intellectual aspirant and human being, I owe an extraordinary amount to you. You were one of the most effective of those who encouraged my nascent strivings. You were the first of those who supported the theory of relativity.[4] You have contributed, in a decisive way, to providing me with support and working conditions available to only a fortunate few. In all matters of action and direction, there was agreement and consensus between us. I often had the opportunity to admire the objectivity with which you always served higher goals—free of influences from personal or political gains.

Your thoughts will continue to resound so long as physics exists, and I hope no less so the example which you have given through your life to those who will follow us.[5]

With these thoughts and feelings, I accept with pleasure, if also with humility, the Planck medal from your hands.

13. Message to the Initiative Committee for the Expansion of the Jewish Agency in Germany

"Schreiben von Prof. Einstein"
[*Einstein 1929mm*]

PUBLISHED 28 June 1929

IN: *Jüdische Rundschau* 34, no. 50 (28 June 1929): 318.

Distinguished Gentlemen and dear Friends:

Unfortunately, I am unable to attend your meeting,[1] one which is as gratifying as it is important. I send you my best wishes for the success of the common Palestine project.

Judaism owes a great deal to Zionism, but most of all, a new flourishing of the feeling of dignity and community among the whole Jewish people. The latter has slowly grown aware of the debt of gratitude it owes to Herzl[2] and his loyal band, who did the first and hardest part of the work. Today, all Jews acknowledge the common task. May they prove through their achievements that in Judaism the world has sufficient moral power and devotion to complete the great communal project.

If you really want something, it will happen.

A. Einstein

14. To the Bund der Freunde der Sowjet-Union[1]

[Berlin,] 30 June 1929

Although I strongly disapprove of the constant criticism of the Soviet Union, as well as the militarism in our country, I cannot sign your appeal, since it also involves taking a political position that does not correspond to my conviction.[2]

With best regards,

15. To the Pittsburgh Conference of Jewish Women's Organizations[1]

Berlin, 30 June 1929

To your inquiry,[2] I would like to make the following reply:

The modern Jews are undoubtedly a people in the sense of common descent and certain shared traditions.[3] However, among the Jews of the present, one cannot speak of a unified culture as was still exhibited by the Jews of antiquity. The answer to your question also depends upon whether a hall of your building should showcase: (a) personalities and their contributions to culture in general; (b) individual national cultures in their distinctiveness. It is clear that the Jews could well claim a hall if viewpoint (a) is decisive. But if viewpoint (b) is decisive, then the creation of a Jewish hall is only justified if other nations of the past (for example, the Greeks of antiquity) also receive special halls.

Respectfully yours,

A. Einstein

16. To Mileva Einstein-Marić

[Caputh,] 4 July 1929

Dear Mileva,

I am entirely of your opinion. Because of the tax, it's better if you take over the mortgage debt on the house. I would also find out if the 6% interest rate isn't too high in the Swiss context. If it isn't, then I would agree to it, and if it is, we could try to sell the mortgage to another creditor.[1]

I am doing decidedly better, healthwise, so that I can sail again.[2] I am now living in Caputh in the countryside all summer. If Tetel comes, he will also find it very beautiful and relaxing.[3] The house is not yet finished, but we are living in an interim apartment right on the lake.[4] The new boat, which I was given by wealthy financiers for my fiftieth birthday, is wonderful, so beautiful that I'm a little scared because of the responsibility.[5]

I'm working a lot, but the evil spirit is making me turn around and around in circles, so I still don't know whether my new theory of electricity is any good or not.[6] My best regards to you and Tetl, from your

Albert

Rudi and Toni[7] were very pleased to receive Tetl's letters.

I hope the graduation exam is not stressing him too much.[8] I would think it very good if he left home soon in order to learn to find his way in life. I tell you this because I can imagine that you are surely strongly opposed to that. But think about Albert's fate. Things might not have gone as they have, had he become acquainted with the world earlier.[9] It would be especially good if he were to go to England for a time to learn the language. What do you think about that?

17. To Semen L. Frank

Berlin, 7 July 1929

Dear Prof. Frank,

I do not, in the least, maintain that the judgment of the three engineers was justified.[1] Moreover, I have never made any secret of the fact that I do not agree with the system of violent political action. But since, in my view, proof of judicial murder has not been presented in this case, I cannot speak out against it. This is all the more imperative because I know very well that false statements about Russia are systematically made in order to prevent any feelings of solidarity for the Soviets from arising in the working population of the Western countries.[2] Your claim that I harbor some kind of prejudices in favor of the Russian rulers is not true. On the other hand, I am informed by my best friends and colleagues about the admirable spirit that prevails in the world of Russian students and scholars.[3] Therefore, it cannot be as dismal as you imagine it to be. But hearing that can only be a pleasure for you.

Best regards, your

A. Einstein

18. To Ivan Vasilyevich Obreimov[1]

[Caputh,] 11 July 1929

Dear Colleague,

It was with great interest that I received your letter of July 4, in which you inform me that you would like to confer about "the unified field theory."[2] Overwork and my poor state of health, unfortunately, make it impossible for me to respond to your request appropriately.[3] However, I would like to briefly express my current view of this subject in the following words:

I truly believe that the hypothesis of teleparallelism is worthy of serious consideration, because it could possibly permit us to comprehend the structure of physical space[time]. On the other hand, I am presently rather convinced that my interpretation of the electromagnetic field thus far is not correct. The situation is progressing only very slowly, since I have, up to now, not been able to overcome certain mathematical difficulties.[4] I believe that it would be expedient to postpone the discussion of the whole subject until greater clarity can be achieved about the formal connections. At the moment, so much is still lacking that I have not yet been able to reach a firm conclusion regarding both the choice of the field equations and the terms of success of the whole project.

Yours very respectfully,

19. To Gustav Stresemann[1]

[Berlin,] 11 July 1929

Distinguished Minister,

After mature consideration, in my capacity as a member of the League of Nations' Committee for International Cooperation,[2] I take the liberty of making the following proposal:

Up to this point, the Institute for International Intellectual Cooperation, founded by France in Paris, has been primarily subsidized only by states that are politically close to France. In particular, up to now, the institute has not been subsidized by Germany or England.[3] Even though it is my firm conviction that the director of the institute, Monsieur Louchaire,[4] sincerely intends to lead, in a purely international way, the work done in the service of the International Committee on Intellectual Cooperation, the financial non-participation of Germany and England very naturally produces a preponderant French influence that certainly does not foster a beneficial impact on the institute or, furthermore, on the committee, from the point of view of international solidarity.[5]

I am convinced that Germany could immediately get a German hired as subdirector of the institute, should it decide to make an annual contribution on the order of 40,000 marks, and if a prominent person could be found for this position, someone who could win trust and influence through his activity and his interest in increasing understanding, he could have a major influence on the leadership of the institute. A subdirector of English nationality is already available, but all those in the know are convinced that he is not a sufficiently competent man.[6]

It has often been said that Germany should not participate until England does. I am not of that opinion. For one thing, it will be easier to gain influence in the institute if Germany's participation precedes England's. On the other hand, given the current distribution of political power in England, the latter could hardly fail to follow Germany's example.

Respectfully yours,

20. To Chaim Herman Müntz

[Caputh, 13 July 1929][1]

Dear Mr. Müntz,

The proof of compatibility[2] can be carried out, e.g., as follows:
Consider that the equations

$$\overline{\overline{\mathfrak{G}}}^{\mu\alpha} = 0; \qquad \varphi_\alpha = 0 \qquad S^\alpha = 0$$

are compatible.

Proof: The 12 equations

$$\overline{\overline{\mathfrak{G}}}^{11} = 0 \quad \overline{\overline{\mathfrak{G}}}^{22} = 0 \quad \overline{\overline{\mathfrak{G}}}^{33} = 0$$

$$\overline{\overline{\mathfrak{G}}}^{12} = 0 \quad \overline{\overline{\mathfrak{G}}}^{23} = 0$$

$$\overline{\overline{\mathfrak{G}}}^{13} = 0$$

$$S^1 = S^2 = S^3 = \varphi_1 = \varphi_2 = \varphi_3 = 0$$

can certainly be fulfilled.

From this, it initially follows that $S^4 = 0$ and $\varphi_4 = 0$.

Then we have $\overline{\mathfrak{G}} = 0$ if this is satisfied for only one x_4.

Now, since $D_\mu(\dot{\mathfrak{G}}^{\mu\alpha}) = 0$, it also follows that

$$D_\mu(\overline{\overline{\mathfrak{G}}}^{\mu\alpha}) = 0 .$$

If for one section x_4, $\overline{\overline{\mathfrak{G}}}^{4\alpha} = 0$, then the $\overline{\overline{\mathfrak{G}}}^{4\alpha}$ also vanish everywhere. Thus, all of the equations are fulfilled.

That the temporal continuation is also fulfilled by these equations would, in my opinion, require a separate proof.

Best regards, your

A. Einstein

21. From Joseph Sauter

Wabern, 13 July 1929

Dear Mr Einstein,

I would like to suggest that you send the enclosed document with your signature back to me.[1]

Following a teaching method often used for children, I have attempted today to learn the first page (3) of your article on Riemannian geometry by heart.[2] In this way, I was forced to become accustomed to your manner of thinking, and that gave me a great deal of childish joy. But I also found some joy in all the formulas of the whole article, in particular, when I noticed the following:

The tensor $\Lambda^{\nu}_{\alpha\beta}$ will represent the entire field, the components $\nu = \beta$ the electromagnetic partial field, and the components $\nu \neq \beta$ the gravitational partial field;

The right-hand side of (7a) has the same structure as the first term or the right-hand side of (8), below;[3]

In the latter formula, $\dfrac{\partial g_{\mu\alpha}}{\partial x^{\sigma}}$ represents the gradient of the pseudoscalar $g_{\underline{\mu\alpha}}$, and $\dfrac{\partial g_{\sigma\alpha}}{\partial x^{\mu}} - \dfrac{\partial g_{\mu\sigma}}{\partial x^{\alpha}}$ represents the curl of the pseudovector $g_{\underline{\sigma}i}$, where the underlined indices are to be considered as constant;

Your quantity $\Lambda^{\nu}_{\alpha\beta}$ is, according to the next-to-last formula on page 6,[4] the product of a first vector with the curl of a second vector, which is contragradient with respect to the first.

Most respectful greetings from yours sincerely,

Sauter.

P.S. What pleases me most is that everything, even the $g_{\mu\nu}$, can be derived from the h^{ν}_{α}!

..................*Le cose tutte quante*
Hann' ordine tra loro; e questo è forma,
Che l'universo a Dio fa simigliante.

Even Dante would have taken his pleasure in this![5]

⟨Wabern, 13 July 1929⟩[6]
Berlin, July 1929

To an incorrigible,

May the following serve as an answer to your letter of :

(1) As a disciple of Kirchhoff, I do not distinguish between a "physical description" and a "physical theory." By mapping new concepts onto those that are known, a description of novel things can emerge from comparison against those known ones. There is, therefore, nothing to be objected to in the first sentence of my article.[7]

(2) You forget that every point P is the origin of a local coordinate system and thus is the seat of an "n-Bein" [axis system] in the sense that I mean; and only the point $x^{\nu} = 0$ is the origin of the Gaussian coordinate system. A confusion of my ∞^{n} n-Beine with the single n-Bein of the basis vectors $\mathfrak{e}_{\nu}$ at the point $x^{\nu} = 0$ is therefore excluded.

(3) With respect to an orthogonal coordinate system, no distinction between covariant and contravariant need be made; thus, we can let all the indices α be lower indices, as was traditionally done. By the way, you would not be happy with your suggestion $A^{\nu} = h^{\nu\alpha}A_{\alpha}$ for very long, since, in my formula (2) you would be forced to sum over only lower indices α four lines later!

(4) For all I care, you can replace "becomes" with "may become" in the first sentence of §2, and replace "thus, they are to be considered as equal and parallel" with "so they may be considered to be not only equal, but also parallel" in the second.

(5) I do not mind if you delete "which can be easily verified" in the third-to-last line on page 5.[8]

22. To Cornel Lanczos[1]

[Caputh,] 27 July 1929

Dear Mr. Lanczos,

Using the old method of applying the divergence, I have now found a system of 20 equations with 8 identities. I start from the identities

$$\mathfrak{P}^{\mu\alpha}{}_{/\alpha} = (L^{\alpha}_{\underline{\mu}\underline{\nu}/\nu} - L^{\sigma}_{\underline{\mu}\underline{\tau}}\Lambda^{\alpha}_{\sigma\tau})_{/\alpha} \equiv L^{\alpha}_{\underline{\mu}\underline{\nu}/\alpha/\nu} = \mathfrak{f}^{\mu\nu}{}_{/\nu} \qquad (1)$$

$$\mathfrak{Q}^{\mu\alpha}{}_{/\mu} = \left(L^{\alpha}_{\underline{\mu}\underline{\nu}/\nu} - \frac{1}{2}L^{\alpha}_{\underline{\sigma}\underline{\tau}}\Lambda^{\mu}_{\sigma\tau}\right)_{/\mu} \equiv 0\,, \text{[2]} \qquad (2)$$

where the $\mathfrak{P}^{\mu\alpha}$, $\mathfrak{Q}^{\mu\alpha}$, and $\mathfrak{f}^{\mu\nu}$ are abbreviations whose meanings you can guess from (1) and (2). I then write these identities as follows: (2′)

$$\mathfrak{Q}^{\mu\alpha}{}_{/\mu} \equiv 0 \qquad (2')$$

$$\mathfrak{Q}^{\mu\alpha}{}_{/\alpha} \equiv \mathfrak{f}^{\mu\alpha}{}_{/\alpha} + (\mathfrak{Q}^{\mu\alpha} - \mathfrak{P}^{\mu\alpha})_{/\alpha} \qquad (1')$$

Now, I formulate the equations

$$\mathfrak{Q}^{\mu\alpha} = 0 \qquad (I)$$

$$\mathfrak{f}^{\mu\alpha}{}_{/\alpha} + (\mathfrak{Q}^{\mu\alpha} - \mathfrak{P}^{\mu\alpha})_{/\alpha} = 0 \qquad (II)$$

These are the 16 gravitation equations plus four Maxwell equations, among which the above 8 identities hold. In this case, we have

$$\mathfrak{Q}^{\mu\alpha} - \mathfrak{P}^{\mu\alpha} = -\frac{1}{2}\mathfrak{S}^{\mu}_{\underline{\sigma}\underline{\tau}}\Lambda^{\alpha}_{\sigma\tau}, \qquad (3)$$

where $$\mathfrak{S}_{\underline{\sigma}\underline{\tau}} = L^{\mu}_{\underline{\sigma}\underline{\tau}} + L^{\sigma}_{\underline{\tau}\underline{\mu}} + L^{\tau}_{\underline{\mu}\underline{\sigma}},$$

which is thus a quadratic form of remarkably simple structure.

That these transformations are possible is based, first, on the fact that the commutation relations here have the character of divergences; and, second, on the simplicity of (3) with respect to the symmetric character of the quadratic term of (2) in the indices μ and α.

(Now, it is only still necessary to see whether applying the other divergences does not lead to something analogous.)[3]

With friendly greetings, yours truly,

A. Einstein.

Criticism and continuation on July 29.

23. To Chaim Herman Müntz

Caputh. Saturday [between June and August 1929][1]

Dear Mr. Müntz,

I have found something that clears up the situation with a single stroke, if it turns out correct. It is, however, something quite strange.

Let $\mathfrak{f}^{\alpha}_{\underline{\mu}\underline{\nu}}$ be a contravariant tensor density that is antisymmetric in all three indices. The divergence commutation theorem[2] then gives:

$$\mathfrak{f}^{\alpha}_{\underline{\mu}\underline{\nu}/\nu/\mu} = -\frac{1}{2}(\mathfrak{f}^{\alpha}_{\underline{\sigma}\underline{\tau}}\Lambda^{\rho}_{\sigma\tau})_{/\rho} \equiv 0$$

For $x_4 = (x_4)_\nu$ (cross-section Θ), let all four $\mathfrak{f}^{\alpha}_{\underline{\mu\nu}}$ be equal to zero. Furthermore, let $\mathfrak{f}^{1}_{\underline{23}}$ and two of the components $\mathfrak{f}^{4}_{\underline{12}}, \mathfrak{f}^{4}_{23}, \mathfrak{f}^{4}_{31}$ be zero everywhere.

Proposition:

Then, the third component also vanishes everywhere.

Proof: We write down the identity for $\alpha = 4$. Then μ and ν are spacelike. Within the cross-section Θ, the first term vanishes completely. Then, carrying out the operations /μ and /ν, only the $\mathfrak{f}$ and their derivatives with respect to x_1, x_2, x_3 occur, as factors in each term. For similar reasons, all the elements of the second term for which $\rho \neq 4$ also vanish. Then, within Θ, second term contributes only

$$-\frac{1}{2}\mathfrak{f}^{4}_{\underline{\sigma\tau},4}\Lambda^{4}_{\sigma\tau},$$

where the sum is taken over the spacelike σ, τ. Within Θ, the expressions

$$\mathfrak{f}^{4}_{\underline{12},4}\Lambda^{4}_{12} + \mathfrak{f}^{4}_{\underline{23},4}\Lambda^{4}_{23} + \mathfrak{f}^{4}_{\underline{31},4}\Lambda^{4}_{31}$$

thus vanish. If $\mathfrak{f}^{4}_{23}$ and $\mathfrak{f}^{4}_{31}$ vanish everywhere ⟨i.e., also within Θ⟩, then the product

$$\mathfrak{f}^{4}_{\underline{12},4}\Lambda^{4}_{12}$$

also vanishes within Θ.

Since we may assume that Λ^{4}_{12} does not vanish over any finite portion of Θ, it thus follows that $\mathfrak{f}^{4}_{\underline{12},4}$ vanishes within Θ, or $\mathfrak{f}^{4}_{\underline{12}}$ itsel f vanishes within the sheet $\Theta^{\lambda}((x_4)_\nu + dx_4)$.

By repeating this chain of reasoning, it follows that $\mathfrak{f}^{4}_{\underline{12}}$ vanishes altogether.

From the above I conclude from that the equations $\mathfrak{S}^{\alpha}_{\underline{\mu\nu}} = 0$ [3] are equivalent to only three independent equations, as are the equations

$$\varphi_i \delta^{iklm} = 0 \text{ [4]}$$

or

$$\varphi_i = 0.$$

For our problem, this means the following:

If I begin with the equations

$$L^{\alpha}_{\underline{\mu\nu}/\nu} - \frac{1}{2}L^{\alpha}_{\underline{\sigma\tau}}\Lambda^{\mu}_{\sigma\tau} = 0 = \mathfrak{G}^{\mu\alpha} \quad (\mathfrak{G}^{\mu\alpha}{}_{/\mu} = 0), \text{[5]}$$

whose antisymmetric equations are given by

$$\mathfrak{S}^{\alpha}_{\mu\nu/\nu} - L^{\ \ \nu}_{\underline{\mu}\underline{\alpha}/\nu} = 0$$

or, respectively,
$$\mathfrak{S}^{\alpha}_{\underline{\mu}\underline{\nu}/\nu} - h(\varphi_{\underline{\mu};\underline{\alpha}} - \varphi_{\underline{\alpha};\underline{\nu}}) = 0 \,,$$

then, without overdetermination, this is compatible with

$$\mathfrak{S}^{\alpha}_{\underline{\mu}\underline{\nu}} = 0 \qquad \varphi_{\mu} = 0 \,.$$

Then these 8 equations are equivalent to only 6 equations, according to the above proof.

Dear Mr. Müntz! All this appears to be rather curious. See if you can find an error in it. If not, then the problem is indeed solved. Show or send this letter also to Mr. Lanczos (Nikolassee, Teutonenstr. 1)[6]

Hearty greetings, your

Potsdamer Str. 35, Kaputh[7]

24. To Chaim Herman Müntz

[Caputh, between June and August 1929][1]

Dear Mr. Müntz,

I have given myself the task of finding which are the simplest manifolds with teleparallelism and, in particular, without taking any account of their physical applicability.[2] I have thus returned to the earlier method, which was based only upon the identities rather than on Hamilton's principle.[3]

I start with the commutation relations[4]

$$\left(\mathfrak{B}^{\sigma}_{\underline{\alpha}\underline{\beta}/\beta} - \frac{1}{2}\mathfrak{B}^{\sigma}_{\underline{\rho}\underline{\tau}}\Lambda^{\alpha}_{\rho\tau}\right)_{/\alpha} \equiv 0 \ldots \qquad (1)$$

or, compactly,
$$\langle \mathfrak{G}^{\sigma\beta} \equiv \rangle \ \mathfrak{G}^{\alpha\sigma}{}_{/\alpha} \equiv 0 \,,$$ [5]

where $\mathfrak{G}^{\alpha\sigma}$ is defined by (1).

One can easily prove that $\mathfrak{G}^{\alpha\sigma}$ becomes symmetric[6] when the following two relations hold:[7]

$$\varphi_{\alpha} = 0 \ldots \qquad (2)$$

$$S^{\alpha}_{\underline{\mu}\underline{\nu}} = 0 \ldots \qquad (3)$$

since, the second term in $\mathfrak{G}^{\alpha\sigma}$ is symmetric because of (2), and the first term becomes symmetric due to (3). In the case that

$$\mathfrak{B}_{\underline{\alpha}\underline{\beta}}^{\sigma} + \mathfrak{B}_{\underline{\sigma}\underline{\alpha}}^{\beta} + \mathfrak{B}_{\underline{\beta}\underline{\sigma}}^{\alpha} = 0 \,,$$

after applying the divergence $/\beta$ and taking $\mathfrak{B}_{\sigma\alpha/\beta}^{\beta} \equiv 0$ into consideration, we namely obtain:

$$\mathfrak{B}_{\underline{\alpha}\underline{\beta}/\beta}^{\sigma} = \mathfrak{B}_{\underline{\sigma}\underline{\beta}/\beta}^{\alpha} \,.$$

However, when combined with equations (2) and (3), $\mathfrak{G}^{\alpha\sigma} = 0\ldots$ (4) are compatible with each other. Equations (2) and (3) indeed each comprise three independent individual equations, while equations (4) include 6 independent equations; that is all together, hence 12 independent conditions, as required.

Whether or not this has a physical significance is for now completely unclear to me. But formally, it is indeed quite remarkable.

All this can also be somewhat generalized. In fact, it is sufficient to make the more restricted assumption that, instead of (2),

$$(\varphi_{\tau}\mathfrak{B}\Lambda_{\underline{\alpha}\underline{\sigma}}^{\tau})_{/\sigma} = 0\ldots, \qquad (2a)$$

as can easily be proven.

What do you think?

Best regards, your

A. Einstein.

25. To Chaim Herman Müntz

[Caputh?, after July 1929][1]

Dear Mr. Müntz,

Our letters crossed in the mail. It is now clear that the equations[2]

$$L_{\underline{\mu}\underline{\nu}/\nu}^{\alpha} - \frac{1}{2}L_{\underline{\sigma}\underline{\tau}}^{\alpha}\Lambda_{\sigma\tau}^{\mu} = 0 \qquad (1)$$

are compatible with the equations

$$S_{\mu\nu}^{\alpha} = 0 \text{ and } \varphi_{\mu} = 0. \qquad (2)$$

Then, according to your considerations, at most 4 of the latter 8 equations are independent of the others.[3]

If all 8 were independent of each other, then we would have 2 equations—namely, $(10+8)$—more than required, compared to only 4 identities. It would be sufficient to reduce those 8 to 6 in order to achieve compatibility, which *my* considerations clearly accomplish.

Now, one might think that the problem would be underdetermined. However, that is impossible, since equations (1) alone already represent 16 individual equations with only 4 identities, so there is already no underdetermination here. Equations (2) thus provide an (admissible) overdetermination together with (1).

In the first approximation, these equations yield everything that we are seeking except for the tensor character of the electromagnetic field, which is perhaps not even present at all. *The all-important question is now whether there is a strict centro-symmetric solution with two constants that we can interpret as charge and mass.*

The lovely aspect of your identity is that it is linear and homogeneous in φ and S.[4] It is admirable that you were able to come up with it.

Best wishes, yours sincerely,

A. Einstein.

26. Greeting to the 16th Zionist Congress

[*Einstein 1929oo*]

DATED 28 July 1929

PUBLISHED 1929

IN: *Protokoll der Verhandlungen des XVI. Zionistenkongresses und der konstituirenden Tagung des Council der Jewish Agency für Palästina, Zürich, 28. Juli bis 14 August 1929.* London: Zentralbureau der Zionistischen Organisation, 1929, p. 28.

[Not selected for translation.]

27. From Eduard Einstein

Zurich, 28 July 1929

Dear Papa,

I'm writing on the day I turn nineteen. On the table, there's a birthday cake with candles, a camera,—which has long been ours, it is true, but which now expressly passes into my possession—then the fountain pen with which I'm writing here, stuff for tennis, ties, books, and the usual accessories. On such a day, one can probably become a little pensive.

I must now ask you to be lenient if I don't write very often. Right now, so much is going on inside; I mean in my brain. When I look back on the last few years, I see that I sank, stage by stage, to an ever-higher consciousness. I can observe this step by step in my essays, verses, and memories,[1] in the very consistent series of books that I loved. Who could persuade me that these processes lasted only four, five, at most six years? It's astonishing, how many worldviews I had in that time.[2] That is sad for the mind; it's really not long enough for life; should we not specialize it, harness it, it exhausts all greater possibilities in a few years. I have a friend who even went through the whole range in a single year.

My own case was extreme: without religion, without fatherland, without race, often alone during my childhood, often sick in bed next to the bookshelves:[3] so it happened that the mental monologuing that others toil for a long time to make their second nature came naturally to me from the outset. At the age of seven or eight, I had a quantity of mental contents with an immediacy that I have never experienced since; it was a real shock for me when I began going to school, so much outside world. Adaptation, dismantling; with a certain effort, I grew into it.[4] Since then, the conflict between mind and life has continued throughout my whole life, waxing and waning. By a stroke of good luck in high school, I was in a class with interesting fellow students, friends with whom I could associate during the vacations; letters went back and forth.[5] I still regret not having been a Scout, even more than before. And also, that I can't serve in the military because of an ear disorder.[6]

Now I'm often lost in thought, but under certain circumstances, I'm in the mood for tea, cigarettes, and company in convivial contact in the evening.

28. To Eduard Einstein

Caputh, 30 July [1929]

Dear Tetel,

I'm coming to Zurich on the 9th. If Albert[1] has already left and you are at home, I'll stay with you. I have to go there to attend to Zionist business.[2] Since it will be much too turbulent there for my frail condition, I will stay only a few days. It would be great if you came with me. I'm staying alone with Else in Caputh (Potsdamerstr. 35,) in a house on a lake. My splendid new boat is now ready, and I sail every day. It's wonderful for relaxing. I think that, given the prevailing cool weather, it would also be better for you than the mountains.

Your letter depresses me a little, because it shows so little spirit of enterprise and joy in life.[3] How beneficial an occupation would be for you! Even a genius like Schopenhauer[4] would be demoralized by the lack of an occupation. He didn't know that, but it was true. Besides, these days, insuring life by money is a utopia, and that is good. One must have the feeling that, through honest work, one is giving back to others what one has received. Otherwise, remorse will never fail to appear. It will be good when you are with me for a longer time. You'll see.

See you soon (I hope).

Best wishes to you both, your

Papa

29. To Anatoly Lunacharsky[1]

Albert Einstein. Confidential. Copy.
Berlin, V. 31/VII. 29

Dear Mr. Lunatscharsky,

I take the liberty of writing to you regarding an important matter. Some time ago, the distinguished and respected Russian scholar Vladimir Beneshevich from Leningrad was sentenced to deportation, a punishment which, according to the judgment of local scholars who know the circumstances better, the person concerned would hardly be able to survive.[2] The local Academy of Sciences, which values Mr. Beneshevich for purely academic reasons, has already tried to have him pardoned without success.[3] So, I am trying once again, here, in an unofficial way.

Mr. Beneshevich has come into his misfortune as a result of the fact that he procured scholarly material through relations with a prelate of the Vatican.[4] With these, however, he only pursued purely academic purposes. Not only from the standpoint of pure scholarship, but also from that of prestige and sympathy toward the Soviet government, a pardoning of the scholar would be highly gratifying. My request is that you would personally support this pardon.

With warm regards and best wishes, your

A. Einstein

30. To Paul Levi[1]

Berlin, 8 August 1929

Dear Paul Levi,

You have been granted the greatest boon: You have enchanted even the flippant Berliner Tageblatt. See the morning edition for 8 August.[2] But jokes aside. It's uplifting to see how you, through love of justice and sheer acumen as an independent individual, have boldly cleared the air, like a marvelous pendant to Zola.[3] In the finest of us Jews, something of the social justice of the Old Testament lives on.

Best regards and good wishes for the holidays, your

A. Einstein

P.S. I have heard nothing more about the German-Belgian matter.[4]

31. To Elsa Einstein

Sunday [Zurich, 11 August 1929]

Dear Else,

So far, everything has gone well. Tetel[1] and Mileva are in good health. However, they are not coming to Berlin in the fall, but going instead to Florence, Tete visiting Maja. That was my idea.[2] Yesterday morning was quiet. I wasn't able to reach Zangger.[3] In the afternoon, I visited Grossmann, whom I found in a deplorable state (multiple sclerosis).[4] He had a crying fit. It was terrible. Then, I went to Weizmann's home, where there was a little conversation. He must be suffering a great deal, apparently from his kidneys, not from sepsis. A doctor in Paris seems to have given the right diagnosis. After losing a lot of weight, he has put on a few pounds again.[5] In the evening, I went to see the Karrs,[6] where it was very pleasant, as if there were no horror. Despite all this moving about, I slept well. The sessions begin today at noon.[7] I am actually glad about that. I want to avoid the socializing. Husmann and little Micha[8] have already gotten in touch; I had written to the latter. This morning, I am staying home, as I will in the evening. Here, there is immense wealth; everything is swimming in fat and pomade. But I look forward to Caputh.

Greetings and kisses, your

Albert

32. Address at the Constituent Meeting of the Jewish Agency Council

"Die konstituierende Tagung der Jewish Agency für Palästina"
[*Einstein 1929pp*]

DATED 11 August 1929

PUBLISHED 1930

IN: *The Jewish Agency for Palestine: Constituent Meeting of the Council held at Zurich, August 11th–14th, 1929. Official Report.* London: The Jewish Agency for Palestine, 1930, p. 15.

PROF. EINSTEIN'S SPEECH.

Professor Albert Einstein: We all feel that this is a great day that we are living through, and I wish to say nothing more than just wherein it is that I see the greatness of this day. [1] I see the tragedy of the modern Jew in the fact that though he represents a nation, it is a nation decomposed into atoms. [2] The individual Jew is isolated, and suffering from the misery of isolation. That misery has turned into a tragic situation. But how should a remedy be created, save by the establishment of one's own home? He who saw it clearly, with absolute clearness, was Herzl. (*Tumultuous, ever-renewed applause.*) What he saw was so simple, and still we none of us understood him; [3] for only a small band of men gathered round him. He saw that a common work was only possible if the people could itself create that community which was able to break down the individual's isolation. He also saw with the sure instinct of political genius that that work could be nothing else but the upbuilding of Palestine. Nearly all of us failed to see it. Nearly all Jews opposed him, from fear of losing or jeopardising their modest position in the countries of their domicile. Herzl knew that it was not so, and with a small band of faithful ones he began the difficult work, which at first seemed impossible from a political point of view. But in these days we see that he succeeded. We see that all Jews have realised that he brought the redemption. And let us recognize it with a full heart, and do our homage to him. (*Tumultuous applause.*) Equally let us recognize that our great leader Weizmann [4] was destined to overcome a great part of the difficulties that towered in the path indicated by Herzl. [5] Nor must we forget that the upbuilding of Palestine signifies not only a great task for all Jews, but that this task also means a great gift to us. We owe it not only to the two great leaders, but also to the brave and enthusiastic minority who call themselves Zionists. We others must remember the fact that we owe to those men our national solidarity, and I believe that also we should not forget that they have a moral right, in future, too, to exercise a most important influence on the labours which we desire to perform. (*Tumultuous and enthusiastic applause.*)

33. Statement on the Jewish Agency

[Zurich,] 11 August 1929

On this day, the seed of Herzl[1] and Weizmann has matured in a wonderful way.[2] None of those present remain unmoved.

A. Einstein[3]

34. From Max Born

Göttingen, 12 August 1929

Dear Einstein,

Some time ago, there appeared here a young Russian who brought along a six-dimensional theory of relativity. Since I had already experienced feelings of fear at the various five-dimensional theories and had little confidence that anything beautiful could come of these approaches, I was *very* skeptical.[1] But the man spoke intelligently and soon convinced me that there is something to his thoughts.

Although I understand less than ε of these matters, I presented his work to the Göttingen Academy, and I am sending to you, enclosed here, a preprint with the *urgent* request that you read and evaluate it.[2] The man, whose name is Rumer,[3] left Russia because relativity theorists are badly treated there (seriously!); the theory of relativity is held to contradict the official "materialistic" philosophy, and those who support it are persecuted. Joffe[4] had already told me that earlier. Mr. Rumer came to Germany and somehow got a chance to study at a technical school in Oldenburg, and now he wants to take his final technical examinations. He will then try to establish himself here or, failing that, emigrate to South America.

If you have a *good* impression of his work, I would ask you to do something to help this man. He is quite thoroughly acquainted with the mathematical literature on Riemannian geometry etc. up to modern times, and would likely be an ideal assistant for you. He is personally very pleasant and makes an extremely intelligent impression. Whether he is really a Russian or a Jew, I do not know; the latter is more probable. His address is: Georg Rumer, Oldenburg, Am Festungsgraben 8.

I am still not very well. I spent eight days with my children in Waldeck, but it was very loud and busy there. My nerves are in a bad state. Next week, I will travel alone to Lake Lucerne, where an acquaintance of mine (a Swiss lawyer) has a small house and a motorboat in Kehrsiten-Bürgenstock. (My address there: Hotel Schiller). I saw you in the most recent Illustrierten in a sailboat, evidently well tanned.[5]

Hedi[6] has bowel problems and is staying on a strict diet. With warm wishes to your wife and Margot,[7] yours truly,

Max Born.

Rumer wants to write a book, and I am enclosing the planned table of contents.[8]

35. From Georg Rumer

Göttingen 14 August 1929

[Not selected for translation.]

36. To Chaim Weizmann

Berlin, 15 August 1929

Dear Mr. Weizmann,

It gives me particular pleasure to send a belated farewell in writing, which I probably could not have delivered to the overworked master in person. My joy over the success of your ardent efforts over the years was great,[1] and so was my admiration for the way in which you proceeded and brought together almost all those who were recalcitrant. My best wishes for the restoration of your health as well,[2] which is already so auspiciously returning and is so important for your great work.

One more thing: The good F. Warburg ought not to be pestered because of Magnes.[3] Over there, an awareness of the necessity of more knowledgeable and better leadership will gradually emerge all by itself. In the meantime, we must be patient, even if it is also painful. We can do that without making sordid concessions.

Heartfelt greetings with best wishes, your

A. Einstein

37. To Maja Winteler-Einstein

[Caputh, 19 August 1929][1]

Dear Maja,

Here, I sit in my hermitage with the wonderful sailboat, without telephone, far enough from Berlin. This peace and quiet has already helped me make a scientific discovery that finally rounds out what I've been working on for a year.[2] Tetel is a very nice fellow, sensitive and good-hearted. You are certain to enjoy him.[3] I have only reluctantly given him up for the fall vacation, but I have the feeling that you two must become acquainted and will learn a lot from each other. I myself can't come, because I couldn't remain hidden, and uncomfortable situations would be inevitable. Something important for the Palestine project happened in Zurich.[4] The world's main Jewish institutions have officially joined the project, which, up to now, they had not had the courage to do for political reasons. This was a major step forward, and no eye remained dry.

Life in the countryside suits me wonderfully well (even if we are half-broke because of the expensive house, thanks to the city of Berlin).[5] An excess of impressions just pesters and numbs. My heart is not bothering me anymore, so that I almost feel as I used to.[6] There is also something to be said for the late semesters and their tranquility.

Best wishes from your

Albert

38. On the Jewish Agency

"Die Gründung der Jewish Agency"
[*Einstein 1929qq*]

DATED 19 August 1929[1]
PUBLISHED 24 August 1929

IN: *Das Tagebuch* 10, no. 34 (24 August 1929): 1389.

The material and moral difficulty of Jewish existence consists in the fact that Jews represent an ethnic minority that is not, or not sufficiently, consolidated.[2] This situation brings with it not only an exacerbation of the individual's economic struggle, but also a peculiar psychological endangerment. Abandoned to the influence of the majority and their values, the Jew so often loses the natural *assurance of the feeling of being alive* and *being valued for his ethnicity*, his tradition, and his ancestors, without which harmonious personalities cannot develop.[3] The recognition of this plight allowed the Viennese writer Theodor Herzl[4] to bring Zionism into being. The colonization of Palestine through the cooperation of all Jews is supposed to give new life to Jews' feeling of community and national self-awareness. Subsequent development has proven that Herzl's psychological acumen marked out the right path.[5]

At first, only a small but devoted minority responded to Herzl's call, whereas the rest of the Jews opposed the enterprise; some mockingly, some hostile, and some with indifference. But the younger generation became increasingly involved in the project, which was also helped along from outside by England's cooperation.[6]

But until recently, there was strong opposition among Jews to the Zionists, which was nourished by the latter's occasionally *excessive emphasis on their nationalist convictions*.[7]

Yet the success of the colonization project, combined with the acknowledgment of its beneficial moral effects, caused the resistance gradually to wane, especially after the project found active moral support on the non-Jewish side as well. We owe it to these circumstances and to the exceptionally intelligent, patient, and energetic leadership of the president of the Zionist World Organization, Prof. Chaim Weizmann,[8] that, in recent years, almost all the world's Jewish organizations have *united in active support for the colonization project*.

That is the significance of the expanded Jewish Agency that was just formed in Zurich.

39. To Ulrich von Beckerath[1]

[Caputh,] 19 August 1929

Dear Sir,

The path taken by the "Supranational Republic"[2] seems to me not the right one. The state does not exist solely to wage wars, but designates a local organization for solving highly important economic and cultural tasks. Thus, if it were really possible to separate the majority of the inhabitants of a region from the state, that would mean either an annihilation of higher organizing values, or else a complete disenfranchisement of the majority. It looks to me, more or less, as if you are recommending universal castration as a means of fighting hereditary diseases.

Best regards,

40. To Otto Blumenthal[1]

[Caputh,] 19 August 1929

Dear Colleague,

I would like to publish an already completed review article on the mathematical apparatus of general field theory in the Mathematische Annalen (about fifteen printed pages).[2] There is, however, a certain difficulty: Prof. Cartan from the Sorbonne, whose earlier works have many points in common with my investigations,[3] has written by my request an article in French (around twelve pages long) on the prehistory of the problem,[4] which is to be published immediately after my article. I would thus ask you to answer the following two questions:

(1) Can Mr. Cartan's article be published in the Annalen in French ? (2) How long would it take before the articles could appear?

With friendly greetings and best wishes for the vacation period, your

41. From Eduard Einstein

[Zurich, between 20 August and 11 September 1929][1]

Dear Papa,

Is your theory already finished?[2] My graduation exam isn't.[3] The happiest surprise was that physics isn't part of it. Yesterday, we had a discussion about "the natural-scientific worldview." Edgar Meyer's son[4] gave a masterful speech. He spoke fluently about difficult questions bordering on epistemology. It was particularly amusing when he described the worldview that would emerge if the eyes were accidentally adjusted for slightly different vibrations. A fountain pen, he said, would appear transparent. Toward the end, he gave his natural-scientific outpouring free rein and presented the usual evolutionary view focused on cognition and to which all natural scientists subscribe, so long as they're among fellow natural scientists: how living beings start from protozoa and clamber up to ever-higher cognitive functions, and, who knows, maybe things will continue in this way toward an increasingly complete understanding of the world. Then, it was fun to see how, in the discussion, he had to eat humble pie: The drive for knowledge is only one among a hundred other drives. The level of the discussion was almost always high, and there were even arguments about the origin of the basic rules of logic. Fritz Wolgensinger[5] led the debate from the high pedestal: rejected knowledge; life is entirely subjective, knowledge is objective, he said with such fervor that the improbable occurred: I rose and struck a blow for the natural sciences.

How are you? Will your villa soon be finished?[6] You will surely entertain yourself with your theory. Theories are basically healthier as an occupation than subjective creations; somehow, one always runs into a bulwark and must not, like the artist, hearken inward, "eyes inward," Gottfried Benn says;[7] à propos, have you already written on behalf of the poor man?

I have forty-three hours of medicine per week, and, for the exam, I have to dissect a body cavity.[8]

You can learn more about me from the letter to Frau Mendel.[9] I couldn't properly address it to you because I probably already told you one or two things last spring.[10]

Best regards from your

Teddy

42. From Chaim Herman Müntz

[Berlin,] 23 August 1929

Dear and honored Professor,

To add to my other worries, when I intended to go out to visit you on Wednesday, I again clumsily sprained my leg; so I am at least writing to you in answer to your kind recent communication from my bed, where I will presumably have to stay for several days more; to be sure, my answer comes rather late.

Have I now understood you correctly?

You obtain the identity:

$$2\overline{G}^{\mu\alpha} \equiv A^{\mu\alpha} - \delta^{\nu}_{\mu\underline{\alpha},\nu} - \delta^{\sigma}_{\mu\underline{\alpha}}\Lambda^{\nu}_{\sigma\nu}; \tag{2}$$

then $\delta^{\nu}_{\mu\underline{\alpha},\nu}$ should be a tensor?

But we also have:

$$(2^*) \qquad \delta^{\nu}_{\mu\underline{\alpha};\nu} = \delta^{\nu}_{\underline{\alpha}\mu,\nu} + \Delta^{\mu}_{\sigma\nu}\delta^{\nu}_{\sigma\underline{\alpha}} + \Delta^{\alpha}_{\sigma\nu}\delta^{\nu}_{\mu\underline{\sigma}} + \Delta^{\nu}_{\sigma\nu}\delta^{\sigma}_{\mu\underline{\alpha}};$$

$$\Delta^{\mu}_{\sigma\nu}\delta^{\nu}_{\sigma\underline{\alpha}} = \Delta^{\mu}_{\sigma\nu}\Lambda^{\nu}_{\sigma\underline{\alpha}} + \Delta^{\mu}_{\sigma\nu}\Lambda^{\sigma}_{\alpha\underline{\nu}} + \Delta^{\mu}_{\sigma\nu}\Lambda^{\alpha}_{\nu\underline{\sigma}} =$$

$$= \Delta^{\mu}_{\sigma\nu}\Lambda^{\nu}_{\sigma\underline{\alpha}} + \Delta^{\mu}_{\nu\sigma}\Lambda^{\nu}_{\alpha\underline{\sigma}} + \frac{1}{2}\Delta^{\mu}_{\sigma\nu}\Lambda^{\alpha}_{\nu\underline{\sigma}} + \frac{1}{2}\Delta^{\mu}_{\nu\sigma}\Lambda^{\alpha}_{\sigma\underline{\nu}} =$$

$$= \Lambda^{\mu}_{\sigma\nu}\Lambda^{\nu}_{\sigma\underline{\alpha}} + \frac{1}{2}\Lambda^{\mu}_{\sigma\nu}\Lambda^{\alpha}_{\nu\underline{\sigma}}; \text{ similarly, } \Delta^{\alpha}_{\sigma\nu}\delta^{\nu}_{\mu\underline{\sigma}} \equiv \text{tensor} \;.$$

It should therefore follow from (2) and (2*) that $\Delta^{\nu}_{\sigma\nu}\delta^{\sigma}_{\mu\underline{\alpha}}$ should be a tensor, which, however, is not really possible.

Would you be so kind (and not angry) as to let me know briefly, before I continue working (by telephone if convenient?), whether you have no doubts about identity (2)?

$$\text{NB: } 2\overline{G}^{\mu\alpha} = G^{\mu\alpha} - G^{\alpha\mu};$$

$$G^{\mu\alpha} = \Lambda^{\alpha}_{\mu\underline{\nu};\nu} - \Lambda^{\sigma}_{\mu\underline{\tau}}\Lambda^{\alpha}_{\sigma\tau}; \; A^{\mu\nu} = \Lambda^{\sigma}_{\mu\underline{\nu};\sigma}.$$

We would be sincerely happy to hear that your dear wife's health has improved; we would really like to take advantage of your friendly invitation to go on a sailing party with you.[1]

In the meantime, heartiest greetings from our house to your house!

Yours faithfully,

Müntz

43. To Élie Cartan

Berlin, 25 August 1929

Dear Mr. Cartan,

Please forgive my long silence.[1] This has been caused by many doubts as to the correctness of the course I have adopted. But now, I have come to the point that I am persuaded I have found the simplest legitimate characterization of a Riemann metric with distant parallelism that can occur in physics.[2] I am now writing up the work for the Mathematische Annalen and would like to be permitted to add yours to mine, as we have agreed.[3] The publication should appear in the Mathematische Annalen because,[4] for the present, only the mathematical implications are explored, not their application to physics.[5]

Kind regards, your

A. Einstein.

Translators' note: Translation by Jules Leroy and Jim Ritter in *Elie Cartan—Albert Einstein: Letters on Absolute Parallelism, 1929–1932*, ed. Robert Debever (Princeton: Princeton University Press, 1979), p. 19.

44. To Chaim Herman Müntz

Sunday, [25 August 1929][1]

Dear Mr. Müntz,

There was evidently an error in my letter.[2] The identity is given by

$$2\overline{G}^{\mu\alpha} \equiv A^{\mu\alpha} - S^{\nu}_{\underline{\mu}\underline{\alpha},\,\nu} - S^{\sigma}_{\underline{\mu}\underline{\alpha}}\Delta^{\ \nu}_{\sigma\ \nu}.$$

The last term contains Δ, not Λ. The two last terms together have tensor character:

$$\Delta^{\nu}_{\sigma\nu} \equiv \Lambda^{\nu}_{\sigma\nu} + \frac{\partial \lg h}{\partial x^4}$$

$$S^{\nu}_{\underline{\mu}\underline{\alpha},\,\nu} + S^{\sigma}_{\mu\alpha}\Delta^{\nu}_{\sigma\nu} \equiv \left(S^{\nu}_{\underline{\mu}\underline{\alpha},\nu} + \frac{\partial \lg h}{\partial x^{\sigma}} S^{\sigma}_{\underline{\mu}\underline{\alpha}}\right) + \varphi_{\sigma} S^{\sigma}_{\underline{\mu}\underline{\alpha}}$$

$$\equiv \frac{1}{h}\left(\frac{\partial (h S^{\nu}_{\underline{\mu}\underline{\alpha}})}{\partial x^{\nu}}\right) + \varphi_{\sigma} S^{\sigma}_{\underline{\mu}\underline{\nu}}.$$

Both terms on the right-hand side have tensor character.

I am very sorry that you are once again ill, and so shortly before your departure.[3] I hope that we can go sailing once again with your wife.[4] Write to me beforehand, so that I will be there for sure.

The theory is completely in order. I have checked everything several times and also found a direct proof for the compatibility of the equations.

The field equations:

$$\Lambda^{\alpha}_{\underline{\mu}\underline{\nu};\nu} - \Lambda^{\sigma}_{\underline{\mu}\underline{\tau}}\,\Lambda^{\alpha}_{\sigma\tau} = 0$$

$$\Lambda^{\nu}_{\underline{\mu}\underline{\alpha};\nu} = 0\,.$$

The third system indeed says nothing new. If twelve of these equations are fulfilled everywhere and *all* of them are fulfilled on a section $x^4 = \xi$, then they are all fulfilled everywhere.[5] I am very happy that I have now been able to overcome the difficulties after all; my confidence in the uniqueness of these equations, with their far-reaching overdetermination, is complete. If they don't appear to be valid in their consequences, then I will no longer believe in teleparallelism.

Best regards, your

A. E.

45. From Robert A. Millikan[1]

Pasadena, 26 August 1929

[See documentary edition for English text.]

46. To Georg Rumer

[Caputh], 30 August 1929

Dear Colleague,

Unfortunately, neither a research assistant nor the means to pay for an assistant, are at my disposal, so I cannot consider working with you at the present time.[1] For Mr. Lanczos, we had obtained a one-time fellowship from the Notgemeinschaft deutscher Wissenschaft.[2]

As to what concerns the problem that is of mutual interest to us: I have stopped working with the multi-dimensional view altogether.[3] Considering the five-dimensional interpretation of Kaluza, apart from the cylinder condition, mainly the non-interpretability of the potential component disturbs me.[4] In contrast, the theory of teleparallelism in four dimensions, in light of my most recent, not yet published results, does appear to me to be extremely promising.[5]

I am very sorry that at the moment I cannot be of direct assistance to your career. But I am quite willing to recommend you, whenever you wish to apply for any eligible position.

Kind regards, your

47. Message on the Recent Violence in Palestine for Zionist Solidarity Rally

"Einsteins Brief"
[*Einstein 1929rr*]

DATED 31 August 1929
PUBLISHED 3 September 1929

IN: *Jüdische Rundschau*, 3 September 1929, p. 447.

Berlin, 31 August 1929

To the Chairman of the Jewish Rally

Dear Mr. Chairman,

To my regret, I am unable to participate in person in today's rally,[1] but I ask you to convey to the assembly the expression of my sympathy.

Shocked by the tragic catastrophe in Palestine,[2] Jews must now show that they are really up to the great task that they have undertaken.

First of all, I declare the obvious fact that such setbacks will not cause our devotion to the great project, nor our determination to continue the peaceful work of construction, to waver in the slightest.

But what must happen in order to prevent the repetition of such sad events in the future?

The first and most important thing is the *creation of a way of living with the Arab people, in which the—perhaps inevitable—frictions are overcome by organized cooperation*, so that the causes of conflict cannot build up in a menacing fashion. As a result of the lack of natural contact in everyday life, there arises an atmosphere of mutual fear and distrust that fosters such deplorable venting of passion.[3] Above all, we Jews must show that we have gained so much understanding and so much psychological experience through the difficult times in our past that we are capable of coping with this psychological and organizational problem, especially since no insurmountable contradictions prevent Jews and Arabs from living together in Palestine. Therefore, above all, let us be wary of blinding chauvinism of every kind and not believe that reason and understanding can be replaced by English police bayonets.

However, we must absolutely demand one thing from the mandatory government to which the well-being of the country is entrusted: Institutions for the security of peaceful working men must be found, institutions that do justice to the scattered distribution of Jewish settlements and are, at the same time, capable of acting to bridge national contradictions. Hence, it goes without saying that a corresponding participation of Jews in the security force must be secured.[4] The mandatory government cannot be spared the reproach that it has failed to fulfill this obligation in sufficient measure, setting aside the fact that the authorities were mistaken regarding the real situation in the country.

The greatest danger in the present situation is that blind chauvinism has been able to gain ground in our ranks.[5] Even as we are responsible for the defense of our lives and property, we must never forget for a minute that our national task is supranational at its core, and that the power of the whole movement is based on its ethical value, on which it stands or falls.[6]

May all friends of the Jewish Palestine project be steadfast in these convictions and, in this spirit, multiply their efforts and lend their best strengths to the service of the great and holy cause. May the present meetings of the Jewish community in Berlin also contribute to the realization of this goal.

Respectfully yours,

A. Einstein

48. To Sir Horace Rumbold[1]

Berlin, 31 August 1929

[See documentary edition for English text.]

49. Calculations

[Berlin, after 4 September 1930][1]

[Not selected for translation.]

50. To Konrad Baetz[1]

[Berlin, after 4 September 1930][2]

[Not selected for translation.]

51. To Marie Curie-Skłodowska[1]

Caputh, 6 September 1929

Dear Mrs. Curie,

I wholeheartedly wish you the best of luck for your trip to Dollaria.[2] Hopefully, we will nevertheless see each other?

I would be happy to contribute to the reorganization of the Commission, if possible.[3] The only possibility would be for me to negotiate privately with a leading personality from the League of Nations. It seems to me that the Commission would be able to work successfully only if it is removed from the political arena. This would only be the case should it be possible that (1) only very reliable people are selected as members in the commission and (2) the choice of new members of the Commission is carried out by the Commission itself, and the council of the League of Nations would simply confirm them.[4] If you agree on this, we could try to launch an unofficial effort in that direction with Langevin, Mrs. Bonnevie, and some suitable English person (Russell or Eddington?)[5] through a leading personality of the League of Nations. What do you think of that?

In the matter of Mr. Rosenblum,[6] I have already undertaken some steps and will give you a report as soon as possible.

He, and his work, made an excellent impression on me as well.

With kind regards, your

52. To Liga gegen Imperialismus[1]

Caputh, 6 September 1929

Your resolution opposing Jewish construction work in Palestine makes it impossible for me to continue to associate in any way with the League against Imperialism.[2] I therefore resign my position as the honorary president of the League and ask you to communicate this to the central office, seeing to it that (1) my resignation and the reason for it are communicated to the members of the League, (2) and—in addition—that my name will no longer appear on any publications, circulars, etc., whether internal or external.

Respectfully yours,

53. To Weltliga für Sexualreform[1]

[Caputh,] 6 September 1929

I have insufficient human experience at my disposal to justify publicly expressing my opinion on these difficult social questions. I feel a degree of certainty only on the following points:

Abortion, up to a certain stage in pregnancy, should be permitted at the wish of the woman concerned.[2] Homosexuality should be exempt from punishment, except for the necessary protection of the young.[3] Regarding sex education: no secretmongering.[4]

Respectfully yours,

54. Fragment of draft of Doc. 55

[before 13 September 1929][1]

[Not selected for translation.]

55. "Unified Field Theory Based on Riemannian Metric and Distant Parallelism"

"Auf die Riemann-Metrik und den Fernparallelismus gegründete einheitliche Feldtheorie"
[*Einstein 1930m*]

COMPLETED 13 September 1929
RECEIVED [19 August 1929]
PUBLISHED after 20 February 1930

IN: *Mathematische Annalen* 102 (1930): 685–697.

In the following article, I wish to describe the theory that I have been developing for the past year in such a way that it can be readily understood by anyone who is familiar with the general theory of relativity. This presentation is needed because reading the previously-published works on this subject[1] would entail an unnecessary expenditure of time due to more recently discovered connections and improvements not contained in those works. The subject is presented here in such a way as to make it possible for the reader to comprehend it most expediently. In particular, through Mr. Weitzenböck and Mr. Cartan,[2] I have learned that the mathematical treatment of the continua of the relevant type is, in itself, not new. Mr. Cartan very courteously wrote an article on the history of the mathematical object used here, which provides a complement to the content of my description; his article is published in this journal, immediately following this one.[3] I would like to express my sincere thanks to him for his valuable contribution. The most important aspect of the present treatment that is, in any case, new is the discovery of the simplest field laws compatible with a Riemannian manifold with teleparallelism. I only a briefly discuss the physical significance of this theory.[4]

§1. The Structure of the Continuum

Since the dimensionality plays no role in the considerations that follow, we can assume an n-dimensional continuum. In order to conform to the fact of the metric and of gravitation, we assume the existence of a Riemannian metric. In nature, however, electromagnetic fields also exist, and they cannot be represented by a Riemannian metric. We are thus confronted by the question of how we can ascribe an additional structure to our Riemannian space in such a manner that the whole has a unified character.

In the neighborhood of every point P, the continuum is (pseudo) Euclidean. At every such point, there is a local Cartesian coordinate system (i.e., an orthogonal n-Bein) for which the Pythagorean theorem holds. The orientation of these n-Beine plays no role in a Riemannian manifold. Now, we shall assume that there is a directional relationship between these elementary local Euclidean spaces. We also assume, owing to the spatial structure of Euclidean geometry, that it is meaningful to speak of a parallel orientation of all the local n-Beine (which would be meaningless in a space with *only* a metric structure). In the following, we will always imagine the orthogonal n-Beine to be in a parallel orientation. The orientation of a local n-Bein at *one* point P, which is intrinsically arbitrary, then unequivocally determines the orientation of all the local n-Beine at every point in the continuum. Our task now is, in the first instance, to describe such a continuum mathematically and then establish the simplest limiting laws which govern such a continuum. We do this in the hope that we can thus derive the general laws of nature, just as the earlier theory of general relativity attempted to do for gravitation, under the assumption of a merely metric spatial structure.

§2. Mathematical Description of the Spatial Structure

The local n-Bein consists of n mutually perpendicular unit vectors, whose components with respect to an arbitrary Gaussian coordinate system are denoted as $h_s^{\ \nu}$. Here, as always, a lower Roman index denumerates the individual unit vectors of the local n-Bein, while a Greek index, depending on whether it is raised or lowered, expresses the contravariant or covariant transformation character of the quantity with respect to a variation in the Gaussian coordinate system.[5] The general transformation properties of the $h_s^{\ \nu}$ are as follows: If all the local systems (the n-Beine) are rotated in the same manner, which is allowed, and if, at the same time, a new Gaussian coordinate system is introduced, then, between the new and the old $h_s^{\ \nu}$, the following transformation law holds:

$$h_s^{\ \nu'} = \alpha_{st}\frac{\partial x^{\nu'}}{\partial x^{\alpha}}h_t^{\ \alpha}, \tag{1}$$

where the constant coefficients α_{st} themselves form an orthogonal system:

$$\alpha_{sa}\alpha_{sb} = \alpha_{as}\alpha_{bs} = \delta_{ab} = \begin{cases} 1 \text{ if } a = b, \\ 0 \text{ if } a \neq b. \end{cases} \tag{2}$$

The transformation law (1) can be readily generalized to objects whose components have arbitrarily many local and coordinate indices. Such objects are called tensors. The algebraic laws for tensors (addition, multiplication, contraction with respect to Roman and Greek indices) then result immediately from these definitions. The $h_s^{\ \nu}$ are called the components of the fundamental tensor. If a vector in the local system has the components A_s or, with respect to the Gaussian system, the coordinates A^ν, then, from the definition of the $h_s^{\ \nu}$, we have:

$$A^\nu = h_s^{\ \nu} A_s, \tag{3}$$

or—resolved in terms of the A_s—

$$A_s = h_{s\nu} A^\nu. \tag{4}$$

The tensor character of the normalized subdeterminants $h_{s\nu}$ of the $h_s^{\ \nu}$ follows from (4). The $h_{s\nu}$ are the covariant components of the fundamental tensor. Between the $h_{s\nu}$ and the $h_s^{\ \nu}$, the following relations hold:

$$h_{s\mu} h_s^{\ \nu} = \delta_\mu^{\ \nu} \begin{cases} = 1 \text{ if } \mu = \nu \\ = 0 \text{ if } \mu \neq \nu \end{cases} \tag{5}$$

$$h_{s\mu} h_t^{\mu} = \delta_{st} \ .\text{[6]} \tag{6}$$

Due to the orthogonality of the local system, we find for the magnitude of a vector

$$A^2 = A_s^2 = h_{s\mu} h_{s\nu} A^\mu A^\nu = g_{\mu\nu} A^\mu A^\nu; \tag{6}$$

thus we have for the coefficients of the metric:

$$g_{\mu\nu} = h_{s\mu} h_{s\nu}. \tag{7}$$

The fundamental tensor permits us (cf. (3) and (4)) to change local indices into coordinate indices and vice versa (by multiplication and contraction), so that it is only a formal question as to which type of indices one will use on the tensors with which one operates. It is clear that the following relations also hold:

$$A_\nu = h_{s\nu} A_s, \tag{3a}$$

$$A_s = h_s^{\nu} A_\nu .\text{[7]} \tag{4a}$$

Furthermore, the determinant relation also holds:

$$g = |g_{\sigma\tau}| = |h_{a\sigma}|^2 = h^2, \tag{8}$$

so that the invariants of the volume element $\sqrt{g}d\tau$ take on the form $\sqrt{h}d\tau$.

The particular character of the time is most conveniently taken into account in our four-dimensional spacetime continuum by taking the x^4 coordinate (both locally and generally) to be purely imaginary and, likewise, all tensor components with an odd number of the fourth indices.[8]

§3. Differential Relations

Let us denote the increase in the components of a vector or a tensor on undergoing a "parallel shift" in the sense of Levi-Civita[9] on making the transition to an infinitesimally neighboring point in the continuum with δ; then, from the above considerations,

$$0 = \delta A_s = \delta(h_{s\alpha}A^\alpha) = \delta(h_s^\alpha A_\alpha). \tag{9}$$

Resolving the parentheses leads to

$$h_{s\alpha}\delta A^\alpha + A^\alpha h_{s\alpha,\,\beta}\delta x^\beta = 0,$$
$$h_s^\alpha \delta A_\alpha + A_\alpha h_{s,\,\beta}^\alpha \delta x^\beta = 0,$$

where the comma in the second term denotes ordinary differentiation with respect to x^β. On solving these equations, one obtains

$$\delta A^\sigma = -A^\alpha \Delta_{\alpha\beta}^\sigma \delta x^\beta, \tag{10}$$

$$\delta A_\sigma = A_\alpha \Delta_{\sigma\beta}^\alpha \delta x^\beta, \tag{11}$$

where we have set

$$\Delta_{\alpha\beta}^\sigma = h_s^\sigma h_{s\alpha,\,\beta} = -h_{s\alpha} h_{s,\,\beta}^\sigma. \tag{12}$$

(This last relation is based on (5).)[10]

This law of parallel displacement is, in contrast to Riemannian geometry, not symmetric in general. If it is symmetric, then we are dealing with Euclidean geometry; for in that case, one has

$$\Delta_{\alpha\beta}^\sigma = -\Delta_{\beta\alpha}^\sigma = 0$$

or

$$h_{s\alpha,\,\beta} - h_{s\beta,\,\alpha} = 0.$$

Then, however, we would have

$$h_{s\alpha} = \frac{\partial \psi_s}{\partial x_\alpha}.$$ [11]

If we choose the ψ_s to be new variables x_s', then we have

$$h_{s\alpha} = \delta_{s\alpha}, \tag{13}$$

which proves the conjecture.[12]

Covariant differentiation. The local components A_s of a vector are invariant with respect to an arbitrary coordinate transformation. From this, the tensor character of the derivative

$$A_{s,\,\alpha} \tag{14}$$

follows immediately.
If we replace this quantity by

$$(h_s{}^{\sigma} A_{\alpha})_{,\alpha},$$

making use of (4a), then we can recognize the tensor character of

$$h_s{}^{\sigma} A_{\sigma,\,\alpha} + A_{\sigma} h_{s,\,\alpha}^{\sigma}$$

and (after multiplication by $h_{s\tau}$) also of

$$A_{\tau,\,\alpha} + A_{\sigma} h_{s,\,\alpha}^{\sigma} h_{s\tau}$$

and

$$A_{\tau,\,\alpha} - A_{\sigma} h^{\sigma} h_{s\tau,\,\alpha},$$

or, considering (16), also of

$$A_{\tau,\,\alpha} - A_{\sigma} \Delta_{\tau\alpha}^{\sigma}.$$[13]

We refer to this as the covariant derivative ($A_{\tau;\alpha}$) of A_{τ}.

We have thus obtained the law of covariant differentiation,

$$A_{\sigma;\tau} = A_{\sigma,\,\tau} - A_{\alpha} \Delta_{\sigma\tau}^{\alpha}. \tag{15}$$

Analogously, it follows from (3) that

$$A^{\alpha}{}_{;\tau} = A_{,\tau} + A^{\alpha} \Delta_{\alpha\tau}^{\sigma}. \tag{16}$$

Now, analogously, we find the law of covariant differentiation for arbitrary tensors.[14] We explain it by giving an example:

$$A_{\alpha\tau;\rho}^{\sigma} = A_{\alpha\tau,\,\rho}^{\sigma} + A_{\alpha\tau}^{\alpha} \Delta_{\alpha\rho}^{\sigma} - A_{\alpha\alpha}^{\sigma} \Delta_{\tau\rho}^{\alpha}. \tag{17}$$

Since, by making use of the fundamental tensor $h_s{}^{\alpha}$, we can convert local (Roman) indices into coordinate (Greek) indices, we are free to privilege either the local or the coordinate indices when formulating any sort of tensor relations. The former method was preferred by our Italian colleagues (Levi-Civita, Palatini),[15] while I have mainly used the coordinate indices.[16]

Divergence. By contracting the covariant derivative, one obtains the divergence, just as in the absolute differential calculus based only on the metric. For example, from (21),[17] by contraction with respect to the indices σ and ρ, we obtain the tensor

$$A_{\alpha\tau} = A_{\alpha\tau;\sigma}^{\sigma}.$$

In earlier works,[18] I introduced other divergence operators, but I no longer ascribe any particular significance to those operators.

Covariant derivatives of the fundamental tensor. From the formulas obtained above, one readily determines that the covariant derivatives and the divergences of the fundamental tensor vanish. For example, we find

$$
\begin{aligned}
h_{s;\tau}^{\ \nu} &\equiv h_{s,\tau}^{\ \nu} + h_s^{\ \alpha}\Delta_{\alpha\tau}^{\nu} \equiv \delta_{st}(h_{t,\tau}^{\ \nu} + h_t^{\ \alpha}\Delta_{\alpha,t}^{\nu}) \\
&\equiv h_s^{\ \alpha}(h_{t\alpha}h_{t,\tau}^{\ \nu} + \Delta_{\alpha\tau}^{\nu}) \equiv h_s^{\ \alpha}(-\Delta_{\alpha\tau}^{\nu} + \Delta_{\alpha\tau}^{\nu}) \equiv 0.
\end{aligned} \tag{18}
$$

Analogously, one can also prove that[19]

$$h_{s;\tau}^{\ \nu} \equiv g^{\mu\nu}{}_{;\tau} \equiv g_{\mu\nu;\tau} \equiv 0\,. \tag{18a}$$

Similarly, of course, the divergences of $h_{s;\nu}^{\ \nu}$ and $g^{\mu\nu}{}_{;\nu}$ also vanish.[20]

Differentiation of tensor products. As in the ordinary differential calculus, the covariant[21] derivative of a tensor product can be expressed as the product of the derivatives of the factors. If $\overset{\circ}{S}_{\circ}$ and $\overset{\circ}{T}_{\circ}$ are tensors of arbitrary index character, then we have

$$(\overset{\circ}{S}_{\circ}\,\overset{\circ}{T}_{\circ})_{;\alpha} = \overset{\circ}{S}_{\circ;\alpha}\overset{\circ}{T}_{\circ} + \overset{\circ}{T}_{\circ,\alpha}\overset{\circ}{S}_{\circ}\,. \tag{19}$$

From this and from the vanishing of the covariant derivatives of the fundamental tensor, it follows that it can be arbitrarily commuted with (;), the differentiation symbol.

"*Curvature.*" From the hypothesis of "teleparallelism," or from equation (9), it follows that the displacement law (10) or (11) is integrable. From this, we find

$$0 \equiv -\Delta_{\kappa\lambda;\mu}^{\iota} \equiv -\Delta_{\kappa\lambda,\mu}^{\iota} + \Delta_{\kappa\mu,\lambda}^{\iota} + \Delta_{\sigma\lambda}^{\iota}\Delta_{\kappa\mu}^{\sigma} - \Delta_{\sigma\mu}^{\iota}\Delta_{\kappa\lambda}^{\sigma}\,.\text{[22]} \tag{20}$$

These conditions must be fulfilled by the Δ's, in order for them to be expressible in terms of the quantities h as in (12). One can see from (20) that the laws which characterize a manifold of the kind considered here must be quite different from those of the earlier theory. To be sure, all the tensors of the earlier theory also exist according to the the new theory, in particular, the Riemannian curvature tensor, derived from the Christoffel symbols. But in the new theory, simpler and more intuitively apparent tensor relations are found, and they can be used to formulate the field laws.[23]

The tensor Λ. If we differentiate a scalar ψ twice covariantly, then, according to (15), we obtain the tensor[24]

$$\varphi_{,\sigma,\tau} - \varphi_{,\alpha}\Delta_{\sigma\tau}^{\alpha}\,.$$

By commutation of σ and τ, we obtain a new tensor, and subtracting the two tensors gives

$$\frac{\partial\varphi}{\partial x_a}(\Delta_{\sigma\tau}^{\alpha} - \Delta_{\tau\sigma}^{\alpha})\,.$$

From this, we can immediately recognize the tensor character of

$$\Lambda^{\alpha}_{\sigma\tau} = \Delta^{\alpha}_{\sigma\tau} - \Delta^{\alpha}_{\tau\sigma}.^{[25]} \qquad (21)$$

Thus, according to this theory, a tensor exists which contains only the components $h_{s\alpha}$ of the fundamental tensor and its first derivatives. That its vanishing is a sufficient condition for the validity of Euclidean geometry was proven earlier (cf. eq. (13)).[26] A determination of such a continuum compatible with the natural laws will thus consist of conditions for this tensor.

By contraction of the tensor Λ, we obtain the vector

$$\varphi_{\sigma} = \Lambda^{\alpha}_{\sigma\alpha}, \qquad (22)$$

which I previously assumed[27] would play the role of the electromagnetic potential in this theory.[28] But recently, I have rescinded that view.[29]

Commutation rule for differentiation. If one differentiates an arbitrary tensor $T^{\circ}_{\circ}$ covariantly twice, then the important commutation rule holds:

$$T^{\circ}_{\circ;\sigma;\tau} - T^{\circ}_{\circ;\tau;\sigma} \equiv -T^{\circ}_{\circ;\alpha}\,\Lambda^{\alpha}_{\sigma\tau}.^{[30]} \qquad (23)$$

Proof. If T is a scalar (a tensor without a Greek index), then the above theorem follows readily from (15). We will base the general proof on this special case.

First of all, we note that, according to the theory that we are dealing with here, there are parallel vector fields.[31] These are vector fields which have the same components in all the local systems. If (a^{α}) or (a_{α}) is such a vector field, then it must fulfill the condition

$$a^{\alpha}{}_{;\sigma} = 0 \ \text{ or } \ a_{\alpha;\sigma} = 0,$$

as can easily be proven.

Making use of such parallel vector fields, the commutation rule can be related back to the rule for a scalar without difficulty. To simplify the notation, we show the proof for a tensor T^{λ} with only one index. If φ is a scalar, then, from the defining equations (16) and (21), it follows that

$$\varphi_{;\sigma;\tau} - \varphi_{;\tau;\sigma} \equiv -\varphi_{;\alpha}\Lambda^{\alpha}_{\sigma\tau}.$$

If we insert the scalar $a_{\lambda}T^{\lambda}$ into this equation for φ, where a_{λ} is a parallel vector field, then, in every covariant differentiation, a_{λ} can be commuted with the differentiation symbol so that a_{λ} appears as a factor in all the terms. One thus obtains

$$[T^{\lambda}{}_{;\sigma;\tau} - T^{\lambda}{}_{;\tau;\sigma} + T^{\lambda}{}_{;\alpha}\Lambda^{\alpha}_{\sigma\tau}]a_{\lambda} \equiv 0.$$

Since this identity must hold at any particular place for an arbitrary choice of a_λ, then the expression in square brackets vanishes generally, which completes the proof. The generalization to tensors with many arbitrary Greek indices is evident.

Identities for the tensor Λ. If we add the three identities which result from (20)[32] by cyclic permutation of the indices κ, λ, μ, then, after a suitable combination of the terms, taking (21) into account, we obtain the identity

$$0 \equiv (\Lambda^{\iota}_{\kappa\lambda,\,\mu} + \Lambda^{\iota}_{\lambda\mu,\,\kappa} + \Lambda^{\iota}_{\mu\kappa,\,\lambda}) + (\Delta^{\iota}_{\sigma\kappa}\Lambda^{\sigma}_{\lambda\mu} + \Delta^{\iota}_{\sigma\lambda}\Lambda^{\sigma}_{\mu\kappa} + \Delta^{\iota}_{\sigma\mu}\Lambda^{\sigma}_{\kappa\lambda}).$$

This identity can be rearranged by replacing the ordinary derivatives of the tensor Λ by its covariant derivatives (as in (17)); then we obtain the identity

$$0 \equiv (\Lambda^{\iota}_{\kappa\lambda;\mu} + \Lambda^{\iota}_{\lambda\mu;\kappa} + \Lambda^{\iota}_{\mu\kappa;\lambda}) + (\Lambda^{\iota}_{\kappa\alpha}\Lambda^{\alpha}_{\lambda\mu} + \Lambda^{\iota}_{\lambda\alpha}\Lambda^{\alpha}_{\mu\kappa} + \Lambda^{\iota}_{\mu\alpha}\Lambda^{\alpha}_{\kappa\lambda}). \quad (24)$$

This is the condition which permits the Λ's to be expressed in terms of the h's in the manner described above.

By contracting this equation with respect to the indices ι and μ, we obtain the additional identity

$$0 \equiv \Lambda^{\alpha}_{\kappa\lambda;\alpha} + \varphi_{\lambda;\kappa} - \varphi_{\kappa;\lambda} - \varphi_{\alpha}\Lambda^{\alpha}_{\kappa\mu}$$

or

$$\Lambda^{\alpha}_{\kappa\lambda;\alpha} \equiv \varphi_{\kappa,\,\lambda} - \varphi_{\lambda,\,\kappa}, \quad (25)$$

where φ_λ is the abbreviation for $\Lambda^{\alpha}_{\lambda\alpha}$ (cf. eq. (23)).

§4. The Field Equations

The simplest field equations that we are seeking will entail conditions which apply to the tensor $\Lambda^{\alpha}_{\mu\nu}$. Since the number of components of h is n^2, and of these, n must remain undetermined due to the general covariance, the number of independent field equations must be $n^2 - n$. On the other hand, it is clear that a theory is all the more satisfying the more it narrows down the possibilities (without contradicting experience). The number Z of field equations should therefore be as great as possible.[33] If $\bar{Z}$ is the number of identities which relate the quantities in the field equations, then $Z - \bar{Z}$ must be equal to $n^2 - n$.

According to the commutation relation for differentiation, we have

$$\Lambda^{\alpha}_{\underline{\mu\nu};\nu\alpha} - \Lambda^{\alpha}_{\underline{\mu\nu};\alpha;\nu} - \Lambda^{\sigma}_{\underline{\mu\tau};\alpha}\,\Lambda^{\alpha}_{\sigma\tau} \equiv 0 \quad . \quad (26)$$

Here, an underlined index means that it can be "raised" or "lowered,"[34] for example,

$$\Lambda^{\alpha}_{\mu\underline{\nu}} \equiv \Lambda^{\alpha}_{\beta\gamma} g^{\mu\beta} g^{\nu\gamma} ,$$
$$\Lambda^{\underline{\alpha}}_{\mu\nu} \equiv \Lambda^{\beta}_{\mu\nu} g_{\alpha\beta} .$$

We can now write the identity (26) in the form

$$G^{\mu\alpha}{}_{;\alpha} - F^{\mu\nu}{}_{;\nu} + \Lambda^{\sigma}_{\underline{\mu\tau}} F_{\sigma\tau} \equiv 0 , \tag{26a}$$

where we have set

$$G^{\mu\alpha}{}_{;\alpha} \equiv \Lambda^{\alpha}_{\underline{\mu\nu};\nu} - \Lambda^{\sigma}_{\underline{\mu\tau}} \Lambda^{\alpha}_{\sigma\tau} \tag{27}$$

$$F^{\mu\nu} \equiv \Lambda^{\alpha}_{\underline{\mu\nu};\alpha} . \tag{28}$$

Now we take the *field equations* to be:

$$G^{\mu\alpha} = 0 \tag{29}$$

$$F^{\mu\alpha} = 0 .^{[35]} \tag{30}$$

These equations would seem to contain an impermissible overdetermination; for their number is $n^2 + \frac{n(n-1)}{2}$, while we in the first instance know only about them that they obey the n identities in (26a).

From the identity (25), together with (30), we however see that the φ_κ can be derived from a potential. We therefore set

$$F_\kappa \equiv \varphi_\kappa - \frac{\partial \lg \psi}{\partial x^\kappa} = 0 .^{[36]} \tag{31}$$

Equation (31) is completely equivalent to (30). The equations (29), (31) together comprise $n^2 + n$ equations for the $n^2 + 1$ functions $h_{s\nu}$ and ψ. In addition to (26a), there is another system of identities which connect the equations, and we will now proceed to deriving them.[37]

Let $\underline{G}^{\mu\alpha}$ denote the antisymmetric part of $G^{\mu\alpha}$; then by direct calculation from (27), we obtain

$$2\underline{G}^{\mu\alpha} \equiv -S^{\nu}_{\mu\alpha;\nu} + \frac{1}{2} S^{\mu}_{\underline{\sigma\tau}} \Lambda^{\alpha}_{\sigma\tau} - \frac{1}{2} S^{\mu}_{\underline{\sigma\tau}} \Lambda^{\mu}_{\sigma\tau} + F^{\mu\alpha} ,^{[38]} \tag{32}$$

where we have introduced as an abbreviation a tensor which is antisymmetric in all of its indices,

$$S^{\alpha}_{\underline{\mu\nu}} = \Lambda^{\alpha}_{\underline{\mu\nu}} + \Lambda^{\nu}_{\underline{\alpha\mu}} + \Lambda^{\mu}_{\underline{\nu\alpha}} \,. \tag{33}$$

Computation of the first term in (32) yields

$$2\underline{G}^{\mu\alpha} \equiv -S^{\nu}_{\underline{\mu\alpha},\,\nu} - S^{\sigma}_{\underline{\mu\alpha}}\Lambda^{\nu}_{\underline{\sigma\nu}} + F^{\mu\alpha} \,. \tag{34}$$ [39]

Now, however, taking the definition (31) of F_k into account, we have

$$\Delta^{\nu}_{\sigma\nu} - \Delta^{\nu}_{\nu\sigma} \equiv \Lambda^{\nu}_{\sigma\nu} \equiv \varphi_\sigma \equiv F_\sigma + \frac{\partial \lg\psi}{\partial x^\sigma}$$

or

$$\Delta^{\nu}_{\sigma\nu} = \frac{\partial(\lg\psi h)}{\partial x^\sigma} + F_\sigma \,. \tag{35}$$ [40]

Therefore, (34) takes on the form

$$h\psi(2\underline{G}^{\mu\alpha} - F^{\mu\alpha} + S^{\sigma}_{\underline{\mu\alpha}}F_\sigma) \equiv -\frac{\partial(h\psi S^{\sigma}_{\underline{\mu\alpha}})}{\partial x^\sigma} \,. \tag{34b}$$

Due to the antisymmetry, the system of identity equations that we have been seeking then follow:

$$\frac{\partial}{\partial x^\alpha}[h\psi(2\underline{G}^{\mu\alpha} - F^{\mu\alpha} + S^{\sigma}_{\underline{\mu\alpha}}F_\sigma)] \equiv 0 \,. \tag{36}$$ [41]

It consists of n identities, but of those, only $n-1$ are independent, since due to the antisymmetry, $[\ ]_{,\alpha,\,\mu} \equiv 0$ holds independently of which $G^{\mu\alpha}$ and F_μ are inserted.

In the identities (26a) and (36), $F^{\mu\alpha}$ is to be thought of as being expressed by F_μ, according to the relation

$$F_{\mu\alpha} \equiv F_{\mu,\,\alpha} - F_{\alpha,\,\mu} \,, \tag{31a}$$

which follows from (31). We are now in a position to prove the compatibility of the field equations (29), (30), or (29), (31).

First, we must show that the number of field equations minus the number of (independent) identities is smaller by n than the number of field variables. We have:

Number of the equations (29), (31):	$n^2 + n$
Number of (independent) identities:	$n + n - 1$
Number of field variables:	$n^2 + 1$

$$(n^2 + n) - (n + n - 1) = (n^2 + 1) - n \,.$$

The number of identities is thus exactly right. We will however not be content with that, but in addition we prove the following:

Theorem. If, within a section $x^n = const.$, *all the differential equations are fulfilled, and furthermore, everywhere* $(n^2+1)-n$ *of them (suitably chosen) are fulfilled, then all* n^2-n *equations are fulfilled everywhere a priori.*

Proof: Suppose that all the equations are fulfilled within the section $x^n = a$, and furthermore, those equations are fulfilled *everywhere* that corresponds to setting the following equal to zero:

$$\begin{array}{ll} F_1 & \ldots F_{n-1} F_n \\ G^{1\,1} & \ldots G^{1\,n-1} \\ & \ldots \\ G^{n-1\,1} & \ldots G^{n-1\,n-1} \end{array}$$

It follows from (31a), in the first place, then that the $F^{\mu\alpha}$ vanish everywhere. Now, from (36), we can conclude that, on the neighboring sections $x^n = a + da$, the antisymmetric $\underline{G}^{\mu\alpha}$ also must vanish for $\alpha = n$.[1)] Furthermore, it then follows in analogy to (27a) that, in addition, also the symmetric $\underline{\underline{G}}^{\mu\alpha}$ must vanish on the neighboring section $x^n = a + da$ for $a = n$. Repeating this mode of inference verifies the above assertion.

§5. The First Approximation

We now consider a field which differs only infinitesimally from a Euclidean field with ordinary parallelism. Then we can set

$$h_{s\nu} = \delta_{s\nu} + \bar{h}_{s\nu}, \tag{37}$$

where the $\bar{h}_{s\nu}$ are infinitesimally small in first order, and all small quantities of higher order can be neglected. Then, according to (5) or (6), we can set

$$h^{\nu}{}_{s} = \delta_{s\nu} - \bar{h}_{\nu s}. \tag{38}$$

The field equations (29) and (30) are, in this first approximation,[42] given by

$$\bar{h}_{\alpha\mu,\,\nu,\,\nu} - \bar{h}_{\alpha\nu,\,\nu,\,\mu} = 0, \tag{39}$$

1) For $x^n = a$, the $\dfrac{\partial \underline{G}^{\mu n}}{\partial x^n}$ vanish.

$$\bar{h}_{\alpha\mu,\,\alpha,\,\nu} - \bar{h}_{\alpha\nu,\,\alpha,\,\mu} = 0\,.^{[43]} \tag{40}$$

We replace equation (40) by

$$\bar{h}_{\alpha\nu,\,\alpha} = \chi_{,\nu}\,. \tag{40a}$$

We now assert that there exists an infinitesimal coordinate transformation $x^{\nu'} = x^{\nu} - \xi^{\nu}$ which will cause all the quantities $\bar{h}_{\alpha\nu,\,\nu}$ and $\bar{h}_{\alpha\nu,\,\alpha}$ to vanish.

Proof: We first prove that

$$\overline{h'}_{\mu\nu} = \bar{h}_{\mu\nu} + \xi^{\mu}{}_{,\nu}\,. \tag{41}$$

From this, we have

$$\overline{h'}_{\alpha\nu,\,\nu} = \bar{h}_{\alpha\nu,\,\nu} + \xi^{\alpha}{}_{,\nu,\,\nu}\,,$$
$$\overline{h'}_{\alpha\nu\alpha} = \bar{h}_{\alpha\nu\alpha} + \xi^{\alpha}{}_{,\alpha,\,\nu}\,.$$

Taking (40a) into account, the right-hand sides of the above equations vanish when the following equations are fulfilled:

$$\xi^{\mu}{}_{,\nu,\,\nu} = -\bar{h}_{\alpha\nu,\,\nu}\,, \tag{42}$$
$$\xi^{\alpha}{}_{,\alpha} = -\chi\,.$$

These $n+1$ equations for the n quantities ξ^{α} are, however, compatible, since according to (40a),

$$(-\bar{h}_{\alpha\nu,\,\nu})_{,\alpha} - (-\chi)_{,\nu,\,\nu} = 0\,.$$

with our new choice of coordinates, the field equations become:

$$\bar{h}_{\alpha\mu,\,\nu\nu} = 0,$$
$$\bar{h}_{\alpha\mu,\,\alpha} = 0,$$
$$\bar{h}_{\alpha\mu,\,\mu} = 0.$$

If we now split up the $\bar{h}_{\alpha\mu}$ according to the equations

$$\bar{h}_{\alpha\mu} + \bar{h}_{\mu\alpha} = \bar{g}_{\alpha\mu}\,,$$
$$\bar{h}_{\alpha\mu} - \bar{h}_{\mu\alpha} = a_{\alpha\mu}\,,$$

where $\delta_{\alpha\mu} + \bar{g}_{\alpha\mu}$ $(= g_{\mu\nu})$ determines the metric to first order, then the field equations take on the clear and simple form

$$\bar{g}_{\alpha\mu,\,\sigma,\,\sigma} = 0\,, \tag{44}$$
$$\bar{g}_{\alpha\mu,\,\mu} = 0\,, \tag{45}$$
$$a_{\alpha\mu,\,\sigma,\,\sigma} = 0\,, \tag{46}$$

$$a_{\alpha\mu,\,\mu} = 0 \,. \tag{47}$$

The notion suggests itself that, in the first approximation, the $\bar{g}_{\alpha\mu}$ represent the gravitational field and the $a_{\alpha\mu}$ represent the electromagnetic field. Equations (44), (45) would then correspond to the Poisson equation, and (46), (47) to the Maxwell equations for empty space. It is interesting that the field laws of gravitation appear separated from those of the electromagnetic field, which correspond to the experience of the independence of those two fields. Strictly speaking, however, in this theory neither field has a separate existence.

Concerning the covariance of equations (44) to (47), the following holds: the $h_{s\mu}$ obey a general transformation law

$$h'_{s\mu} = \alpha_{st}\frac{\partial x^{\sigma} h_{t\alpha}}{\partial x^{\mu'}} \,.$$

If we choose the coordinate transformation to be linear and orthogonal, as well as conforming to the rotation of the local systems, that is

$$x^{\mu'} = \alpha_{\mu\sigma}x^{\sigma} \,, \tag{48}$$

then we find for the transformation law

$$h'_{s\mu} = \alpha_{st}\alpha_{\mu\sigma}h_{t\sigma} \,. \tag{49}$$

This is precisely the same as for tensors in the special theory of relativity. Since, due to (48), the same transformation law holds for the $\delta_{s\mu}$, it must also hold for the quantities $\bar{h}_{\alpha\mu}$, $\bar{g}_{\alpha\mu}$, and $a_{\alpha\mu}$. With respect to such transformations, the equations (44) through (47) are covariant.

Concluding Remarks

The great appeal of the theory explained here lies for me in its unity and in its high degree of (allowed) overdetermination of the field variables. I have also been able to demonstrate that the field equations lead in first approximation to equations that correspond to the Newton-Poisson theory of gravitation and to the Maxwell theory of the electromagnetic field. Nevertheless, I am still far from being able to assert the physical validity of the derived equations. The reason for this is that I have not yet succeeded in deriving the laws of motion for particles.[44][45]

56. To Otto Blumenthal

[Berlin,] 13 September 1929

Dear Mr. Blumenthal,

Enclosed, you will find the two articles.[1] I am very happy that Cartan's supplement will appear in the original language.[2] I can understand your reasons for why the articles can be published only a half-year from now.[3] It is, however, unfortunate, since collaboration with our colleagues on this fundamental and, after the most recent results, really promising problem will be correspondingly delayed. Physics simply has a different rhythm than mathematics.

Kind regards, your

57. To Élie Cartan

Berlin, 13 September 1929

Dear Colleague,

Today, both our articles go off to the *Mathematische Annalen*.[1] It has taken so long because the editor, Prof. Blumenthal, was on holiday and therefore did not reply.[2] He was especially delighted to receive an original work of yours. Unfortunately, he writes that it will take another half year until the papers are in print.

In the hope of seeing you in Paris in early November,[3] I am, with best wishes, your

A. Einstein

58. To Raschko Zaycoff[1]

[Berlin,] 13 September 1929

Dear Mr. Zaycoff,

It is unacceptable that you always bother me, a very busy person, with your private affairs. You must see that you yourself must bring your matters to a close.[2] Further, I must warn you that, in my opinion, you are not sufficiently discerning in your scientific publications.[3] It is in any case very difficult to find attentive readers. Many have deprived themselves of every scientific credit through frivolous publications and are no longer read. Be careful!

Kind regards, your

59. From Jacques Hadamard[1]

[Paris,] 16 September 1929

[Not selected for translation.]

60. From Georg Count von Arco[1]

Berlin, 17 September 1929

[Not selected for translation.]

61. To Robert Andrews Millikan

Berlin, 18 September [1929][1]

Dear Colleague,

Your friendly invitation was truly a real temptation for me.[2] You have made me an almost shameful offer, much more than I deserve. But upon closer consideration, I see that I unfortunately cannot accept. Then people in other places would rightly be annoyed with me if I only favored this invitation. If I also were to accept the others, as I could not well do otherwise, I would not come through it alive, given my state of health.

Also, I must say that my formal studies of the last years, as interesting as they may be in themselves, have meant that I have only incompletely followed the tumultuous developments of physical theory. On the other hand, my own studies have not blossomed, so far that I could be certain of their physical fecundity. It is only a gamble for the future. So you will not be losing much if I cannot come.

With heartfelt thanks to you and the mutual friend,[3] who has made the generous offer, and with best wishes to all who are there, I am your

A. Einstein.

62. To Hagop-Krikor Djirdjirian[1]

[Caputh,] 20 September 1929

Dear Sir,

My investigations into the unified field theory have not yet matured to the extent that they would justify a review publication.[2] Recently, I submitted a new paper on this subject to the Mathematische Annalen, in which I made not-insignificant changes to the fundamental equations.[3] But also, at the current state of things, I am far from having a clear picture of the physical meaning of the theory. It will still require considerable time before a review article can be considered.

Respectfully yours,

63. To Albert Reich[1]

[Caputh,] 20 September 1929

Dear Sir,

Giving an answer to your question, and indeed to cosmological questions in general, suffers from the difficulty that one does not know which initial state one should assume; quite apart from the fact that our knowledge of the relevant natural laws over very long times is, without doubt, still very incomplete. It is certain that, according to our present knowledge, no arbitrarily formed mass which was initially not rotating can be induced to rotate through any sort of internal processes, including radiation phenomena. The origin of rotational motion can indeed be comprehended only through the eccentric [coalescence] of two or more masses. Translational motion can likewise be thought of as having been present from the very beginning, but it could also have been produced later by attractive forces or radiation pressure. I believe that, given the present state of our knowledge, we can hardly hope to find certain answers to these last cosmological questions.

Respectfully yours,

64. To Reuters Ltd.

Caputh, 20 September 1929

As sympathetic as I am to the battle for freedom of the Indian people, as a foreigner, it would be impossible for me to meddle in such matters.[1] I have never sent such a telegram to Macdonald[2] or any other British politician.

Respectfully yours,

65. From Vladimir A. Fock[1]

Leningrad, 20 September 1929

Dear Professor,

Some time ago, I sent to you one of my articles on quantum theory, which, it seems to me, is intimately connected with your new field theory.[2] I would now like to send you an additional article,[3] which continues the development of the ideas expressed in the previous article.

The subject of my latter article is not actually the field theory, but is rather the establishment of the quantum-mechanical wave equation. In this context, certain relations between the (electrical and gravitational) field quantities occur, so to speak, automatically. I wish to call your attention to these relations, since they could be of interest for the formulation of a unified field theory.

My approach differs from yours, fundamentally, in that I do not refer to any teleparallelism. The h_{va} serve only to formulate the wave equation; the latter proves to be invariant with respect to the choice of the "tetrads." The electromagnetic potential takes its place within Riemannian geometry, and it occurs along with the Ricci coefficients (alongside the quantities which you have identified as potentials).

I would be very grateful to hear your opinion of the ideas developed in these works.

Respectfully yours,

V. Fock

66. From Max Born

Göttingen, 23 September 1929

Dear Einstein,

Hearty thanks for your card.[1] The assertion in my lecture about the impossibility of reducing the statistical laws of quantum mechanics to a causal theory refers to a chapter of the book by Jordan and myself.[2] Unfortunately, the production of the book has been delayed by some months, since Jordan and I have been alternately unable to work. But I have already received the proofs of that chapter 6 and am sending them to you. There, in paragraph 59, you will find our considerations. To be sure, you should read the whole chapter, and I don't know how understandable it will be if one has not read the previous chapters, in particular, our mathematical terminology. I will therefore tell you briefly what seems to us to be the kernel of the matter. The basis is given by a theorem of your Joh. von Neumann:[3] There are statistical manifolds, which are termed "pure states," with the property that they can by no means be represented as mixtures of other manifolds.[4] These pure states, now, do not exhibit any deterministic behavior. On the other hand, the reduction to a causal theory would mean that every manifold could be represented as a mixture of causally determined manifolds. The existence of the pure states thus contradicts this possibility. If we want to keep this possibility open, we would have to decide to abandon the whole of quantum mechanics formalism, and I don't believe that is permissible. Naturally, we would be very grateful for your criticisms.

Your report on the teleparalellism theory was very interesting to me, in particular, the fact that the new field equations are uniquely simple.[5] Thus far, namely, I was bothered by the fact that, alongside the enormously simple and transparent geometry, the field theory looked so extremely complicated.

My assistant Heitler[6] has told me that you have contacted Mr. Rumer.[7] I will think over whether one should send him with a Rockefeller fellowship to you or to Weyl.[8]

My wife is unfortunately in a rather miserable and nervous state and is now being sent to a sanatorium here in the neighborhood.[9] She sends her warmest greetings to you, and also to your wife and to Margot.[10] I add my own best wishes, yours sincerely,

Max Born

Hedi would very much like to know where Miss Margot is (town and address).[11]

67. To Paul Ehrenfest

Caputh, 24 September [1929]

Dear Ehrenfest,

I have written to H.,[1] and I find this matter to be quite reasonable. I also wanted to come, but let myself be talked out of it. I cannot allow my health to be endangered again after I have spent so much effort to arrive at a bearable state. I would, however, very much like to acquaint you with the [tele]parallelism field theory. The newest results are so beautiful that I am quite sure that I have found the natural field equations of such a manifold.[2] It is not improbable that the future will prove them to be correct.

To that end, it would be good if either you were to come to visit us, so that I can show you everything in detail, or else that you send someone who can communicate the whole matter to you later (Tanja?).[3] Whoever it might be, they can stay with us, either here in Caputh (Waldstr. 7) or, from the end of October on, at Haberlandstr. 5.

Best wishes to all of you from your

A. E.

68. To Jacques Hadamard

[Caputh,] 24 September 1929

Dear Mr. Hadamard,

I was extraordinarily pleased by your letter,[1] first because it came from *you*, but second because I see from it the great seriousness with which you yourself grapple with these problems that are so serious for our Europe. I answer you hesitantly because I know very well that, in such human matters, my feeling is stronger than my intellect. But regardless, I want to dare to *justify* my standpoint as well.

First, a qualification. I would not dare to preach to a tribe in Africa in an analogous way without qualification, for there, the patient would be long dead before the cure could be of use. The situation in Europe is, however,—despite Mussolini—a different one.

My first thesis is this: In a Europe systematically prepared, morally and materially, for war, the nonviolent League of Nations will also be morally powerless in the hour of national psychosis.[2] Every citizen will declare his country as the attacked and, indeed, in good faith. Your Boncourt[3] may be a friend of peace, but he is certainly a miserable psychologist if he is a friend of peace. One cannot systematically educate a nation for war and, at the same time, instill in it that war is a heinous crime.

Second, I admit that, at first, the nondefending nations take a heavy sacrifice on themselves, but it is a sacrifice in the service of the wider community and higher development. But without sacrifice, no actual progress can be achieved.

Third: These sacrifices are indeed great, but bearable. If nothing worse has happened to Germany than was actually the case after that, which it has done for four years in a state of complete defenselessness, then a European country which, from the beginning, does not participate in war cannot fare so badly after all.[4]

Fourth: As long as war is systematically prepared for, fear, mistrust, and the ambition and selfishness of individuals will consistently give cause for wars.

Fifth: One cannot wait until the leading strata of states voluntarily renounce their sovereignty: the lust for power will prevent it.

Sixth: If absolute renunciation of involvement in warlike and general military actions is publicly advocated by those who have authority with everyday people, the bellicose spirit would be quite effectively counteracted.

Seventh: Warfare means taking innocent lives and allowing oneself to be killed innocently. (The scoundrel Berchthold[5] is still living, but ten million mostly innocent people were slaughtered.) Can an independent and decent person participate in such business? Would you commit perjury if your state demanded it? Certainly not, but killing innocents?

Frankly speaking, the last argument is, for me, the strongest—at least in its impact on me. For me, humanity stands above fatherland, indeed, above everything else.

I am looking forward to seeing you in November when I come to Paris.[6] There, I will report to you and some colleagues on the "Unified Field Theory," which indeed still seems to be very problematic but extremely interesting.[7]

Warm regards, your

A. Einstein

69. From Rudolf Goldschmidt[1]

Westend, 25 September 1929

Dear Mrs. Einstein, dear Professor,

I was already eager to hear how you are doing, considering how long it has been since we last spoke. It was with all the more pleasure that I received your lovely card. That you, dear Professor, are doing nothing there, I cannot believe; but in the interest of your rest and recuperation, I wish it were true! We have several *choses d'affaire* to discuss. On the subject of magnetostriction, I had a conversation some time ago with Prof. Gerdin,[2] director of the Siemens research laboratory. He was very interested in the "labilizer"[3] and would apparently be happy to carry out the experiments. He has already performed extensive experiments on the behavior of various iron alloys with respect to magnetostriction, and he asserts that only certain alloys are usable, which made me hesitant about proceeding on our own without Siemens. But I would like to discuss that with you. Siemens has quite recently begun to construct underwater acoustic signal devices making use of magnetostriction. These devices were developed in the Siemens research laboratory.

The Patent Office has only formal objections to our patent, no citations.[4] This is also a discussion point for our next meeting! An additional one is the very new development in the loudspeaker matter, which I would like to suggest, and which we should work out together. I look forward to discussing these matters with you and am eager to hear your reactions. Best wishes to both of you, from your friend

Rudolf Goldschmidt

70. To Hugo Bergmann[1]

Berlin, 27 September 1929

Dear Dr. Bergmann,

I had completely forgotten to give you my approval for the reproduction of my manuscript on the theory of relativity.[2] I am rectifying that herewith, in the event that the project has not been dropped in the meantime.

The events in Palestine appear to me to have again demonstrated how necessary it is to establish a kind of genuine symbiosis between Jews and Arabs in Palestine.[3] By this, I mean the existence of continuously functioning, mixed administrative, economic, and social organizations. Separate coexistence must lead to dangerous tensions from time to time. Furthermore, Jewish children must all learn Arabic.

Kind regards, your

A. Einstein

71. To Asis Domet[1]

Caputh, 27 September 1929

Dear Mr. Asis Domet,

Of course I remember you very well and also have read with great interest your article which appeared in the Berliner Tageblatt at the end of August.[2] I am, on the whole, completely of your opinion that the creation of an Arabic-Jewish community must be worked on which, through the exclusion of nationalist fanatics, brings the two tribally related peoples closer together.[3]

I will study your drama in the near future and then send you further news.[4]

Kind regards, your

72. To Leo Kohn[1]

[Caputh,] 27 September 1929

Dear Mr. Kohn,

I have developed concerns about the content of our article, which make its publication in the planned form impossible.[2] We maintained that the events are caused by the incitement of the lowest social class, while the majority of the Palestinian Arabs disavow the issue. This view, however, cannot be correct. A national movement of Arabs against the Jews that has much deeper roots must exist. Apart from the fact that, as I've been told from multiple sides, Jews in Arab areas and Arabs in Jewish areas feel threatened, I find at the end of the J.T.A. of 25 September a report of a boycott movement directed against Jewish goods, which demonstrates a systematic mobilization from the Arab side.[3]

Under no circumstances can I lend my name to a disputable interpretation of the situation. In light of this, a reworking along the following lines is unavoidable: Colonization only in amicable agreement with the Arabs, not in opposition to the Arabs under the protection of English military force. A wave of Arab nationalism which opposes the work of settlement cannot be denied out of existence.[4]

Kind regards, your

A. E.

73. To Unknown

[Caputh, before 3 October 1929][1]

[Not selected for translation.]

74. To Cornel Lanczos

[Caputh,] 4 October 1929

Dear Mr. Lanczos,

It has now become clear to me how we can deal with the notorious behavior of those points which exhibit a singularity in the first approximation, but which correspond to a strict solution that is free of singularities.

We start from the identity[1]

$$G^{\mu\alpha}{}_{;\alpha} \equiv 0\,, \qquad (1)$$

where $$G^{\mu\alpha} = \Lambda^{\alpha}_{\underline{\mu}\underline{\nu};\nu} - \Lambda^{\sigma}_{\underline{\mu}\underline{\tau}}\Lambda^{\alpha}_{\sigma\tau}.$$

Let $$h_{s\alpha} = \delta_{s\alpha} + \bar{h}_{s\alpha}\,.$$ [2]

Correspondingly, $$G^{\mu\alpha} = \overline{G}^{\mu\alpha} + \overline{\overline{G}}^{\mu\alpha} + \ldots = \overline{G}^{\mu\alpha} + \overset{x}{G}^{\mu\alpha}$$

where $\overset{x}{G}^{\mu\alpha}$ combines all the terms that follow the first term.

From (1), we can derive the identity

$$\overline{G}^{\mu\alpha}{}_{,\alpha} \equiv 0 \ \ldots \qquad (2) \text{ (ordinary differentiation!)}$$

From this, and from the field equations $G^{\mu\alpha} = 0$ (3), it follows that:

$$\overset{x}{G}{}^{\mu\alpha}{}_{,\alpha} = 0 \ldots \qquad (4)$$

This is a strict equation in the form of a pure divergence, which is however not an identity, but rather only a result of the field equations (3). (There is no analog for the other identities, since the equations corresponding to (4) are then true identities which hold independently of whether or not the field equations are fulfilled.)

From (4), in the static case, we find for the area enclosing a singularity

$$\int \overset{x}{G}{}^{\mu\alpha} \cos(\alpha n) df = 0 \ldots \qquad (5)$$

This equation must initially be fulfilled for the singularity without an external field; otherwise, a strict, singularity-free solution could not exist. But it must also be fulfilled for the singularity with an external field.

In applying (5), it must be considered satisfactory in a first approximation to insert for $h_{s\alpha}$ solutions of the linear equations

$$\overline{G}{}^{\mu\alpha} = 0$$

$$\overline{F}{}^{\mu\nu} = \overline{\Lambda}{}^{\alpha}_{\mu\nu;\alpha} = 0,$$

since these are the only terms of second order.

Now the question arises as to whether there are singularities in the solutions in first approximation which can be interpreted according to their behavior in fields as ponderable or electrical masses.[3]

75. From Paul Ehrenfest

Leiden, 4 October 1929

Dear Einstein,

Your card disturbed me, in that it banished our hope that we would see you once again here at our home.[1] On the other hand, your suggestion that someone from here could visit you in order to be thoroughly informed about your investigations on teleparallelism gave me great pleasure.[2] I myself have no possibility of traveling just now. The student who would be most suited for this mission: my pupil Casimier, is in Copenhagen, working with Bohr as his private assistant.[3] Otherwise, only Tanitschka occurs to me as a candidate.[4] At the moment, she indeed knows practically nothing about your work. She also naturally has, for that reason and for other reasons, all possible "buts" against the idea that she of all people should be entrusted with this "Einstein mission." I myself, however, believe that she is the first to be considered after Casimier. In particular, because she has an excellent training in differential geometry, and a clear and critical mind. I would even think it possible that she could be helpful to you in carrying out certain intermediate calculations. I have the impression that she, in fact, would be very happy to visit you, but that she is ashamed, especially of her lack of knowledge of "normal physics." For, in "crazy physics," she presently has an almost unpleasant special proficiency, because, in the past months, she has (living in Göttingen), as Born's private assistant, been working continuously on the final editing of the Born-Jordan book on quantum mechanics,[5] and, indeed, she is solely responsible for making sure that the various contributions of all the various people will fit logically cohere TO SOME EXTENT.[6]

To be sure, she couldn't stay in Berlin for very long, since she will have to return to Göttingen soon after. In fact, even during her stay in Berlin, she would have to continue to take care of the corrections to Born's book.

Please don't be startled if some of those corrections should suddenly be delivered to your address for Tanitschka. That would just mean that Tanitschka herself would arrive soon afterwards (depending on the date when she travels, either to Caputh or to Haberlandstr.). In that case, let the corrections simply wait at your place.

I certainly hope that she decides to visit you. But she would have to write to Born about it beforehand, in order to know when he would need her again in Göttingen.

One more point: How should we pay the ADDITIONAL COSTS of her "Berlin" detour on the way to Göttingen? That will naturally be a small amount, less than 50 marks. I would suggest that those 50 marks be paid from the "theoretical physics account" which is jointly at the disposal of you, Fokker, and myself.[7] I would also suggest that, if the amount turns out to be higher, an additional 25 marks be made available. If *you* are in agreement, I am practically certain that Fokker will also agree. Whether Tanitschka will agree remains questionable, to be sure.

I would ask you to answer, with a few words, whether you agree to my suggestion, and, at the same time, to inform me more exactly as to when you will be moving from Caputh to Haberlandstrasse.

By way of explanation, I would furthermore like to say the following: In the case that Casimier is sent to you, I would apply for his travel costs to be paid by the Lorentz Fund.[8] But for Tanitschka, I wouldn't do that, since she is not a genuine theoretical physicist. But the small fund that is administered by you, Fokker, and me is indeed an honest source for cases like this one.

Now, let us send you our most heartfelt greetings! At the end of November, I hope to be able to visit you in Berlin.

How would it be if Margot[9] "suddenly" came, for once, to Leiden to have a look at Holland while staying at our house. We would be VERY, VERY happy at that. Besides me, especially Galinka[10] would enjoy that enormously. Not least also for music, for example, singing canons! [He]y, dear Margotl, answer, or come on short notice instead of an answer!

However: don't come to Holland right now without renewing your vaccination, because of the alastrim epidemic which is rapidly spreading.[11] We have all had fresh vaccinations.

76. On Palestine

[Caputh, before 7 October 1929][1]

[Not selected for translation.]

77. Protest against the British Press's Reaction to Recent Arab Attacks on Jews in Palestine

[*Einstein 1929ss*]

DATED 7 October 1929

PUBLISHED 12 October 1929

IN: *The Manchester Guardian Weekly* 21, Saturday, 12 October 1929, pp. 13–14.

[See documentary edition for English text.]

78. To Ilse Kayser-Einstein

[Caputh, between 8 and 31 October 1929][1]

Dear Ilse,

Thank you for the kind gift, which is patiently waiting for me in the cellar. We are having a wonderful time here and are looking forward to Rudi's[2] visit. I'm still going barefoot in good weather and, otherwise, I'm happy as a clam and go sailing alone, and with Mama, and all the women friends that God has created. You would love it here. Mama and Herta[3] whimper to their heart's content.[4]

I hope things go well, with best regards from your

Albert

79. To Willy Hellpach[1]

Caputh Potsdam Waldstr. 7, 8 October 1929

Dear Mr. Hellpach,

I read your article about Zionism and the Zurich conference, and I very much feel the need to answer you, if only briefly, as someone who is very devoted to the idea of Zionism.[2]

The Jews are a community of blood and tradition, in which religion is by no means the only binding factor. This is demonstrated by the attitude of other people toward the Jews.[3] I only discovered that I was a Jew when I came to Germany fifteen years ago, and this discovery was communicated more by non-Jews than by Jews.[4]

The tragedy of the Jews lies in that they are people of a certain type of development which lacks the support of a community that connects them. The uncertainty of the individual, which can increase to the point of moral instability, is the result. I recognized that a recovery of this people was only possible if all the Jews of the Earth were bound into a living community to which the individual joyfully belonged and which would make bearable the hate and setbacks which he has to endure from all sides around him.

I saw the undignified mimicry of valuable Jews, the sight of which made my heart bleed.[5] I saw how schools, satirical magazines, and countless cultural factors of the non-Jewish majority undermined the self-esteem of even the best of my tribesmen, and I felt that it cannot be allowed to continue like this.

I recognized that only a common effort that is dear to the hearts of all the world's Jews could allow this people to heal. It was a great deed by Herzl[6] that he recognized and demonstrated with great energy that, given the existing traditional stance of the Jews, the creation of a homestead—or, to express it more correctly, central place—in Palestine was the task on which one could concentrate their energies.

You call all of that nationalism, and not totally unjustly. But a striving for community, without which we cannot live and die in this hostile world, can always be labeled with this ugly name. In any case, it is a nationalism that strives not for power, but for dignity and recovery. If we did not have to live amongst intolerant, narrow-minded, violent people, I would be the first to reject every form of nationalism in favor of universal humanity![7]

The allegation that we Jews could not be proper citizens of, for example, the German state if we want to be a "nation" equates to a misunderstanding of the nature of the state, which itself arises from the intolerance of the national majority. (Switzerland!) We will never be protected from this intolerance, whether we call ourselves a "people," or a "nation," or not.

To be brief, I have laid this out nakedly and brutally, but I know you, from your writings, as someone who is mindful not of the form, but the meaning.

With kind regards, your

80. To the Jewish Telegraphic Agency

[Caputh,] 8 October 1929

To your query from 2 October, I offer the following comment: I do not doubt that the Mufti's remark regarding a statement by me in which the analogy of body and head played a role is a deliberate falsification.[1] Your correction mirrors the truth.

Respectfully yours,

81. To Leonhard Sklarz[1]

Caputh Potsdam Waldstr. 7, 8 October 1929

Dear Mr. Sklarz,

When I became aware of the League's resolution against imperialism, I immediately broke off my relationship with the League.[2] It seems to me that the deeper reasons for the League's strange position are based on the dependence of its leadership on Moscow.[3] Justice and objectivity may therefore hardly be expected from that side.

Respectfully yours,

82. From Georg Wolfgang Kellner[1]

Charlottenburg 5, Riehlstr. 12 IV, Berlin, 8 October 1929

Dear Professor,

As I heard, you recently made roughly the following objections to my remarks on "Causality in Quantum Mechanics":[2]

Take a potato and lay it in the cellar at a certain place. Then, leave the cellar and close the door. After a certain period of time, open the door and search for the potato. You will find it in its original position [---] Now, one could not imagine that the potato would have spread out, in the meantime, over the whole cellar and only contracted and returned to its original position when one opened the cellar door.

To this, on the contrary, I would answer that an electron is indeed not a potato. The essential difference appears to me to be that, within the potato, there are forces which hold it together, but there are none within the electron! That a position measurement of the electron is preceded by a collapse of its wave function can perhaps be seen from the following example: Let a homogeneous monochromatic electron wave which is, according to the quantum-mechanical equations, infinitely extended impinge on an atomic nucleus of charge Ze, with a very large Z. If the process that follows leads to capture of the electron (which can depend only upon phase relations), then, once quantum electrodynamics is available in its completed form, we will be able to follow precisely via calculations just *how* the electron is captured, and thus *how* the plane wave contracts into the Laguerre function which characterizes the ground state of the atom. And this will be all the more surely a position measurement, as, with increasing atomic number Z, this Laguerre function approaches, ever more closely, a Dirac delta function; and that is precisely, in my opinion, a position measurement.

And there is something else which could perhaps be suited to supporting my viewpoint. If one rejects my viewpoint, then one must, in principle, reject uniting quantum mechanics with the law of causality. One could give it a try by introducing some parameters or others which determine the orbit of the electron; and one would then have to require that the statistical predictions of quantum mechanics would result when one averaged over those parameters. That, however, is precisely not possible, in principle, as follows from the existence of the uncertainty relations.

Finally, in order to maintain the law of causality, one could reject all of quantum mechanics. And I seem to recall that you have sought a way out of the dilemma in essentially that direction. I recently heard that you made the following objection against quantum mechanics:[3]

Let an electron wave fall onto a screen which contains a slit. A part of the electron wave will emerge through that slit, to be sure, in various directions. Behind the slit, there is a hemispherical screen which captures the electron wave. Now, of course, at some particular place on that hemisphere, a corpuscular electron will be detected. Thus, it is *completely determined* that the particle cannot be detected at any other place on the screen, including all those positions where the wave *simultaneously* arrives. This is, so to speak, the causal connection of spatially separated events with a superluminary velocity and would thus contradict the theory of relativity unless the particle already, before, e.g., on passing the slit, *knows* to which point it should fly.

However, I see, in this objection against quantum mechanics, only an indication of the correctness of my view of the theory. If it is indeed true that the norm of Schrödinger's wave function can be interpreted as the electron density,[4] and if a position measurement is possible only after the "collapse," then something must have caused the electron to collapse before it hit the detection screen, since its position is, in fact, then measured upon hitting the screen. *This could indeed only be the potential of the screen*, and it is self-evident that it reaches out to infinity, at least to the location of the slit. My view of quantum mechanics does not, in any case, contradict the theory of relativity; I indeed believe that, also, those who support the statistical interpretation see no contradiction with the theory of relativity. If that should nevertheless be the case, then I would see that only as a positive point in favor of my viewpoint.

I hope that I have been able to show that, thus far, no substantive objection against my interpretation of quantum mechanics has been raised (the remarks of Mr. Infeld in the most recent issue of the Z[eitschrift] f[ür] Phys[ik][5] certainly do not count as such). And I therefore consider myself to be justified to believe that the consistent implementation of that interpretation will contribute essentially to the solution of the causality problem.

I hardly have a greater wish than that these remarks to you, revered Professor, will have convinced you, or, at the very least, will have made your disapproving standpoint less determined.

I send you best wishes with my utmost respect, yours truly,

G. W. Kellner

83. To Georg Wolfgang Kellner

[Caputh?, after 8 October 1929][1]

Answered

Kasten mit reflekt. Wänden [2]

9 Schieber [3]

„Kartoffel" [4]

Soll die Kartoffel nach dem Zerfliessen Ihres Wellenpaketes durch den Schieber lokalisiert werden? E. [5]

[2]"Crate with reflecting walls"
[3]"Damper"
[4]"„Potato""
[5]"Would the potato be localized after the wave packet is dissolved? E."

84. From Eduard Einstein

Florence (Colonnata,) [9 October 1929][1]

Dear Papa,

The grapes are hanging on the house.[2] In the garden, the cute peasant girl who helps on the farm makes herself rope for a swing; if I could speak Italian, I would go and play with her. The cypresses lend the impression, even the one withered palm—the country house with the iron door, with the chicken coop and flower beds—anyone who has ever frozen in their life, today, they can't imagine it. In fact, it is warm, but not as warm as for the guest from Germany,[3] the lady who today had a temperature of 40.5, so that the doctor came and said that it could perhaps be typhus; but he could only say with certainty in seven days. In my mind, there is a wondrous blurriness; it comes to no clear conclusions: one has a state of mind practically made for a poem. When the cute peasant girl is in sight, I always look over and smile at her. For this reason, she now looks the entire time to see if I am looking over. Earlier, Aunt Maja brought out a drawer full of photographs; in them, I saw, to my delight, that I have a whole host of ancestors. Almost enough for a family tree. I always start from Colonnata with the tram to Florence.[4] Two of my school friends are staying in a boarding house there.[5] They get on so well with one another that they always run off with one another. I often meet up with them, and then we go to the museums. I really enjoy seeing the paintings. My favorites are a pair of paintings by Botticelli.[6] I wander through the streets. I ate lunch in a restaurant; I rustled up a waiter who spoke a little German, ate spaghetti, ordered fish. At first, the waiter said it would come in five minutes, but it only came in half an hour; a thin slice. I drank wine, and that unhinged me. Wine turns you inside out, while tea, for example, only boosts you. That which one calls "I" is distorted by wine to dubiousness, by tea to a clearer point of concentration. By the way, I drank little wine, or almost none at all. As a result, the world changed for me. I lodged myself there between the Dome and the Campanile. There, the street children gallivanted around. One of them pranced around with a coin, held it in front of everyone's nose amongst excited debate. Then he went away. Came back with a cigarette in his mouth. I easily guessed that he had exchanged the money for it. When one of the others pulled the cigarette from his mouth and threw it on the ground, it almost came to a brawl, especially because he hadn't even lit the cigarette yet. Now, I stood up and climbed into the tram to Fiesole.[7] In Fiesole, I escaped the busy monks, the artworks, the museum, appeased the beggar women with a few coins, and took a wonderful walk down and back while the evening drew near. Was glad that no one was there to hold me to account for everything not seen. This morning, one of the school friends said to me worriedly, with a wry look in the Baedeker: "We still have so much to do." In the museum, he only says things like: "Is this Hall D? In Hall F, there's supposed to be a famous painting; The Baedeker says the third saint on the left in this image is St. Bartholomew; In this image from the School of Lippi,[8] Lippi himself is also supposed to have helped paint the heads of these two angels;" and so on. The poignant copyists are everywhere; their copies are really very similar: perhaps a few of the fine details are lost, a little of the color changed, the face of the Madonna a touch coarser, looks now perhaps a bit more like a cheap three-color print. In other news, my *Matura* was very good; oral mathematics and German were both first-rate; the only let-down was written history. I had the second-best result in the class; admittedly miles away from the top spot.[9] Here with Uncle[10] and Aunt, we are very happy together, but you see, yourself, that I am not currently predestined to write.

Many warm wishes, your

Teddy

85. From Maja Winteler-Einstein

Wednesday [9 October 1929][1]

Dear Albert,

We are so happy to meet Tetel.[2] I like him even better than Hans Jo, and that says a lot.[3] I have never before met a boy of that age who has seriously reflected over so much, is so open for all things human, artistic, and spiritual, and yet is so unpretentious in every relationship. And how musical he is! I am just sorry that we currently have two ladies as guests, so I cannot dedicate myself to him as much as I would like. Of all the bad luck, one of them became ill; but today, she is feeling better, so that the danger of typhus can be considered eliminated. Miza has not come along;[4] however, she wrote me very nicely, and I have answered her likewise. I prefer it that she is in Zurich; otherwise, Tetel would be too divided in his feelings. Just a pity that Tetel's holidays are so short. In his overeagerness, he wants to be in Zurich right at the beginning of the semester.[5] So his stay is much too short for us, as well as for Florence. We have already had lively discussions and played music with great enjoyment (at least for my part). Today, it is a week that he has been here.

And how is it in your house?[6] Write to me again. Hopefully, you all have had just as beautiful and warm an October as us. Best wishes to you all, but especially to you, from your

Maja

What is going on with Toni?[7]

86. Stresemann's Mission

"Stresemanns Mission"
[*Einstein 1929uu*]

DATED 12 October 1929
PUBLISHED November 1929

IN: *Nord und Süd: Monatsschrift für internationale Zusammenarbeit* 52, no. 11 (November 1929): 953–954.

STRESEMANN'S MISSION [1]

by

DR. ALBERT EINSTEIN

IF WE are still to credit the saying of the Greek sage that no man can be esteemed happy before his death, we may now esteem Stresemann a happy man. For it was given to him to live successfully in the service of a great idea and to die in that service—an idea, in the realization of which the more progressive spirits of our generation can no longer doubt.

For the fulfillment of his great mission he was most happily equipped. He was a strong character hardened by a laborious struggle for existence in his youth, which his comparatively humble origin made inevitable. An [2] atmosphere of struggle and conflict was natural and indeed necessary to him, even when his constitution had been undermined by suffering. But nature had endowed him likewise with a fine sense of the beauties of speech and literature that lent a more delicate quality to his gift of eloquence. Indeed, his oratory, sustained as it was by his consciousness of his exalted task and a certain wholesome optimism, cast a spell which it was difficult to withstand. Therewith came a capacity of grasping unfamiliar ways of thought and spheres of feeling, that is very seldom met with in our public servants.

His greatest achievement is, to my mind, his ability to induce a number of large political groups, against their own political instincts, to give their support to a comprehensive campaign of European conciliation. His success in this matter depends on a none too obvious psychical phenomenon. The men who possessed insight and courage enough to withstand the war psychosis, tell their countrymen the truth, and confront them with the evil consequences of a war, especially one so intensified and so prolonged, were, not merely during the War but even after it and until the present day, regarded with distrust and even hatred. On the other hand those who from indifference, weakness, or even criminal purpose, shared the guilty responsibility for the outbreak of the great war, are utterly excused. No one throws a stone at Berchtold, but Forster and Muhlon are objects of hatred. [3] [4]

During the War Stresemann was warlike enough to be able, later on, to hold the confidence of those whom he wanted to win over to his own wise and noble purposes. He had to contend against the resentment of men defeated in war, injured in their pride, and robbed of their accustomed privileges; and, unlike his comrade-in-arms Briand, he was not backed by a victorious and exultant soldiery. [5] [6]

Stresemann had one characteristic that is always found in great leaders. He did not exert his influence as the representative of a caste, a profession, or a nation, nor indeed as a type of any kind, but directly as a man of mind and the bearer of an idea. He was as different from politicians of the usual stamp as a genius differs from an expert. Herein lay the magic and the strength of his personality.

87. To Sozialistischer Studentenbund Dresden[1]

Caputh, 18 October 1929

Folly and ungainliness (as opposed to wisdom and grace) find themselves in the happy situation of more easily getting the majority on their side.

Here as well. What may be "nonacademic behavior" is decided by the academic majority; an objective refutation is therefore impossible. The only thing that I do not grasp is the following: Why does the Dresden Socialist Students' League place any value in regaining recognition? Does it believe that, in doing so, it would achieve some form of higher existence? I do not care about this recognition!

Respectfully yours,

88. From Paul Ehrenfest

Leiden, 20 October 1929

Dear Einstein,

I presume that Tanitschka is already with you. If that is not the case, she will certainly arrive in Berlin by the beginning of November.[1]

Now, I would like to permit myself to pose a question, which (in somewhat different form!) repeats a question that my wife[2] asked me in relation to your theory of gravitation and which I was unable to answer. She wanted to ask it of you herself, but since she had to hurry off to Russia, she couldn't go to see you in Caputh.[3]

I would be very grateful if you would be willing to discuss this question with Tanitschka in detail. She can then communicate your answer to me. That will, in any case, save you from having to write it all out.

I would thus like to formulate the question as follows:

1. Let, in your spatially closed world, nothing at all be present except for two very heavy mass *points* and arbitrarily many light signals. What, then, does the phrase "naturally measured distance of these two masses from one another" mean, and can this distance assume different values?

2. The same question, in the case that one also assumes that, in that world, some "measuring rods" and "clocks" are present (whose masses are completely negligible compared to the masses of the two mass points.) (and how should one for example such

3. Question 1 repeated, but under the assumption that not two, but rather three, or ten, or quadrillions of mass POINTS are present in the spatially closed world.

4. In the same way, question 2 repeated.

(*Vague* explanation concerning the *goal* of these questions:

If only two points exist in the world, then "you don't know if they are far apart or near to each other;" compared to what? And thus their distance "makes no sense," and one can't see how they could prescribe to the measuring rods which are dictatorially dominated by the two masses (the sole source of the g_{mn} are the two masses) to determine this or that distance between the two dictator masses, since they themselves don't know whether they are far apart or near to each other. (This is, naturally, atrociously formulated.)

5. Can one formulate your cosmological gravitation equations in such a way that no coefficients occur in them "whose numerical value still depends on the chosen mass units"? I ask this, since relying on not carefully defined measurement scales is very disturbing, in particular, for the clarification of the above questions.

The above questions are NOT the questions that my wife posed, but they would help me to continue the discussion with her if you could answer them via Tanitschka (to save you the effort of writing!).

I hope that your answers will make it clear to me how the last paragraph of your talk on the ether in Leiden[4] fits into your cosmological considerations, especially with the idea that the g-field "originates" completely from the masses. (Review the paragraph!!)

Excuse the disturbance. But I believe that it would be really important if you would clarify the whole thing!

With my heartfelt greetings!

Greetings to your wife, and to Ilse and Margot.[5] And please do tell Margot that she should come to Leiden and have a look at Holland. Galinka[6] and I would be very happy to have her here for some time!!!

I am playing quite often now many organ works of Bach (see in particular volume 1!!)[7] with Galerl (she unashamedly rasps *prima vista*)[8] and, from various cantatas and oratorios, the soprano, alto, and tenor parts. Every Sunday afternoon, we listen to a usually excellent performance of a Bach cantata.[9] Now, every Sunday for half a year. Sometimes, a cantata will be repeated after three or four weeks. Pawlik[10] despises nearly all music except for that of Bach, Mozart, and Brahms. Unfortunately, he can't play any instrument. He has a fine ear. And intense pleasure from the really musical within the music.

89. Message to Thomas A. Edison[1]

[Caputh, on or before 21 October 1929][2]

The historians who come after us will consider our time to have been highly significant, due to the deep changes in human society brought about by the enormous technical developments which are taking place. Human and animal muscles have been replaced by machines, and the surface of our planet has been fused into a single economic unit by improvements in transportation and logistics, as well as communications. These developments have supplied the preconditions for the political unification of humanity, which, however, can proceed only very slowly because of the backwardness of human traditions and values.

Among the heroes of this global-historical process, you, honored Mr. Edison, have played a preeminent role by contributing greatly to putting the electrical force at the service of humanity. Accept the calls of thanks and of joyous admiration from your contemporaries, as one whose life has been rendered priceless by his work.

90. Calculations

[Caputh, on or before 21 October 1929][1]

[Not selected for translation.]

91. To Eduard Einstein

[Caputh, 21 October 1929][1]

Dear Tetel,

I was very happy about your two cheerful letters, and also especially that it went so beautifully in Florence for you and my sister.[2] Then you will also perhaps go there sometimes if, in doing so, the strict father may not also be totally forgotten. When I see you again, you will already be a well-appointed defiler of corpses and *studiosus*.[3] I am now living in the new cottage in Caputh (until the end of the month),[4] which brings me great joy despite the impoverishment connected to it. One has a wonderful view, and the sail boat is beyond any comparison. Today, I went out on the water with Ehrenfest's daughter, Tanja, who is a great light and must spout my new theories in Göttingen and Leiden.[5] This evening, I congratulated Edison by overseas radio,[6] which made a great fuss. I still don't know yet whether my theory is true. It has, however, grown quite stately and round under the general distrust of the herd of colleagues.

The solitude of rural living pleases me magnificently. I believe that we will move out here completely and, by and by, give up city living. The solitary walks make me indescribably happy.

So, my dear *studiosus*. Get well acquainted with the severe muse, and best wishes to you and Mama[7] from your

Papa

92. To Maja Winteler-Einstein

[Caputh, 21 October 1929][1]

[Not selected for translation.]

93. To Paul Ehrenfest

[Caputh, 22 October 1929][1]

[Not selected for translation.]

94. To Thomas A. Edison

[Caputh, 22 October 1929][1]

Dear Mr. Edison,

I take singular pleasure in being able to send you these heartfelt good wishes from so far away.[2] These days, intellectual workers, and especially physicists and technicians around the world, commemorate you and your work with admiration and congeniality. You have been equally successful as a pioneer, a developer, and an organizer.

Your construction of the electric light bulb has made a huge contribution to the development of an enormous electrical industry.[3]

The great creators of technology, of which you are one of the most successful, in the course of a century, have brought mankind into a completely new situation, to which it is far from being adapted. Today, man needs hardly any muscular work to produce the goods necessary for life; he no longer needs to be a motor or a slave.

The great creators of coming generations will probably be organization men who will have to see to it that the powers of technology no longer serve war, but rather people's economic security and liberation.[4]

Albert Einstein

95. From Thomas A. Edison

[Dearborn, 22 October 1929][1]

[See documentary edition for English text.]

96. To Paul Ehrenfest

[Caputh, after 22 October 1929][1]

Dear Ehrenfest,

I had already discussed the questions with Tanja, but I couldn't really understand the reasoning behind them.[2] Now I will try to answer your individual questions.[3]

(1 and 2) If the world were considered to be quasi-Euclidean,[4] i.e., not as closed, and the two masses rest against each other, then the distance can be measured with practically massless immobile rods and unit measuring sticks. If measuring rods and clocks are not allowed because they might disturb the field, then one cannot measure the distances *directly*. Then the metric expresses itself only indirectly (coincidences of a different kind), but this is not a very essential point.

A closed world with only two masses has not been possible to construct.[5] Whether it would be possible according to the theory (*statically, it is certainly not possible*), we don't know. In nonstatic cases, the *spatial* distance has no significance at all, only the four-dimensional $ds = \sqrt{g_{\mu\nu}dx^{\mu}dx^{\nu}}$.[6]

(3) In this case, if the metric is quasi-static and measuring rods are allowed, the spatial distance can be defined and measured.[7]

I can see that these answers do not lead to anything.[8] I believe that this whole series of questions arises from the attempt to see only the masses as real, and not the four-dimensional spacetime continuum; but instead to replace the latter through the concept of distance. That is, however, not a possible way to see the problem. I believe, finally, that the matter dissolves in the continuum and not the reverse.[9]

As to the cosmological term, it has the dimensions $(l(\text{length}))^{-2}$, and thus of the numerical order of the hydrogen atom's diameter. But I consider the cosmological matter to be a presumption about which the Lord God is merely laughing. The whole theory is not *so* true as to be able to bear such an extrapolation—even though it was fun to consider it just once!

97. To Nicholas Roerich[1]

[before 24 October 1929][2]

[See documentary edition for English text.]

98. On Art Films

[*Einstein 1929tt*]

PUBLISHED 25 October 1929

IN: *The New York Times*, 25 October 1929, p. 26.

[See documentary edition for English text.]

99. To Eduard Büsching[1]

Berlin, 25 October 1929

Dear Sir,

I regard your book to be a genuine book of religious edification.[2] Its title, however, seems misleading to me. It should be called something like "There is no *personal* God." We followers of Spinoza[3] see our God in the wonderful order and lawfulness of all that exists and in its soul as it reveals itself in man and animal.

It is a different question whether belief in a personal God should be contested. Freud endorsed this view in his latest publication.[4] I myself would never engage in such a task. For such a belief seems to me preferable to the lack of any transcendental outlook of life, and I wonder whether one can ever successfully render to the majority of mankind a more sublime means in order to satisfy its metaphysical needs.

Respectfully yours,

100. To *Die Weltbühne*[1]

[Berlin,] 25 October 1929

Dear Editors,

Perhaps it may interest you to know that the press for "political" reasons gave a truncated account of my short speech to Edison.[2] When one's own self in even a trivial matter has anything to do with the daily press, one sees how distorted by bias the reports are, which the public—in Switzerland, jokingly called the "Sovereign"—receives of events, hardly different than in the early days of the personified sovereign. I enclose a transcript of my remarks here.

Respectfully yours,

101. To Caesar Koch[1]

[Caputh, 26 October 1929][2]

[Not selected for translation.]

102. On the Hedwig Wangel Foundation[1]

"Über die Hedwig Wangel Stiftung"

[*Einstein 1929zz*]

DATED 1 November 1929

IN: *Das Tor der Hoffnung. 4. und 5. Jahresbericht der Hedwig Wangel-Hilfe e.v.* 1 April 1928–31 December 1929, p. 2.

[Not selected for translation.]

103. From Elisabeth, Queen of the Belgians

Laeken, 1 November 1929

Dear Mr. Einstein,

I hear from Madame Barjansky[1] that you will be in Paris on the 6th.[2]

It would please us greatly if you were to come to Brussels afterward and visit us in Laeken.—Bring your violin with you again as well.[3] So, *hopefully* see you again soon!

With warmest regards, your

Elisabeth

104. To Michele Besso

[Berlin,] 4 September [November] 1929[1]

[Not selected for translation.]

105. Statement on Euthanasia

"Stellungnahme zur Sterbehilfe"

[*Einstein 1929vv*]

DATED 6 November 1929

PUBLISHED 7 November 1929

IN: *The Richmond Times Dispatch*, 7 November 1929, p. 3.

[See documentary edition for English text.]

106. To Hugo Bergmann

Berlin, 6 November 1929

Dear Mr. Bergmann,

I'm terribly sorry to have to bother you once again with this matter. I find the spirit of your letter excellent,[1] but I have to object to the form, from the psychological standpoint, that you blame and bemoan instead of simply postulating energetically. That is to say, when one blames and bemoans, one lets the nationalistic reader feel that he still has many allies. So one is then thrust into the opposition. So please turn the positive side toward the outside, and send me the revised letter again, together with the first version.

Incidentally, in a note for the English press, I have already had the opportunity to draw attention to you in connection with the Jewish-Arabic organizational problem.[2]

Kind regards, your

A. Einstein

107. To Martha Steinitz[1]

[Berlin,] 6 November 1929

Dear Mrs. Steinitz,

I too would welcome it if from the Jewish side something were to be contributed so that no death sentences are carried out on the victims of political incitement.[2] I would gladly be ready to join such a campaign myself but, as an individual, would not like to intervene all that much. If you know of a practical, viable way, please follow up.

Kind regards, your

108. To Elsa Einstein

[Paris,] Thursday evening, [7 November 1929][1]

Dear Else,

After a pleasant journey without incident, I have arrived here. Langevin, Haber(!), Ogden, Alice, Botsch, Secr. Kühn[2] were at the station. This evening, I want to go to the Steinhardts right away, because I have time. I have a huge apartment at my disposal. Borel[3] was already there. The ambassador is in the chamber, where a very important meeting is taking place.[4] It is still uncertain when I will see him. I am not accepting private invitations. Instead, there is a tea on Sunday for my colleagues to attend. Langevin is coming midday tomorrow. Tomorrow afternoon is my first lecture.[5] For the time being, the whole business amuses me greatly. I do not believe that I will die from it; but that remains to be seen.

Love and kisses, your

Albert

109. "Unified Theory of the Physical Field"

"Théorie unitaire du champ physique"
[*Einstein 1930u*]

PRESENTED 8 and 12 November 1929
PUBLISHED after 20 March 1930

IN: *Annales de l'Institut Henri Poincaré* 1 (1930): 1–24.

1. The "unified theory of the physical field" proposes to advance the general theory of relativity by combining, within a single discipline, the theories of the electromagnetic field and of the gravitational field. [p. 21]

At the present time, this new theory is nothing more than a mathematical construct, only loosely related by some very tenuous connections to physical reality. It has been discovered by means of entirely formal considerations, and its mathematical consequences have not been sufficiently developed to permit its comparison with experience. Nevertheless, this attempt seems to me to be very interesting for its own sake; it above all offers splendid possibilities for future development, and it is in the hope that the interest of mathematicians will be awakened that I permit myself to present and analyze it here.

2. From the formal point of view, the fundamental idea of the general theory of relativity is the following: the four-dimensional space in which phenomena occur is not amorphous, but rather possesses a structure; the existence of the latter makes itself apparent in the Riemannian metric of that space.

Physically, this indicates the existence of a fundamental quadratic form

$$ds^2 = g_{\mu\nu} dx^\mu dx^\nu$$

characteristic of that space, which expresses its metric, and which, set equal to zero, [p. 2]

$$ds^2 = g_{\mu\nu} dx^\mu dx^\nu = 0$$

defines the law of the propagation of light in that space. This quadratic form is thus intimately related to physical reality; its introduction is not a simple intellectual game, and its use is justified by the correspondence that one can establish between its coefficients $g_{\mu\nu}$ and a system of known phenomena—the phenomena of gravitation.

The structure of space having been defined by the fundamental quadratic form, the problem which then arises is the following: What is the *simplest* law which can be imposed on the coefficients $g_{\mu\nu}$? The answer is given by the tensor theory of Riemann. Starting from the $g_{\mu\nu}$ and their derivatives, a tensor $R^{i}{}_{k,lm}$,[1] called Riemann's curvature tensor, can be constructed. By contracting it with respect to its indices i and m, one obtains another tensor of second rank, R_{kl}. The simplest law to which one can subject the $g_{\mu\nu}$ is expressed by the equation

$$R_{kl} = 0 .$$

This theory would be the ideal physical theory if it could describe the force fields which really exist in nature, that is to say, the ensemble of the gravitational field and the electromagnetic field. But the equation $R_{kl} = 0$, which seems to describe the gravitational phenomena, absolutely fails to take into account the electromagnetic phenomena. The metric alone does not suffice to describe this ensemble.

In order to characterize space completely, one thus tries to find a linear form $\varphi_{\mu}dx^{\mu}$ in addition to the fundamental form $g_{\mu\nu}dx^{\mu}dx^{\nu}$; its coefficients φ_{μ} would then be the components of the electromagnetic vector potential. The complete field equations would then take on the form $R_{kl} + T_{kl} = 0$, T_{kl} being a term dependent upon the potentials, for example, the electromagnetic tensor of Maxwell, or something analogous. However, this procedure is not satisfactory. In fact, the equation $R_{kl} + T_{kl} = 0$ comprises two *independent* terms; logically, one could change one of them without affecting the other. One thus introduces, in this manner, two inde- [p. 3] pendent elements into the theory, corresponding to *two* "states" of space. Nature would then present a lack of unity in which our minds absolutely refuse to believe. On the contrary, it would seem to be more satisfying to attribute this fault to the imperfection of the theory and to search for a means to complete it, to enrich it, to attain the unity to which our minds tend indomitably.

The unified theory of a physical field thus begins by ascertaining that the metric alone cannot suffice to describe the phenomena. However, it furnishes us with at least a part of the truth; it certainly possesses a physical substratum. The problem is thus reduced to that of finding the element which will complement the metric and which will permit the description of space without leaving anything out.

3. To that end, let us search for the sense of the notion of the Riemannian metric and which representations can be imputed to it.

We consider a continuum of n dimensions which presents a Riemannian structure. Such a continuum is characterized by the fact that Euclidean geometry is valid within an infinitely small domain around a given point. Furthermore, if one considers two points A and B at a finite distance apart, one can compare between them the *lengths* of two linear elements situated at A and at B, but one cannot do the same for their *directions*; in Riemannian geometry, there is no distant parallelism.

4. Let us imagine, in such a space, at a given point, a Cartesian coordinate system, i.e., a system of n orthogonal axes, each one carrying a unit vector. We will call such a system of axes an n-Bein.

The infinitesimal Euclidean domain around a point is completely characterized when an n-Bein is defined at that point. The metric of the space is known if one has attached an n-Bein *to each point* of that space. But inversely, the metric of a Riemannian space does not suffice in order to determine, in an unequivocal manner, an n-Bein for each point. In fact, the metric of the space remains unchanged if one submits all of the n-Beins to arbitrary rotations. When one specifies only the metric, the orientation of the n-Beins is not defined; a certain arbitrariness still remains in the determination of the structure of the space. One thus can see in this manner that the description of spaces in terms of n-Beins is, in a way, richer than their description by means of the fundamental quadratic form. One could suspect that the arbitrariness introduced by this description might provide a means of reconnecting the structure of the space to the cause of electromagnetic phenomena, which had hitherto not found its place within the theory. [p. 4]

This is not the first time that such spaces have been envisioned. From a purely mathematical point of view, they had already been studied previously. M. Cartan[2] was so kind as to write a note for the *Mathematische Annalen* which traces the diverse phases of the formal development of these considerations.

Suppose that we have attached an n-Bein at A; the structure of the space will then be defined if we place an arbitrary n-Bein at every other point, taking them all to be parallel to the first one by definition. One can thus establish, in addition to a relation of magnitude between any two points in the space, also a relation of direction. The notion of distant parallelism now possesses a precise sense, which it lacks in the Riemannian theory. Two vectors which originate at two finitely separated points will be parallel if they have the same components in their respective local systems. When one characterizes the structure of a space by a field of n-Beins, one expresses, *at the same time*, the existence of a Riemannian metric and of a distant parallelism; between two infinitesimal elements of the space, there is now not only a relation of magnitude expressed by the metric, but also a directional relation expressed by the orientation of the n-Beins.

In summary, the only new hypothesis which need be introduced in order to arrive at a geometry more complete than that of Riemann concerns the existence of "directions" in the space and of relations among those directions. This notion of "directions" is not contained either in the notion of the continuum nor in that of space. We thus require a supplementary hypothesis to allow for the possibility that the space contains something like directional relations, expressed by the existence of parallelism at finite distances.

It is, however, easy to see that, even with the hypothesis of distant parallelism tied to that of a Riemannian metric, the field of *n*-Beins is defined only up to a rotation (common to all the *n*-Beins).

[p. 5] 5. Let us introduce a Gaussian system of general coordinates and consider the *n*-Bein attached at the point *P*. Let $h_s^{\ \nu}$ be the components of the unit vectors of the *n*-Bein in the Gaussian coordinate system. In what follows, every Greek index will refer to those coordinates, and every Roman index will refer to the local *n*-Bein. The $h_s^{\ \nu}$ thus represent the νth component of the unit vector corresponding to the *s*-axis of the *n*-Bein. In a four-dimensional space, $n = 4$, we then have sixteen quantities $h_s^{\ \nu}$, which completely describe the structure of the space.

If these quantities are given, one can calculate the components of any vector A in a local system as a function of its components in the Gaussian system. One has

$$A^{\nu} = h_s^{\ \nu} A_s, \tag{1}$$

and, inversely,

$$A_s = h_{s\nu} A^{\nu}, \tag{2}$$

where the $h_{s\nu}$ are the minors of the determinant $h = |h_s^{\ \nu}|$ divided by h. Following this convention, the summation over indices that appear twice needs to be made.

To obtain the metric of this space, we calculate the magnitude of a vector A. In a local system where Euclidean geometry holds, we find:

$$A^2 = \sum A^2 = \sum h_{s\mu} h_{s\nu} A^{\mu} A^{\nu}. \tag{3}$$

The coefficients of the fundamental metric form $g_{\mu\nu} = dx^{\mu} dx^{\nu}$ will thus be given by

$$g_{\mu\nu} = h_{s\mu} h_{s\nu} \tag{4}$$

We can thus see that a field of *n*-Beins, $(h_s^{\ \nu})$ completely determine the metric $(g_{\mu\nu})$, but the inverse is not true.

The quantities $h_s^{\ \nu}$ form the fundamental tensor analogous to the tensor $g_{\mu\nu}$ in the old theory; for the case of $n = 4$, there are sixteen quantities $h_s^{\ \nu}$, but only ten $g_{\mu\nu}$.

The concept of tensor is broadened in this theory. In fact, we have to consider here not only coordinate transformations, but also those which modify the orientation of the *n*-Beins. The *n*-Beins are determined only up to a rotation; the only admissible relations must thus be invariant with respect to such rotations; for [p. 6] example, a change of the coordinate system and at the same time of the orientation of the local systems. Rotations are characterized by the constant coefficients α_{st}, independently of the coordinates and obeying the relations

$$\alpha_{s\mu}\alpha_{s\nu} = \alpha_{\mu s}\alpha_{\nu s} = \delta_{\mu\nu} = \begin{cases} 1 & \mu = 1 \\ 0 & \mu \neq \nu \end{cases} \tag{5}$$

leading to

$$h_s^{\ \nu'} = \alpha_{st}\frac{\partial x^{\nu'}}{\partial x^{\rho}} h_t^{\ \rho} . \tag{6}$$

To each local index, there corresponds a transformation α, and to each Greek index, there corresponds an ordinary transformation.

6. The algebraic laws which govern these tensors are nearly the same as those which govern the tensors of the absolute differential calculus. One can define the sum and the difference of two tensors *T* and *S* with the same indices. The product of two tensors obeys the same transformation law as a tensor of higher rank.

The contraction operation is also applicable, both with respect to Greek and Roman indices. For the former, it is always necessary that an upper index be identical to a lower index. Exchange of indices is possible; in particular, one can replace a Roman index by a Greek index by making use of the fundamental tensor $h_s^{\ \nu}$. For example, consider the tensor $T :_s^{\ \lambda}$. It obeys the relation

$$h_s^{\ \tau} T :_s^{\ \lambda} = T^{\tau\lambda} . \tag{7}$$

One can thus go from the local components to the components of the same tensor in the Gaussian system and vice versa.

Finally, we calculate the volume element in this new theory. This important quantity has the following expression in the theory of general relativity:

$$d\Omega = \sqrt{g} \cdot d\tau$$

where

$$g = |g_{\mu\nu}| \quad \text{and} \quad d\tau = dx^1 \cdot dx^2 \ldots$$

$$g_{\mu\nu} = h_{\mu s} \cdot h_{\nu s} \quad \text{and} \quad g = h^2 ;$$

Now one has [p. 7]

$$d\Omega = h d\tau. \tag{8}$$

The fact that the irrational has disappeared is an additional advantage of the new theory.

7. Let us now consider the parallel displacement of a vector A^{μ}. In a Riemannian manifold, that displacement is given by the formula

$$\delta A^{\mu} = -\Gamma^{\mu}_{\alpha\beta} A^{\alpha} \delta x^{\beta}.$$

The $\Gamma^{\mu}_{\alpha\beta}$ are the Christoffel symbols, which must satisfy two conditions:

1. The translation defined by them has to conserve the metric, that is to say, it leaves the magnitudes of the vectors considered invariant; and
2. The $\Gamma^{\mu}_{\alpha\beta}$ must be symmetric with respect to α and β.

$$\Gamma^{\mu}_{\alpha\beta} = \Gamma^{\mu}_{\beta\alpha}.$$

In this geometry, the parallel displacement is not integrable. If one carries it out along a closed curve, the initial vector does not coincide with the final vector, and their difference is measured by the Riemann tensor $R^{i}_{k,lm}$.

In the new theory, things appear in a different light. The parallel displacement of a vector A is given by an analogous formula:

$$\delta A^{\mu} = \Delta^{\mu}_{\alpha\beta} A^{\alpha} \delta x^{\beta}. \tag{9}$$

But here, the displacement is integrable: if one moves a vector along a closed curve, the initial vector will always coincide with the final vector. The Riemann tensor generated starting from the $\Delta^{\mu}_{\alpha\beta}$ will therefore be identically equal to zero. Furthermore, the $\Delta^{\mu}_{\alpha\beta}$ are not symmetric with respect to the indices α and β. One can readily verify these results by calculating the expression for the $\Delta^{\mu}_{\alpha\beta}$ as a function of the h.

Let us consider two neighboring points x^{β} and $x^{\beta} + dx^{\beta}$; their n-Beins are "parallel" to each other. The vectors A_s and $A_s + \delta A_s$ will be parallel if they have the same components in the two n-Beins. The condition of parallel displacement from x^{β} to $x^{\beta} + dx^{\beta}$ is thus $\delta A_s = 0$.

[p. 8] By expressing A_s as a function of the components of A in the Gaussian system

$$A_s = h_{s\mu} A^{\mu},$$

one obtains

$$\delta(h_{s\mu}A^{\mu}) = 0. \tag{10}$$

Multiplying by $h_s^{\ \sigma}$, one can deduce

$$0 = h_s^{\sigma}\left\{h_{s\mu}\delta A^{\mu} + A^{\alpha}\cdot\frac{\partial h_{s\alpha}}{\partial x^{\beta}}\delta x^{\beta}\right\} \tag{11}$$

where, denoting ordinary differentiation by a comma (,), we write

$$\Delta^{\mu}_{\alpha\beta} = h^{\mu}_{s}h_{s\alpha,\,\beta} \tag{12}$$

and also

$$\Delta^{\mu}_{\alpha\beta} = -h_{s\alpha}h_{s,\beta}^{\ \ \mu}. \tag{13}$$

By the same mechanism as the one utilized in absolute differential calculus, one can generate the operator for covariant differentiation, starting from the $\Delta^{\mu}_{\alpha\beta}$. Denoting this operation by a semicolon (;), we find for a contravariant tensor of first rank

$$A^{\mu}{}_{;\sigma} = A^{\mu}{}_{,\sigma} + A^{\alpha}\Delta^{\mu}_{\alpha\sigma} \tag{14}$$

and, for a covariant tensor of first rank,

$$A_{\mu;\sigma} = A_{\mu,\,\sigma} - A_{\alpha}\Delta^{\alpha}_{\mu\sigma}. \tag{15}$$

One finds analogous formulas for tensors of higher ranks. They are similar to the formulas of the absolute differential calculus based exclusively on the metric, and they can be deduced in the same manner.

One can easily verify that the covariant derivative of the fundamental tensor is identically zero,[3]

$$h_{s;\tau}^{\ \nu} = h_{s\nu;\tau} = g_{\sigma\tau;\rho} = g^{\sigma\tau}{}_{;\rho} \equiv 0. \tag{16}$$

One finds, in fact,

$$h_{s;\tau}^{\ \nu} = h_{s,\tau}^{\ \nu} + h^{\alpha}_{s}\Delta^{\nu}_{\alpha\tau} = \delta_{st}(h_{t,\tau}^{\ \nu} + h^{\alpha}_{t}\Delta^{\nu}_{\alpha\tau}) = h^{\alpha}_{s}(h_{t\alpha}h_{t,\tau}^{\ \nu} + \Delta^{\nu}_{\alpha\tau}) \equiv 0.$$

The covariant derivative of a product of two tensors is obtained by the ordinary rule from differential calculus. For example, consider two tensors of arbitrary rank, T: and S: [p. 8]

$$(T{:}S{:})_{;\tau} = T{:}_{;\tau}S{:} + T{:}S{:}_{;\tau}. \tag{17}$$

Two covariant derivatives do not commute, that is to say that the order of the derivatives is important. Let T:: be an arbitrary tensor. Take its successive covariant derivatives—first in the order σ, τ, then in the order τ, σ—and then take their difference. This leads us to the fundamental formula

$$T{::}_{;\sigma;\tau} - T{::}_{;\tau;\sigma} \equiv -T{::}_{;\alpha}\Lambda^{\alpha}_{\sigma\tau}, \tag{18}$$

where

$$\Lambda^{\alpha}_{\sigma\tau} = \Delta^{\alpha}_{\sigma\tau} - \Delta^{\alpha}_{\tau\sigma}.$$

It is easy to demonstrate this formula for simple cases. Suppose, initially, that *T*:: reduces to a scalar ψ. In this case, the second covariant derivative is the same as the ordinary derivative

$$\psi_{;\sigma} = \psi_{,\sigma},$$

and we have

$$\psi_{;\sigma;\tau} = \psi_{,\sigma,\tau} - \psi_{,\alpha}\Delta^{\alpha}_{\sigma\tau}$$

$$\psi_{;\tau;\sigma} = \psi_{,\tau,\sigma} - \psi_{,\alpha}\Delta^{\alpha}_{\tau\sigma}$$

$$\psi_{;\sigma;\tau} - \psi_{;\tau;\sigma} = -\psi_{,\alpha}(\Delta^{\alpha}_{\sigma\tau} - \Delta^{\alpha}_{\tau\sigma}) = -\psi_{;\alpha}\Lambda^{\alpha}_{\sigma\tau}.$$

The difference indeed has the form mentioned. One can see that $\Lambda^{\alpha}_{\sigma\tau}$ is a tensor.

The case of a vector $T:: = A^{\mu}$ can be reduced to the previous case if we take into account that, in this theory, distant parallelism exists. In fact, the existence of this parallelism involves the possibility of the existence of a uniform field of vectors (parallel fields); it is possible to imagine that, at each point in space, there would be a vector pointing in the same direction as a given vector.

Given this, let us consider a uniform field of *arbitrary* vectors a_{μ}; one can readily show that $a_{\mu;\sigma} = a^{\nu}{}_{;\sigma} = 0$. Using the given vector A_{μ}, we form the scalar

$$\psi = A^{\mu} \cdot a_{\mu}$$

[p. 10] Now we can apply to this scalar the formula given above for the difference *D*. Then, taking into account the rule for the derivative of a product, we find $(A^{\mu}_{\alpha\mu})_{;\sigma} = a_{\mu}A^{\mu}{}_{;\sigma}$. The arbitrary quantities a^{μ} become common factors and can be canceled, and finally, we arrive at a relation of the same form

$$A^{\mu}{}_{;\sigma;\tau} - A^{\mu}{}_{;\tau;\sigma} \equiv -A^{\mu}{}_{;\alpha}\Lambda^{\alpha}_{\sigma\tau},$$

which is easy to generalize to a tensor of any rank.

8. An important difference between the theory presented here and the theory of Riemann is worth our undivided attention. In Riemann's theory, there is no tensor which could be expressed uniquely by means of the first derivatives of the fundamental tensor. In ours, the difference

$$\Lambda^{\alpha}_{\mu\nu} = \Delta^{\alpha}_{\mu\nu} - \Delta^{\alpha}_{\nu\mu} \tag{19}$$

is a tensor which contains only first derivatives. Furthermore, this tensor is remarkable because, in a certain sense, it is the analogue of Riemann's tensor: *if* Λ *is zero, the continuum is Euclidean.*[4]

This result can be readily established. According to the formula given for $\Delta^{\alpha}_{\mu\nu}$, one has

$$\Lambda^{\alpha}_{\mu\nu} = h_s{}^{\alpha}(h_{s\mu,\nu} - h_{s\nu,\mu}) = 0\,.$$

Multiplying by $h_t{}^{\alpha}$, one obtains, due to $h_s{}^{\alpha}h_{t\alpha} = \delta_{st}$,[5] the following result:

$$h_{t\mu,\nu} - h_{t\nu,\mu} = 0$$

$h_{t\mu}$ thus has the form

$$h_{t\mu} = \frac{\partial\psi_t}{\partial x^{\mu}}\,.$$

If we take as the Gaussian coordinates precisely the ψ_t, which is possible; $\psi_t = x^t$, then the expressions

$$h_{t\mu} = \delta_{t\mu} = \begin{cases} 1 & t = \mu \\ 0 & t \neq \mu \end{cases}$$

are constants; they can be ordered into a matrix form where the diagonal terms are 1 and the off-diagonal terms are 0. Since the $h_{s\mu}$ and the $g_{\mu\nu}$ are constants, the continuum is Euclidean.

9. Let us consider the quantity Λ, which will play a fundamental role in the new theory. Altogether, there are $6 \times 4 = 24$ of these quantities Λ; however, there are [p. 11] only sixteen of the h. Thus, there must be a certain number of identities which relate the various Λ among themselves. In order to find them, let us begin with the expression for Λ as a function of the Δs. Since the parallel displacement is integrable, the "curvature" tensor analogous to the Riemann tensor will thus be identically equal to zero. As a result, we have

$$\Delta^{\iota}_{\kappa\lambda,\mu} - \Delta^{\iota}_{\kappa\mu,\lambda} - \Delta^{\iota}_{\sigma\lambda}\cdot\Delta^{\sigma}_{\kappa\mu} + \Delta^{\iota}_{\sigma\mu}\Delta^{\sigma}_{\kappa\lambda} \equiv 0\,. \quad (20)$$

We carry out cyclic permutations of the indices κ, λ, and μ, and take the sum, and then introduce the covariant derivative instead of the ordinary derivative. We thus arrive at the following identity for the Λs:

$$(\Lambda^{\iota}_{\kappa\lambda;\mu} + \Lambda^{\iota}_{\lambda\mu;\kappa} + \Lambda^{\iota}_{\mu\kappa;\lambda}) + (\Lambda^{\iota}_{\kappa\alpha}\Lambda^{\alpha}_{\lambda\mu} + \Lambda^{\iota}_{\lambda\alpha}\Lambda^{\alpha}_{\mu\kappa} + \Lambda^{\iota}_{\mu\alpha}\Lambda^{\alpha}_{\kappa\lambda}) \equiv 0\,. \quad (21)$$

Contracting once with respect to i and μ, and setting $\Lambda^{\alpha}_{\mu\alpha} = \varphi_{\mu}$, we obtain another important identity:

$$\Lambda^{\alpha}_{\mu\nu;\alpha} - \left(\frac{\partial \varphi_\mu}{\partial x^\nu} - \frac{\partial \varphi_\nu}{\partial x^\mu}\right) \equiv 0 . \qquad (22)$$

In order to deduce another identity, we need to apply the rule for permutations of covariant derivatives, which can be expressed as

$$T{:}_{;\sigma;\tau} - T{:}_{;\tau;\sigma} = -T{:}_{;\alpha}\Lambda^{\alpha}_{\sigma\tau} .$$

Now, we introduce a new notation: we agree that an underlined index can change its position, that is to say, it can be raised or lowered. For example, writing $\Lambda^{\alpha}_{\underline{\mu}\underline{\nu}}$ signifies that we take the contravariant components of $\Lambda^{\alpha}_{\mu\nu}$,

$$\Lambda^{\alpha}_{\underline{\mu}\underline{\nu}} = \Lambda^{\alpha}_{\mu\nu} g^{\mu\sigma} g^{\nu\tau} .$$

With this convention, we apply the preceding rule to $\Lambda^{\alpha}_{\underline{\mu}\underline{\nu}}$, taking its derivatives with respect to ν and to α. We then have

$$\Lambda^{\alpha}_{\underline{\mu}\underline{\nu};\nu;\alpha} - \Lambda^{\alpha}_{\underline{\mu}\underline{\nu};\alpha;\nu} \equiv -\Lambda^{\alpha}_{\underline{\mu}\underline{\nu};\sigma}\Lambda^{\sigma}_{\nu\alpha} .$$

The right-hand side can be written as

$$-\Lambda^{\alpha}_{\underline{\mu}\underline{\nu};\sigma}\Lambda^{\sigma}_{\nu\alpha} \equiv -\,(\Lambda^{\alpha}_{\underline{\mu}\underline{\nu}}\Lambda^{\sigma}_{\nu\alpha})_{;\sigma} + \Lambda^{\alpha}_{\underline{\mu}\underline{\nu}}\Lambda^{\sigma}_{\nu\alpha;\sigma} .$$

[p. 12] In the first term of the right-hand side, we change the names of the mute indices σ, α, and ν into α, σ, and τ; that term thus becomes:

$$-(\Lambda^{\sigma}_{\underline{\mu}\underline{\tau}}\Lambda^{\alpha}_{\tau\sigma})_{;\alpha} \equiv +(\Lambda^{\sigma}_{\underline{\mu}\underline{\tau}}\Lambda^{\alpha}_{\sigma\tau})_{;\alpha} .$$

We thus have

$$\Lambda^{\alpha}_{\underline{\mu}\underline{\nu};\nu;\alpha} - (\Lambda^{\sigma}_{\underline{\mu}\underline{\tau}}\Lambda^{\alpha}_{\sigma\tau})_{;\alpha} - \Lambda^{\alpha}_{\underline{\mu}\underline{\nu};\alpha;\nu} - \Lambda^{\alpha}_{\underline{\mu}\underline{\nu}}\Lambda^{\sigma}_{\nu\alpha;\sigma} \equiv 0 ,$$

or

$$(\Lambda^{\alpha}_{\underline{\mu}\underline{\nu};\nu} - \Lambda^{\sigma}_{\underline{\mu}\underline{\tau}}\Lambda^{\alpha}_{\sigma\tau}) - \Lambda^{\alpha}_{\underline{\mu}\underline{\nu};\alpha;\nu} - \Lambda^{\alpha}_{\underline{\mu}\underline{\nu}}\Lambda^{\sigma}_{\nu\alpha;\sigma} \equiv 0 , \qquad (23)$$

which is exactly the identity that we were seeking. We further introduce the notations:

$$G^{\mu\alpha} \equiv \Lambda^{\alpha}_{\underline{\mu}\underline{\nu};\nu} - \Lambda^{\sigma}_{\underline{\mu}\underline{\tau}}\Lambda^{\alpha}_{\sigma\tau}$$

$$F^{\mu\nu} \equiv \Lambda^{\alpha}_{\underline{\mu}\underline{\nu};\alpha} .$$

The identity (23) can then be written as:

$$G^{\mu\alpha}{}_{;\alpha} - F^{\mu\alpha}{}_{;\alpha} - \Lambda^{\alpha}_{\underline{\mu}\underline{\nu}} F_{\nu\alpha} \equiv 0 . \qquad (24)$$

10. Having defined a method for mathematically describing the structure of space, let us now examine the fundamental problem of the theory, which is to establish the field equations. As in the general theory of relativity, this problem consists in finding the most simple conditions which can be imposed upon the elements that define the structure of the space, that is to say the quantities $h_s^{\ \nu}$. It is a question of making a choice among various diverse possibilities; the difficulty of this choice resides in the absence of markers which could guide us. Before writing down the definitive field equations, it seems to me that it would be interesting to indicate the route that I followed in order to discover them.

My point of departure was defined by the identities satisfied by the quantities $\Lambda^{\alpha}_{\mu\nu}$. In a general sense, the study of certain identities can be a great help in making the choice of the field equations, by suggesting to us possible forms of the relations that we are seeking. The study of those identities must then, logically, precede the choice of the system of equations. But one cannot know *a priori* which quantities are those among which one must establish the identities.

A first path marker that appears here would seem to be the following: the relations that we are seeking should presumably contain $\Lambda^{\alpha}_{\mu\nu}$ and its derivatives, this tensor being the only one that can be uniquely expressed in terms of the fundamental tensor. [p. 13]

The simplest condition that we can impose would be

$$\Lambda^{\alpha}_{\mu\nu} = 0 .$$

It is evident that this condition is too restrictive: the space would be Euclidean. Furthermore, *it contains only the first derivatives*, and it is probable that the equations which govern natural phenomena are of the second order, as, for example, Poisson's equation.

Then let us try to set

$$\Lambda^{\alpha}_{\mu\nu;\sigma} = 0 .$$

This relation is also not acceptable, since it is nearly equivalent to the first one; but it is useful because it immediately suggests to us that we try to cancel the divergences that one can form starting from $\Lambda^{\alpha}_{\mu\nu;\sigma}$. Let us therefore start from that covariant derivative and contract it in every possible manner (which is equivalent to taking its divergence). We have two possibilities:

either $$\Lambda^{\alpha}_{\mu\nu;\alpha} = 0 \qquad (25)$$

or
$$\Lambda^{\alpha}_{\underline{\mu\nu};\nu} = 0\,. \tag{26}$$

One sees immediately that the ensemble of these systems would not be suitable, because the number of equations cannot be chosen arbitrarily: Without a special investigation, one cannot guarantee the compatibility of these equations. Now, it is indispensable that the system chosen should be such that the equations be compatible.

11. In general, for a space of n dimensions, there are n^2 variables $h_s^{\ \nu}$. But in a generally covariant theory, the choice of the coordinate system being arbitrary, among the n^2 variables, n can be chosen arbitrarily. In consequence, the number of independent equations will be $n^2 - n$. The number of equations could be even [p. 14] greater than $n^2 - n$, provided that they are related by a suitable number of identities which make the system compatible. In any case, the system must satisfy the rule that the *excess of the number of equations over the number of identities must be equal to the number of variables minus n.*

Consider, for example, the equations of general relativity. We have 10 unknown functions $g_{\mu\nu}$; the coordinate system being arbitrary, we can choose them in such a way that 4 of the functions $g_{\mu\nu}$ are arbitrary. The 6 unknowns then must satisfy 10 equations. But, as we know, we at the same time have 4 identities,

$$(R^{ik} - \tfrac{1}{2}g^{ik}R)_{;k} \equiv 0$$

which reestablish the compatibility.(1)

One can cite other cases in which the number of equations is greater than the number of unknowns without their being incompatible. For example, the Maxwell equations,

$$\text{rot}\,H - \frac{1}{c}\frac{\partial E}{\partial t} = 0 \qquad \text{rot}\,E + \frac{1}{c}\frac{\partial H}{\partial t} = 0$$

$$\text{div}\,E = 0 \qquad \text{div}\,H = 0$$

consist of eight individual equations with six unknowns; the system is however compatible, since the equations are related by two well-known identities.

What, in the end, is signified by the presence of a greater number of equations than of unknowns?

In the chosen example, the two vectorial Maxwell equations canonically determine the problem. If the fields E and H are given at some moment t, all the rest is

(1)The symbol ";" is employed here with a well-known meaning, different from that defined in the rest of this article.[7]

determined. But the other scalar relations make the initial conditions non-arbitrary. Thus, an over-determination of the problem, a number of equations greater than the number of unknowns (with, however, the identities which make them compatible), in this case lifts the arbitrariness of the initial conditions which would otherwise have existed. It is furthermore clear that a theory which is compatible with empirical evidence is all the more satisfying when it limits such arbitrariness most completely. Having said this, let us return to our problem.

12. For a space of 4 dimensions, $n = 4$, we have 16 unknowns $h_s^{\ \nu}$, of which 4 are arbitrary;[6] thus only 12 of them can be determined by the field equations. The number of equations which at first glance would form a suitable system would be 22, namely the 6 equations (25) and the 16 equations (26). We thus require 10 identities, which in this case do not exist. One can understand in this manner how the condition of compatibility allows us to limit, in an efficient way, the arbitrariness in the choice of the field equations. [p. 15]

Let us now examine the identity (24). It suggests that we take as our field equations the following system:

$$G^{\mu\alpha} = 0 \tag{27}$$

$$F^{\mu\alpha} = 0 \tag{28}$$

or, explicitly,

$$\Lambda^{\alpha}_{\mu\nu;\nu} - \Lambda^{\sigma}_{\underline{\mu}\underline{\tau}}\Lambda^{\alpha}_{\sigma\tau} = 0 \tag{27a}$$

$$\Lambda^{\alpha}_{\underline{\mu}\underline{\nu};\alpha} = 0 \tag{28a}$$

This system, only slightly different from the system (25), (26), always contains 22 equations, chosen in such a way that they satisfy 4 identities of the type (24).

Nevertheless, the excess $22 - 4 = 18$ is still greater than the difference $16 - 4 = 12$. In order to ensure that the new system of equations is compatible, there must be an additional six identities which relate them among themselves. We will prove that those necessary identities exist. To show that, let us first put the equations (28) into a different form, guided by the identity (22),

$$\Lambda^{\alpha}_{\mu\nu;\alpha} - (\varphi_{\mu,\,\nu} - \varphi_{\nu,\,\mu}) \equiv 0\,. \tag{22}$$

We have set $\Lambda^{\alpha}_{\mu\nu;\alpha} = 0$; according to (22), this results also in

$$\frac{\partial\varphi_\mu}{\partial x^\nu} - \frac{\partial\varphi_\nu}{\partial x^\mu} = 0\,.$$

Thus, φ_μ is the derivative of a scalar which we can conveniently designate here by $\log\psi$

$$\varphi_\mu = \frac{\partial \log\psi}{\partial x^\mu}.$$

[p. 16] Let us then set

$$F_\mu = \varphi_\mu - \frac{\partial \log\psi}{\partial x^\mu};$$

we thus have $F_\mu = 0$. Then we can replace the equations

$$\Lambda^{\alpha}_{\underline{\mu}\underline{\nu};\alpha} = 0$$

by the equations $F_\mu = 0$ and write our system of equations as follows:

$$G^{\mu\alpha} = 0 \qquad (29)$$

$$F_\mu = 0 \qquad (30)$$

or

$$\Lambda^{\alpha}_{\underline{\mu}\underline{\nu};\nu} - \Lambda^{\sigma}_{\underline{\mu\tau}}\Lambda^{\alpha}_{\sigma\tau} = 0 \qquad (20a)$$

$$\varphi_\mu - \frac{\partial}{\partial x^\mu}\log\psi = 0. \qquad (30a)$$

Now we have sixteen equations (29) and four equations (30), twenty equations in all. We have introduced a new variable, the scalar ψ; there are thus $16+1 = 17$ unknowns, of which four are arbitrary. In order for the system to be compatible, we must have seven identities between the $G^{\mu\alpha}$ and the $F_\mu = 0$

$$20-(17-4) = 7.$$

We have found only four of them, the identities (24). Now, there are some more identities that connect the quantities envisaged, and—one might say miraculously—there are just three of them. I cannot say what the profound reason for their existence is. It derives essentially from the nature of the space that we are considering. This type of space had been studied, by the way, by several mathematicians before me, notably by Weitzenböck, Eisenhart, and Cartan;[8] I hope that they are willing to help us to discover the hidden origin of these new identities.

Whatever the reason may be, they exist; I will tell you how one can arrive at them.

We decompose the tensor $G_{\mu\alpha}$ into its symmetric part $\underline{\underline{G}}^{\mu\alpha}$ and its antisymmetric part $\underline{G}^{\mu\alpha}$. We then have

$$\begin{cases} 2\underline{G}^{\mu\alpha} = (\Lambda^{\alpha}_{\underline{\mu\nu}} - \Lambda^{\mu}_{\underline{\alpha\nu}})_{;\nu} - \Lambda^{\sigma}_{\underline{\mu\tau}}\Lambda^{\alpha}_{\sigma\tau} + \Lambda^{\sigma}_{\underline{\alpha\tau}}\Lambda^{\mu}_{\sigma\tau} \\ \phantom{2\underline{G}^{\mu\alpha}} = (\Lambda^{\alpha}_{\underline{\mu\nu}} - \Lambda^{\mu}_{\nu\alpha})_{;\nu} - \Lambda^{\sigma}_{\underline{\mu\tau}}\Lambda^{\alpha}_{\sigma\tau} + \Lambda^{\sigma}_{\underline{\alpha\tau}}\Lambda^{\mu}_{\sigma\tau}, \end{cases}$$

since the $\Lambda^{\mu}_{\underline{\alpha\nu}}$ are antisymmetric in α, ν.

One can express $2\underline{G}^{\mu\alpha}$ as a function of $F^{\mu\alpha} = \Lambda^{\nu}_{\mu\alpha;\nu}$ and of a tensor which is [p. 17] antisymmetric with respect to an arbitrary pair of the indices α, μ, ν:

$$S^{\alpha}_{\underline{\mu\nu}} = \Lambda^{\alpha}_{\underline{\mu\nu}} + \Lambda^{\nu}_{\underline{\alpha\mu}} + \Lambda^{\mu}_{\underline{\nu\alpha}}. \tag{31}$$

Evidently, we have

$$2\underline{G}^{\mu\alpha} = S^{\nu}_{\underline{\alpha\mu};\nu} + F^{\mu\alpha} + C,$$

where the complementary term C is given by

$$C = \Lambda^{\sigma}_{\underline{\alpha\tau}}\Lambda^{\mu}_{\sigma\tau} - \Lambda^{\sigma}_{\underline{\mu\tau}}\Lambda^{\alpha}_{\sigma\tau}.$$

In order to calculate it, observe that on exchanging the mute indices σ and τ, we obtain

$$\Lambda^{\sigma}_{\underline{\alpha\tau}}\Lambda^{\mu}_{\sigma\tau} = \Lambda^{\tau}_{\underline{\alpha\sigma}}\Lambda^{\mu}_{\tau\sigma} = -\Lambda^{\tau}_{\underline{\alpha\sigma}}\Lambda^{\mu}_{\sigma\tau}$$

and

$$\Lambda^{\sigma}_{\underline{\mu\tau}}\Lambda^{\alpha}_{\sigma\tau} = \Lambda^{\tau}_{\underline{\mu\sigma}}\Lambda^{\alpha}_{\tau\sigma} = -\Lambda^{\sigma}_{\underline{\mu\sigma}}\Lambda^{\alpha}_{\sigma\tau}.$$

Furthermore, we have the equality

$$\Lambda^{\alpha}_{\underline{\tau\sigma}}\Lambda^{\mu}_{\sigma\tau} = \Lambda^{\alpha}_{\underline{\tau\sigma}}\Lambda^{\mu}_{\sigma\tau}$$

due to

$$\Lambda^{\alpha}_{\underline{\tau\sigma}}\Lambda^{\mu}_{\sigma\tau} = \Lambda^{\alpha}_{\beta\gamma}g^{\beta\tau}g^{\gamma\sigma}\Lambda^{\mu}_{\sigma\tau} = \Lambda^{\alpha}_{\beta\gamma}\Lambda^{\mu}_{\underline{\gamma\beta}} = \Lambda^{\alpha}_{\tau\sigma}\Lambda^{\mu}_{\underline{\sigma\tau}}.$$

Thus,

$$\begin{aligned} -C = \Lambda^{\sigma}_{\underline{\tau\alpha}}\Lambda^{\mu}_{\sigma\tau} - \Lambda^{\sigma}_{\underline{\tau\mu}}\Lambda^{\alpha}_{\sigma\tau} &= \frac{1}{2}(\Lambda^{\sigma}_{\underline{\tau\alpha}} + \Lambda^{\tau}_{\underline{\alpha\sigma}} + \Lambda^{\alpha}_{\underline{\sigma\tau}})\Lambda^{\mu}_{\sigma\tau} \\ &\quad -\frac{1}{2}(\Lambda^{\sigma}_{\underline{\tau\mu}} + \Lambda^{\tau}_{\underline{\mu\sigma}} + \Lambda^{\mu}_{\underline{\sigma\tau}})\Lambda^{\alpha}_{\sigma\tau} \\ -C &= \frac{1}{2}S^{\alpha}_{\underline{\sigma\tau}}\Lambda^{\mu}_{\sigma\tau} - \frac{1}{2}S^{\mu}_{\underline{\sigma\tau}}\Lambda^{\alpha}_{\sigma\tau}. \end{aligned}$$

And finally,

$$2\underline{G}^{\mu\alpha} = -S^{\nu}_{\underline{\mu\alpha};\nu} + \frac{1}{2}S^{\mu}_{\underline{\sigma\tau}}\Lambda^{\alpha}_{\sigma\tau} - \frac{1}{2}S^{\alpha}_{\underline{\sigma\tau}}\Lambda^{\mu}_{\sigma\tau} + F^{\mu\alpha}. \qquad (32)$$

Let us develop the covariant derivative (the underlined indices are contravariant). We have

$$-S^{\nu}_{\underline{\mu\alpha};\nu} = S^{\tau}_{\underline{\alpha\mu};\tau} = S^{\tau}_{\underline{\alpha\mu},\tau} + S^{\sigma}_{\underline{\alpha\mu}}\Delta^{\tau}_{\sigma\tau} + S^{\tau}_{\underline{\sigma\mu}}\Delta^{\alpha}_{\sigma\tau} + S^{\tau}_{\underline{\alpha\sigma}}\Delta^{\mu}_{\sigma\tau}.$$

[p. 18] Now, exchanging σ and τ,

$$S^{\tau}_{\underline{\sigma\mu}}\Delta^{\alpha}_{\sigma\tau} = \frac{1}{2}S^{\sigma}_{\underline{\tau\mu}}\Delta^{\alpha}_{\tau\sigma} = \frac{1}{2}(S^{\tau}_{\underline{\sigma\mu}}\Delta^{\alpha}_{\sigma\tau} + S^{\sigma}_{\underline{\tau\mu}}\Delta^{\alpha}_{\tau\sigma})$$

$$= \frac{1}{2}S^{\mu}_{\underline{\sigma\tau}}(\Delta^{\alpha}_{\tau\sigma} - \Delta^{\alpha}_{\sigma\tau}) = -\frac{1}{2}S^{\mu}_{\underline{\sigma\tau}}\Delta^{\alpha}_{\sigma\tau}$$

because of

$$S^{\tau}_{\underline{\sigma\mu}} = S^{\mu}_{\underline{\tau\sigma}} = S^{\sigma}_{\underline{\mu\tau}}$$

and also

$$S^{\tau}_{\underline{\alpha\sigma}}\Delta^{\mu}_{\sigma\tau} = S^{\sigma}_{\underline{\alpha\tau}}\Delta^{\mu}_{\tau\sigma} = \frac{1}{2}S^{\alpha}_{\underline{\sigma\tau}}(\Delta^{\mu}_{\sigma\tau} - \Delta^{\mu}_{\tau\sigma}) = \frac{1}{2}S^{\alpha}_{\underline{\sigma\tau}}\Lambda^{\mu}_{\sigma\tau}.$$

Thus,

$$-S^{\nu}_{\underline{\mu\alpha};\nu} = -S^{\nu}_{\underline{\mu\alpha},\nu} + \frac{1}{2}S^{\alpha}_{\underline{\sigma\tau}}\Lambda^{\mu}_{\sigma\tau} - \frac{1}{2}S^{\mu}_{\underline{\sigma\tau}}\Lambda^{\alpha}_{\sigma\tau} - S^{\sigma}_{\underline{\mu\alpha}}\Delta^{\nu}_{\sigma\nu},$$

and consequently

$$2\underline{G}^{\mu\alpha} = -S^{\nu}_{\underline{\mu\alpha},\nu} - S^{\sigma}_{\underline{\mu\alpha}}\Delta^{\nu}_{\sigma\nu} + F^{\mu\alpha}. \qquad (33)$$

Let us calculate the term $\Delta^{\nu}_{\sigma\nu}$ according to its definition. We find

$$\Lambda^{\tau}_{\sigma\tau} = \Delta^{\tau}_{\sigma\tau} - \Delta^{\tau}_{\tau\sigma}.$$

Now, in general, by definition

$$\Delta^{\mu}_{\alpha\beta} = h_s{}^{\mu}h_{s\alpha,\beta} \qquad \Delta^{\tau}_{\tau\sigma} = \frac{1}{h}\frac{\partial h}{\partial x^{\sigma}} = \frac{\partial \log h}{\partial x^{\sigma}}.$$

On the other hand, we have set

$$\Delta^{\tau}_{\sigma\tau} = \varphi_{\sigma}$$

and

$$F_{\sigma} = \varphi_{\sigma} - \frac{\partial \log \psi}{\partial x^{\sigma}}.$$

Thus,

$$\Delta^{\sigma}_{\sigma\tau} = \varphi_{\sigma} + \frac{\partial}{\partial x^{\sigma}}\log h = F_{\sigma} + \frac{\partial \log(\psi h)}{\partial x^{\sigma}}.$$

Substituting into the preceding equation, after multiplying by ψh, gives

$$\psi h(2\underline{G}^{\mu\alpha} - F^{\mu\alpha}) = \psi h S^{\sigma}_{\underline{\alpha}\underline{\mu},\,\sigma} - \psi h F_{\sigma} S^{\sigma}_{\underline{\mu}\underline{\alpha}} - \psi h \frac{\partial \log(\psi h)}{\partial x^{\sigma}} S^{\sigma}_{\underline{\mu}\underline{\alpha}}.$$

Moving the second term on the right to the left-hand side, we obtain [p. 18]

$$\psi h(2\underline{G}^{\mu\alpha} - F^{\mu\alpha} + S^{\sigma}_{\underline{\mu}\underline{\alpha}} F_{\sigma}) = \frac{\partial}{\partial x^{\sigma}}(\psi h S^{\sigma}_{\underline{\alpha}\underline{\mu}}).$$

Now, if we differentiate the right side with respect to α, it vanishes, and we obtain the identities

$$\frac{\partial}{\partial x^{\alpha}}[\psi h(2\underline{G}^{\mu\alpha} - F^{\mu\alpha} + S^{\sigma}_{\underline{\mu}\underline{\alpha}} \cdot F_{\sigma})] \equiv 0. \tag{34}$$

In fact, the right side can be rewritten, after exchanging the names of the mute indices, as

$$(h\psi S^{\sigma}_{\underline{\mu}\underline{\alpha}})_{,\sigma,\,\alpha} = (h\psi S^{\alpha}_{\underline{\mu}\underline{\sigma}})_{,\alpha,\,\sigma} = -(h\psi S^{\sigma}_{\underline{\mu}\underline{\alpha}})_{,\alpha,\,\sigma},$$

since

$$S^{\alpha}_{\underline{\mu}\underline{\sigma}} = -S^{\sigma}_{\underline{\mu}\underline{\alpha}}.$$

There are only 3 independent identities (34). If $A^{\mu\alpha}$ is an antisymmetric tensor

$$A^{\mu\alpha} = -A^{\alpha\mu}$$

such that

$$(A^{\mu\alpha})_{,\alpha} \equiv 0,$$

we have

$$(A^{\mu\alpha})_{,\alpha,\,\mu} = (A^{\alpha\mu})_{,\mu,\,\alpha} = -(A^{\mu\alpha})_{,\mu,\,\alpha} \equiv 0.$$

This is true whatever $A^{\mu\alpha}$ may be, provided that it is antisymmetric. If we take $A^{\mu\alpha}$ to be the left side of (34), we obtain a relation which is independent of the values assumed by the $G^{\mu\alpha}$ and the F_{σ}, which diminishes by 1 the number of independent identities. Finally, the number of those identities is $4 + 3 = 7$, the number of equations is twenty, and the number of unknowns is seventeen. We thus have

$$20 - 7 = 17 - 4,$$

and the system of equations is thus compatible.

13. One could, by the way, attempt to prove the compatibility of the proposed system of equations directly.[9] To that end, let us suppose that *all* the equations

$$G^{\mu\alpha} = 0 \qquad F_\sigma = 0$$

[p. 20] are satisfied on a section x^4 = constant = a. Let us separate them into two groups: the first containing thirteen equations[(1)]:

$$F_1 = 0 \;\; F_2 = 0 \;\; F_3 = 0 \;\; F_4 = 0$$
$$G^{11} = 0 \;\; G^{12} = 0 \;\; G^{13} = 0$$
$$G^{21} = 0 \;\; G^{22} = 0 \;\; G^{23} = 0$$
$$G^{31} = 0 \;\; G^{32} = 0 \;\; G^{33} = 0$$

and the second containing the other seven. One can easily demonstrate the following proposition: *If all the equations are satisfied within a section* $x^4 = a$, and *if the thirteen equations in the first group are satisfied everywhere in the four-dimensional space, the equations in the other group are also automatically satisfied.*

In fact, one finds that, since

$$F_{\mu\alpha} = F_{\mu,\,\alpha} - F_{\alpha,\,\mu},$$

the F_μ being zero everywhere, then so are the $F_{\mu\alpha}$.

Within the section $x^4 = a$, one finds

$$\frac{\partial \underline{G}^{\mu 4}}{\partial x^4} = 0,$$

as is shown by the identity

$$\frac{\partial}{\partial x^\alpha}[h\psi(2\underline{G}^{\mu\alpha} - F^{\mu\alpha} + S^{\sigma}_{\underline{\mu\alpha}}F_\sigma)] \equiv 0\,.^{[10]} \tag{34}$$

Let us consider an infinitesimally neighboring section $x^4 = a + da$. Since the $F^{\mu\alpha}$ and the F_σ are zero everywhere, one can deduce from the preceding identity that, for $\alpha = 4$, the $G^{\mu\alpha}$ will also be zero in this section. An analogous chain of reasoning, using the identity

$$G^{\mu\alpha}{}_{;\alpha} - F^{\mu\nu}{}_{;\nu} - \Lambda^{\sigma}_{\underline{\mu\tau}}F_{\sigma\tau} \equiv 0 \tag{24}$$

shows us that the symmetric part of $G^{\mu\alpha}$, $\underline{\underline{G}}^{\mu\alpha}$, will also be zero for $\alpha = 4$ in the infinitesimally neighboring section $x^4 = a + da$. The conclusion thus holds for

[(1)]The compatibility of the thirteen equations is not in doubt.

$$G^{\mu\alpha} = \underline{G}^{\mu\alpha} + \underline{\underline{G}}^{\mu\alpha},$$

in a section $x^4 = a + da$, and can be extended little by little to all of the space.

14. Let us now examine, as closely as possible, the physical aspects of the theory. It is difficult to give a physical interpretation of the equations in their generality; one has to limit oneself to a first approximation. [p. 21]

For that purpose, let us imagine a space which differs by an infinitely small amount from a Euclidean space. The latter is characterized by the $h_{s\nu}$ equal to $\delta_{s\nu} = \{ {}^1_0$ (x^4 imaginary), leading us to set

$$h_{s\nu} = \delta_{s\nu} + \bar{h}_{s\nu} \tag{35a}$$

One deduces that it is also necessary to set

$$h_s^{\ \nu} = \delta_{s\nu} - \bar{h}_{\nu s} \tag{35b}$$

We will thus replace the $h_{s\nu}$ by this expression in the given equations and will keep only the first approximation. We then have

$$\Delta^{\mu}_{\alpha\beta} = h_s^{\ \mu} h_{s\alpha,\,\beta} = \bar{h}_{\mu\alpha,\,\beta}$$
$$\Lambda^{\mu}_{\alpha\beta} = \bar{h}_{\mu\alpha,\,\beta} - \bar{h}_{\mu\beta,\,\alpha}.$$

The field equations will then be

$$\bar{h}_{\alpha\mu,\,\nu,\,\nu} - \overline{h}_{\alpha\nu,\,\mu,\,\nu} = 0 \qquad \bar{h}_{\alpha\mu,\,\nu,\,\nu} - \overline{h}_{\alpha\nu,\,\nu,\,\mu} = 0 \tag{36}$$

or

$$\bar{h}_{\alpha\mu,\,\nu,\,\alpha} - \overline{h}_{\alpha\nu,\,\mu,\,\alpha} = 0 \qquad \bar{h}_{\alpha\mu,\,\alpha,\,\nu} - \overline{h}_{\alpha\nu,\,\alpha,\,\mu} = 0 \tag{37}$$

The second equation means simply that one can set

$$\bar{h}_{\alpha\mu,\,\alpha} = \frac{\partial\chi}{\partial x^{\mu}} \tag{38}$$

in such a way that the system can be reduced to

$$\bar{h}_{\alpha\mu,\,\nu,\,\nu} - \bar{h}_{\alpha\nu,\,\nu,\,\mu} = 0, \tag{39}$$

$$\bar{h}_{\alpha\mu,\,\alpha} - \chi_{,\mu} = 0. \tag{40}$$

This form is, however, not very satisfactory, because it, at first view, gives no sufficiently clear information on the field that we are seeking. To arrive at something which is more readily interpretable, we recall that the coordinate system is arbitrary up to a certain point and submit it to an infinitesimal transformation:

$$x^{\mu'} = x^{\mu} - \xi^{\mu}, \tag{41}$$

the ξ^{μ} being infinitely small quantities to first order, which we will choose in a suitable manner to make the system take on a simple form. [p. 22]

Applying the infinitesimal transformation means replacing the $h_{\mu\nu}$ by

$$\overline{h'}_{\mu\nu} = \bar{h}_{\mu\nu} + \xi^{\mu}{}_{,\nu} \tag{42}$$

(an equation that follows the transformation rule for tensors).

One thus has

$$\bar{h}'_{\alpha\nu,\,\nu} = \bar{h}_{\alpha\nu,\,\nu} + \xi^{\alpha}{}_{,\nu,\,\nu}$$

$$\bar{h}'_{\alpha\nu,\,\alpha} = \bar{h}_{\alpha\nu,\,\alpha} + \xi^{\alpha}{}_{,\nu,\,\alpha}.$$

Let us choose the ξ^{μ} in such a way that these two quantities cancel each other in the new coordinate system. I assert that it is sufficient to take

$$\xi^{\alpha}{}_{,\nu,\,\nu} = -\bar{h}_{\alpha\nu,\,\nu} \tag{43}$$

$$\xi^{\alpha}{}_{,\alpha} = -\chi\,. \tag{44}$$

In fact, one initially has

$$\xi^{\alpha}{}_{,\nu,\,\alpha} = \xi^{\alpha}{}_{,\alpha,\,\nu} = -\chi_{,\nu} = -\bar{h}_{\alpha\nu,\,\alpha}\,.$$

Next, the system (43), (44) is compatible, even though it was constituted by five equations for four unknowns; indeed, one finds the identity

$$(-\chi)_{,\nu,\,\nu} - (-\bar{h}_{\alpha\nu,\,\nu})_{,\alpha} \equiv 0\,.$$

Thus the solution of this system furnishes us with the quantities ξ^{μ} such that we have

$$\bar{h}'_{\alpha\nu,\,\nu} = 0\,, \tag{45}$$

$$\bar{h}'_{\alpha\nu,\,\alpha} = 0\,. \tag{46}$$

Let us therefore carry out this coordinate transformation. Our equations become (leaving off the primes):

$$\left.\begin{aligned} \bar{h}_{\alpha\mu,\,\nu,\,\nu} &= 0 \\ \bar{h}_{\alpha\mu,\,\alpha} &= 0 \\ \bar{h}_{\alpha\mu,\,\mu} &= 0 \end{aligned}\right\} \tag{47}$$

[p. 23] If we now decompose the $\bar{h}_{\alpha\mu}$ into a symmetric part $S_{\alpha\mu}$ and an antisymmetric part $A_{\alpha\mu}$, the system separates into two other systems which contain only symmetric or only antisymmetric terms:

$$\left.\begin{array}{lll} S_{\alpha\mu,\,\nu,\,\nu} = 0 & \text{and} & A_{\alpha\mu,\,\mu,\,\nu} = 0 \\ S_{\alpha\mu,\,\mu} = 0 & & A_{\alpha\mu,\,\mu} = 0 \end{array}\right\}. \qquad (48)$$

We have thus arrived at two groups of equations. *The symmetric group provides the laws of the gravitational field, compatible with the Newton-Poisson law*; however, the result is not quite identical to that given by the theory based on Riemannian geometry. *The antisymmetric group supplies the Maxwell equations*, in a more general form. I believe that indeed, the antisymmetric system must be interpreted as giving the general equations of the electromagnetic field (in the first approximation).

We thus find in this case a clean separation between the laws of electromagnetism on the one hand and those of gravitation on the other. But this separation holds *only in the first approximation*; it does not exist in the general case: the field is governed by a unique law.

At the present state of the theory, one however cannot judge whether the *interpretation* of the quantities which represent the field is correct or not. A field is indeed defined in the first instance by the motor actions that it exercises on particles; and, at present, we do not know the law of these actions. Discovering that law would require integrating the field equations, which has not yet been accomplished.

15. In conclusion, we can say the following, condensing the results that we have presented thus far:

The particular structure of space which we have taken as our fundamental hypothesis has led us to certain general field equations which, in the first approximations, reduce to the well-known equations of gravitation and electromagnetism. In spite of this, the results obtained so far do not permit us to experimentally verify the predictions of the theory; in fact, we have not yet been able to deduce the laws of the structures of the particles and their motions in the field, by integration starting from the given field equations. The first step which the theory must take should thus be discovering the integrals—without singularities—which satisfy the field equations, and which would provide us with a correct solution to the problem of [p. 23] the particles and their movement. Only after this has been accomplished will a comparison with experiments become possible.

(Seminar lectures given at the Institut H. Poincaré in November 1929 and edited by Al. Proca).[11]

110. To Hans Albert Einstein

[Paris,] 8 November 1929

Dear Albert,

I'm in Paris right now, in order to receive the honorary doctorate and give a few lectures. On the way back, I'd like to spend a night at your place, if possible, that is, if I'm not completely worn out. In that case, I'll be arriving in Dortmund on November 14 around 6:28 P.M. Then, I can continue my trip on the 15th at 9:10 A.M. Write me your address, which I don't have, here (Legation d'Allemagne Rue de Lille 78.)

For the time being, two things:

(1) The patent expert Seligsohn[1] urges you to visit him the next time you come to Berlin. This is important, because he knows the real paths and not only the official ones.

(2) We have chairs for you, which used to be in our Zurich apartment (seven or so, at the moment I don't know exactly how many), something like this, but a little more slender; maybe you remember

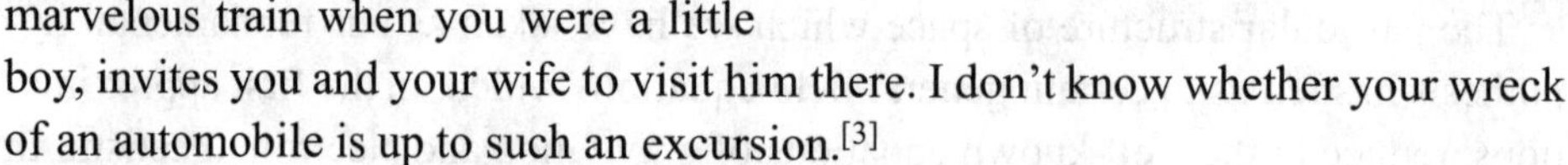

(3) Prof. Hopf[2] at the technical university in Aachen, who once gave you a marvelous train when you were a little boy, invites you and your wife to visit him there. I don't know whether your wreck of an automobile is up to such an excursion.[3]

So now, I have to immediately get on the trapeze and give a talk in French at the Inst. Poincaré on my new theory.[4] It will be a terrible strain. And I have almost a whole week like that to get through! However, it is a great relief to be staying at the embassy because, there, I am not so easy to reach.

If I'm entirely too exhausted next Thursday, I'll travel straight on; but then I won't see you. If you don't hear anything more from me, I'll arrive on the train I mentioned.

In the meantime, best wishes from your

Papa

111. To Elsa Einstein

[Paris,] 12 November 1929

Dear Else,

At least you're doing the right thing in having this matter investigated![1] The symptom can't be traced back to simple fear, though, in your case, I could almost believe it. Hopefully, the results will be good. I will be able to visit you already on Saturday morning, if you haven't already gone home.

My business here is coming to a satisfactory end. The local mathematicians are very interested in the new theory. Yesterday, I visited Langevin, and then Perrin's[2] laboratory. Today has been wonderful: Carnegie Inst. Akademie, Philosoph. Gesellschaft, my second lecture, Deutsche Gesellschaft.[3] Tomorrow, there is still a visit to Briand and the banquet at the embassy.[4] Day after tomorrow, I leave for Dortmund in the morning; I don't want to pass through again[5]

Warm greetings and good luck with the results! Your

Albert

112. From Max Born

Göttingen, Bunsenstr. 9, 13 November 1929

[Not selected for translation.]

113. From Louis de Broglie[1]

94 Rue Perronot, Neuilly/Seine, 14 November 1929

Dear Mr. Einstein,

I thank you with all my heart for the congratulations and kind wishes that you sent me regarding the Nobel prize.[2] Of all those that I've received, these congratulations have touched me the most because of the great admiration that I've had for you for a long time.

Perhaps my idea of the electron waves represents an important enough advancement, but I know how much remains to be done in order to really understand all of that and to get a bit closer to the Truth.[3] We must still do a lot of work, but that's not sad because it is in working that one finds the most pleasure.

Thanking you once again, please be assured, dear Mr. Einstein, of my feelings of respect, friendship, and admiration.

Louis de Broglie

114. From Hugo Bergmann

Jerusalem, 8 October 1929 [between 14 and 25 November 1929][1]

Dear Professor,

I confirm, with heartfelt thanks, the receipt of your letter from 27 September.[2]

The words in the second paragraph of your letter and your assessment of the situation in Palestine pleased me greatly.[3] I have already learned from your letter to the Berlin protest rally that you are completely correct in your assessment of the situation.[4] In fact, your expression "symbiosis" is an apt formulation of the goal that we must set ourselves. In my opinion, the difficulties in achieving this goal also lie in large part on the Jewish side, even if it must also be admitted that the Arab leaders' desire for peace, which perhaps was present a year and a half ago (when they asked for our help with the attainment of a parliament and clearly signified that they would come to terms with the Balfour Declaration),[5] is now no longer perceptible as a result of the Arab movement, which is strengthening daily. At this moment, an agreement is out of the question, but when the commission of inquiry has finished its work, perhaps the basis for negotiations will also be present again, and then, I fear, psychological difficulties on our side will emerge again.[6] During the last twelve years, we have been caught in a dream of a Jewish state, or at least a state in which the Jews would form the decisive majority, which has been shattered by the reality.[7] But we refuse to carry out the psychological adjustment that is necessary in order to see reality as it is. Because of this, your program of "mixed administrative, economic, and social organizations" will run into strong resistance from the Jewish side.[8] I would therefore consider it very meritorious if you, dear Professor, would do your part in order to work toward this; that a program built upon the idea of symbiosis will be *proactively* submitted from us to the English government and the Arabs. We must honestly declare that we will set no obstacles in the way of the political development of the Arab people in Palestine, above all, their demand for a parliament and a parliamentary government, provided that an internationally guaranteed declaration concerning the right of Jewish immigration has been given beforehand, though of course this immigration must not serve as an effort to gain a majority, but is to be adapted to the economic capacity of the country and the funds raised by the Jewish Agency. Furthermore, the workers' organization must strive for an honest trade union policy with respect to the Arabs and renounce the struggle against the employment of the Arab workers (as long as they are not wage squeezers) in any form.[9] It is certainly clear that the unionization and elevation of the Arab worker, as well as the recognition of his equal right to a job during a certain time of transition, will require sacrifices from our side in regards to the level of immigration, because the absorption capacity of the land is limited. But this sacrifice must be made, and in the long run, by elevating the Arab population's standard of living, [it] will only benefit our industry, whose natural market is in the country. What is true of the workers also holds for all other economic organization. One must give the Arabs the opportunity to take part in the concessions we have acquired (electrification, Dead Sea bounties); one would have to agree that the number of Arab village schools will increase; even if this happens at the expense of the Jewish taxpayer, Arab officials must be employed in our business ventures, etc.[10] Until now, we have lived in Palestine like a state within a state and have allowed ourselves to decide all questions based solely on our national interests and have therefore alienated ourselves as much from the government as from the Arabs. We must, as you rightly say, reach a symbiosis. To prepare the way for such an insight is, in my view, the most important requirement of the hour.

With warm regards to you and your wife, your

115. From Chaim Weizmann

Oakwood, 16. Addison Crescent, W. 14, 16 November 1929

Dear Prof. Einstein,

As you can imagine, the events in Palestine following so closely on the conference in Zürich have upset all my plans and have caused me (and are still causing me) great concern.[1] We had to do a great deal to get the commission of inquiry to act decently, at least in part; we are in continuous negotiation with the government, but I do not think anything tangible will happen until the government receives the report of this commission.[2]

In the meantime, I have read your very impressive letter in the Manchester Guardian.[3] I believe it is the best that has yet been written about this unfortunate matter. The situation in the country causes us all great concern, and the terrible confusion prevailing in the Jewish and Zionist world is perhaps the worst that could befall us. In Germany especially, there is so much talk of peace with the Arabs at any price. I am definitely in favor of peace, but the more we speak of it, the less the Arabs will listen.[4] They interpret all these manifestations on our part as weakness and fear. The only people who can serve as intermediaries between us and the Arabs are, first, the English and, second, those Jewish scholars who engage in Arabic studies. I am in contact with Horovitz from Frankfurt on Main, and we intend to call a conference of Arabists at Christmas,[5] at which we will investigate ways and means in detail. Furthermore, we have quietly started certain very cautious moves, but all this is merely a beginning.[6]

In the meantime, friend Magnes, in his well-known dilettante manner, has attempted through the agency of a Mr. Philby (an ex-English official, adventurer, and enemy) to negotiate with the Arab Executive in Palestine in the name of the Jewish people.[7] Of course, he immediately advised our dear Felix[8] to agree to everything, all this without telling us a word. It is a crime which we managed to prevent in time, but all this makes our work extremely difficult.

I hope and trust that you are in good health. I am much better and can once again work vigorously. I would be happy to hear from you. I do not despair. One must work hard and be cautious.

With kind regards, yours ever,

Ch. Weizmann

Translators' note: Translation from *Weizmann 1978*, pp. 84–85

116. From Eduard Einstein

Zurich, 17 November [1929][1]

[Not selected for translation.]

117. To Maja Winteler-Einstein

[Berlin,] 18 November [1929][1]

Dear Maja,

Although I'm completely *meschugge* with letters, telegrams, telephone calls, and visits, I am thinking about you after the secretary told me it was the 18th of November,[2] and I immediately read your letter. Mileva is a fat sow, but I couldn't send you the dear boy any other way, and so here, the end must justify the means.[3] I'm just surprised that she handled it so clumsily—that makes me less inclined to be indulgent.

In Paris, it was wonderful,[4] even if it was also the most extreme strain my rattly old self can still bear. The main thing is that the new theory found lots of understanding there; I am very pleased with its provisional conclusion.

The little house in Caputh is a failure, to be sure, but a very beautiful one, and best of all, the sailboat that High Finance gave me![5] You'll love it when you (hopefully) come next year. In the meantime, best wishes until your next birthday, and warm greetings from your

Albert

Alice and Ogden,[6] whom I saw in Paris, send their greetings. Alice is very nice; I think she owes you a letter. Nonetheless, *do* write her a little note.

118. On Abortion[1]

[*Einstein 1929ww*]

DATED 19 November 1929
PUBLISHED 27 November 1929

IN: *Volksstimme*, 27 November 1929

The right to terminate a pregnancy in the first months must, according to my sense of justice, be granted unconditionally.[2]

A. Einstein

119. Books for a Trip to the Southern Seas

"Mein Bücherkoffer für eine Südseefahrt"
[*Einstein 1930q*]

DATED 19 November 1929
PUBLISHED March 1930

IN: UHU Heft 6, 6. Jahrgang, March 1930, p. 87. Berlin.

[Not selected for translation.]

120. To Heinrich York-Steiner[1]

[Berlin,] 19 November 1929

Dear Mr. York-Steiner,

First of all, I would like to express my great admiration for your wonderful book,[2] which I have read from end to end and whose tendency I totally approve. It is very gratifying that this book has aroused so much interest.

On the occasion of the recent conflicts with the Arabs, I have written various things, so that I am not entirely sure what you are alluding to.[3] You can, however, publish whatever you consider appropriate.

I came to Zionism only after moving to Berlin in 1914, when I was thirty-five years old,[4] before which I lived in an entirely neutral environment. From that time on, it was clear to me that, in order to maintain or to regain a dignified life, we Jews urgently needed to revive a community feeling. I see in Zionism the only aspiration that leads us toward that goal.

It is now time to make sure that this movement avoids the danger of degenerating into a blind nationalism.[5] Above all, we must strive, in my opinion, to ensure that the Arabs meet with psychological understanding and a sincere will for cooperation instead of resentment.

Overcoming this difficulty will be, in my opinion, the touchstone showing that our community has, in the highest sense, the right to live. Unfortunately, I must openly admit that the behavior of our official authorities, as well as the majority of the statements made in this connection, leave, in my opinion, far too much to be desired.

With best regards, respectfully yours,

121. To Hector Laisné[1]

[Berlin,] 20 November 1929

[Not selected for translation.]

122. To Chaim Herman Müntz

[Berlin,] 20 November 1929

Dear Mr. Müntz,

Many thanks for your letter. Please forgive my brevity; I am suffocating under correspondence and obligations. I was pleased to hear of the happy impressions that you experienced.[1]

The theory provides me with a lot of pleasure, and my trust in it is continually growing. I recently lectured on it in Paris.[2] The equations are[3]

$$\Lambda^{\alpha}_{\mu\nu;\alpha} = 0(\ \equiv F_{\mu\nu})\ \left[\varphi_{\mu} = \frac{\partial \lg \psi}{\partial x^{\mu}}\right]\left(F_{\mu} \equiv \varphi_{\mu} - \frac{\partial \lg \psi}{\partial x^{\mu}}\right)$$

$$\Lambda^{\alpha}_{\underline{\mu}\underline{\nu};\nu} - \Lambda^{\sigma}_{\underline{\mu}\underline{\tau}}\Lambda^{\alpha}_{\sigma\tau} = 0\ (\ \equiv G^{\mu\alpha})$$

The corresponding identities are[4]

$$(G^{\mu\alpha} - F^{\mu\alpha})_{;\alpha} - \Lambda^{\sigma}_{\underline{\mu\tau}}\Lambda^{\alpha}_{\sigma\tau;\alpha} \equiv 0 \text{ (commutation relation), and}$$

$$\frac{\partial}{\partial x^{\alpha}}[h\psi(2\underline{G}^{\mu\alpha} - F^{\mu\alpha} + S^{\sigma}_{\underline{\mu\alpha}}F_{\sigma}] \equiv 0\,.$$

From these identities, one can prove directly that: If the quantities

$$\left.\begin{array}{cccc} F_1 & F_1 & F_1 & F_1 \\ G^{11} & \ldots & G^{13} & \\ - & - & - & - \\ - & - & - & - \\ G^{31} & G^{32} & G^{33} & \end{array}\right\} \text{13 equations for the 17 quantities } h_{11}\ldots h_{44},\ \psi$$

vanish everywhere, and *furthermore*, in a section $x^4 = \text{const.}$, the quantities G^{14} G^{24} G^{34} G^{41} G^{42} G^{43} G^{44} and the latter also vanish everywhere. This is the strict proof of the compatibility of the equations.[5]

Best wishes to you and to your wife, your

A. Einstein

The equations in the first approximation, most remarkably, permit a choice of coordinates which fulfill the conditions

$$\bar{h}_{\mu\alpha,\,\alpha} = 0 = \bar{h}_{\alpha,\mu\alpha}\,,$$

whereby they decompose into the equations

$$\text{Newton}\left\{\begin{array}{l} \bar{g}_{\mu\alpha,\,\sigma,\,\sigma} = 0 \\ \bar{g}_{\mu\alpha,\,\alpha} = 0 \end{array}\right. \quad\Bigg|\quad \left.\begin{array}{l} a_{\mu\alpha,\,\sigma,\,\sigma} = 0 \\ a_{\mu\alpha,\,\alpha} = 0 \end{array}\right\}\text{Maxwell}$$

one obtains. ($\bar{g}_{\mu\alpha} = \bar{h}_{\mu\alpha} + \bar{h}_{\alpha\mu}$ $a_{\mu\alpha} = \bar{h}_{\mu\alpha} - \bar{h}_{\alpha\mu}$

123. To Robert Weltsch[1]

[Berlin,] 23 November 1929

Dear Mr. Weltsch,

I have read Bergmann's letter[2] very carefully and find that he is completely right. Not only does practical prudence command us to proceed in such ways, the whole project has no justification for existing if this conception cannot be established among the Jewish people, and especially among its leading men. Then we would have learned nothing from the two thousand year long path of suffering. Which path must we take in order to help reason triumph?

Best regards from your

124. From Georg Rumer

24 November 1929

[Not selected for translation.]

125. To Hugo Bergmann

Berlin, 25 November 1929

Dear Mr. Bergmann,

I am in full agreement with the content and the form of your revised letter,[1] and I authorize you to make any reasonable use of this declaration. Which path is to be taken to make your conception effective is still not clear to me. In particular, I do not know whether the present moment is the right one for the publication of your letter.[2] Brodetsky has recently spoken in a meeting of the Jewish Agency, completely ignoring the serious psychological and economic problems here.[3] If it goes on like this, a catastrophic development is very much to be feared. As soon as I have gained a little clarity, I will write to you again.

Best regards, your

A. Einstein

126. To Martha Steinitz

[Berlin,] 25 November 1929

[Not selected for translation.]

127. To J. Ward

[Berlin,] 25 November 1929

[Not selected for translation.]

128. To Chaim Weizmann

Berlin,W., 25 November 1929

Dear Mr. Weizmann,

I thank you very much for your letter[1] and can well imagine that you are deeply worried. But at the same time, I must tell you, frankly, that the attitude of our leaders makes me uneasy. Brodetsky, in a recent address to the agency,[2] has again been treating the Arab problem with that shallowness and superficiality that brought about the present state of affairs. The economic and psychological problem of the Jewish-Arab symbiosis was not even touched upon, and the conflict was treated as an episode. This was particularly inappropriate, as the more reasonable members of the audience would be fully convinced of the insincerity of such a posture. I am sending you, enclosed herewith, a letter from Hugo Bergmann,[3] which I am convinced goes to the heart of the matter. Unless we find the way to honest cooperation and honest dealings with the Arabs, we have not learned anything on our way of two thousand years' suffering and deserve the fate that is in store for us. In particular, I feel we should beware of relying too much on the English. For unless we succeed in arriving at a genuine cooperation with the leading Arabs, the English will drop us—though not formally, but de facto. And the latter will commiserate with our debacle, piously casting their eyes to heaven in a traditional fashion and proclaiming their innocence without lifting a finger to help us.

With kind regards, your

A. Einstein

129. From Cornel Lanczos

Frankfurt am Main, 27 November 1929

Dear Professor,

I have looked for the missing *Weyl*: *Gruppentheorie und Quantenmechanik*,[1] but I could not find it anywhere. Since the book was sent to me for a review in the Physikalische Zeitschrift, and I was therefore already in possession of it, I do not really believe that I would have borrowed it from you.

On the other hand, I had indeed forgotten to return two books that I had borrowed from you. These are Weyl's *Raum, Zeit, Materie* and Eddington's *Allgemeine Relativitätstheorie*.[2] I sent them off to you today by post as a "small package." I hope that they will arrive in good order. (Please forgive me for this omission!).

There is also a manuscript in my possession by a scholar in Prague, on Galileo's development; I have not yet been able to find it, since my books are not yet ordered. It will certainly turn up in time, and then I will likewise send it back to you with thanks.

Unfortunately, I have not yet been able to start working, since looking for an apartment, moving, etc. have taken up a certain amount of time; but also mainly due to my social responsibilities with regard to my old friends and acquaintances.[3]

I am writing up the investigation of the problem of motion, which I would gladly like to finally finish up. My schedule here is unfortunately much more fragmented than it was in Berlin,[4] where I could retreat from everything.

I would be genuinely happy if the action that Freundlich[5] has set in motion were to be successful; the year in Berlin meant a lot to me, and the interruption of all those relations would be a painful loss. Now—one cannot change one's fate, so I am hoping for the one thing, that Freundlich is a person who does not let himself be brushed off so easily; and I am convinced of his goodwill toward me.

Perhaps you could also put in a good word for me with Laue[6]—it may well be that everything depends on this consultation.

I hope that you have kept many lovely impressions from Paris and returned home with a feeling of satisfaction.[7] And also, I hope, in the best health.

With the warmest wishes to you, and also to your wife and Miss Margot,[8] I remain yours, truly and respectfully,

Lanczos

130. From Chaim Weizmann

Oakwood, 16. Addison Crescent, W. 14, 30 November 1929

Dear Prof. Einstein,

I take up my pen with a heavy heart to reply to your last letter.[1] I do not know how Brodetsky's speech sounded,[2] but what I read in the Rundschau cannot, I believe be objected to. It is possible that B., who lacks the nuances of the German language, was not able to express himself entirely clearly. What really are the great facts. The Arabs have attacked us, have attempted to destroy our development work, have murdered and plundered, and now we are being pressed on all sides to find ways and means of concluding a pact with the Arabs.[3]

Already at the twelfth congress, we solemnly declared how we visualized living together with the Arabs;[4] everything we have done so far in the country is a witness to peace in our actions and thoughts. Why did they attack us? Does anyone believe that it would be possible to negotiate with the Mufti or the Arab Executive[5] today? They would laugh at us if we were to make overtures today, despite what Bergman writes in his articles.[6] The present Arab leaders, murderers and thieves, want but one thing—to drive us into the Mediterranean!

I believe that negotiations with the Arabs are only possible on one condition, namely, if the government were to make clear to the Arabs that they firmly intend to carry out the Mandate.[7] If they do not do so, the Arabs will believe—and rightly so—that we have been abandoned by Britain, and they will not then have much interest in linking themselves with us. They do not yet understand, or do not want to understand, what we are bringing them and what real friendship between us could mean to the world as well. They are too primitive for that and too much under the influence of factors such as Bolshevik, Catholic agitation.[8] The Arab *people* have not yet spoken; the spokesmen of the people today are corrupt Levantines, such as the Mufti and the like. I have clearly explained to Ramsay MacDonald and have also given it to him in writing that we are prepared to sit round a table with the Arabs, but that he—MacDonald—must make it clear to them that the Mandate, National Home, etc. are *choses jugées*.[9] If they accept this, we will also accept a Palestinian parliament,[10] but it must be of a kind that is not a dagger in our back. There are only two alternatives: either we can go to Palestine rightfully, or we are foreigners who slip in through good manners and through money. If the world does not recognize our proper right, that is an error on the world's part and a misfortune for us, and I know the world has misunderstood us for two thousand years. We have no desire to rule over anybody, but we also do not want to bow down before anybody or be ruled by anybody—enough of that! Little Palestine was the one place on Earth where we stood upright. Now come the Magneses,[11] the Bergmans, and break our united front, presenting matters as if *we* do not want peace. Today, you are quoted in the *Near East and India*—a paper that is well-disposed toward us—and a conflict is fabricated between you and us, simply because we do not want to negotiate with the murderers at the graves, still open, of the Hebron and Safed victims.

Because we believe in other tactics.[12] That hypocrite, that Tartuffe, Magnes, lightly abandons the Balfour Declaration.[13] He did not bleed for it; he only gained by it! Believe me, I know the Palestinian Arabs. If we give an inch now, we may as well pack up and must tell ourselves and the people: "The project will not work now, and we shall wait, and tremble, and mourn, and believe in better days and a more just humanity that will some day understand even the Jews!"

I have never been a firebrand, as you know from the whole of my past. I have always preached most unpopular realities to the Zionists, have always been attacked most bitterly. I, too, believe in an agreement with the Arabs; it is necessary for us, for the Arabs, and for the English, but all three parties must assist and—give and take.[14] So far, we have done all the giving, have already explained to the Arabs what we are prepared to do, and that over the fresh graves of our brethren. The Arabs have not even once expressed regret for what has occurred. Naturally, Bergman, and Magnes, and even you (and this grieves my soul!) are now quoted against us, and from the mouths of the murderers, we are told: "Yes, if all Zionists were like Bergman, etc., etc." For centuries, all Jew-haters have used such a phrase and have always divided us by praising one side against the other! We—I—who, all my life, have fought for peace and moderation, are the Hugenberg and Magnes, the Stresemann![15] I hope better times will come, but it is a dark period now, and the thing that makes my life most bitter is the feeling that our faith in ourselves and the justice of our cause has been shaken.

With a heavy heart, yours ever,

Ch. Weizmann

I have written this letter at home and not at the office. Perhaps your secretary will be kind enough to send me a copy.

Ch. W.

Translators' note: Translation from *Weizmann 1978*, pp. 122–124.

131. To Chaim Weizmann

[Berlin, after 30 November 1929][1]

Dear Weizmann,

I was very dismayed by your echo, because it showed me that I have unintentionally wrought havoc with my letter.[2] It never crossed my mind that I was opposing you, and I opened my mouth only once when I was forced to by the envoys of the international association against antisemitism.[3] I am very sorry that you have spent so much time on a letter to me, from which I see that you have misjudged my position. It never occurred to me that we should negotiate with the agitators.

My objection to Brodetsky's speech[4] concerned not what was in it but rather what it *lacked*: a causal-psychological observation of the reasons for the events, and an honest reflection on a preventive policy. Nothing can be achieved with resentment, perseverance, and reliance on England. The enemy has committed crimes, it is true. He was aided and abetted, that is also true. But this analysis is superficial, and it's dangerous to be satisfied with it.

I know that Bergmann's words can easily be turned into accusations against us. But in this case, he seems to be right: without sincere cooperation with the Arabs, there can be no peace and no security.[5] This is for politics in the long term, not for the moment. I really believe that many opportunities have been missed in this respect, but we should reflect on it and not fight with one another. Finally—even if we were not all but defenseless—it would be unworthy of us to seek to maintain a nationalism *à la Prussienne*. Do not answer me now; you need all your strength, unfortunately, now only too much! I will keep quiet, insofar as I can, and not meddle in anything.

I have read your letter very carefully[6] and am sending it back to you because my secretary is already overloaded with work. I hope things will soon go better for you and for Palestine and send you my best regards. Your,

A. Einstein

132. From Georg Rumer

30 November 1929

[Not selected for translation.]

133. From Élie Cartan

Le Chesnay, (S. and W.), 27 ave de Montespan, 3 December 1929

Dear Sir,

It has already been more than three weeks since you left Paris, and I have not yet given you any sign of life. That was not because I didn't think quite a lot as a result of our conversation at Langevin's.[1] I asked you various questions on mathematics and physics which may have seemed somewhat naive to you, but that was because I wanted to start from a secure basis in order to formulate in a precise manner the mathematical problem that you had set yourself. I am now taking the liberty, in this letter, of sending you a summary of the way in which I conceive of this problem. I apologize in advance for considering certain details which will perhaps seem rather useless to you, but that is only to help you understand my point of view;[2] if I am taking the wrong route, I would hope that you will let me know so. I will give you some details of the theory of involuted systems which I developed starting thirty years ago,[3] and which seems to me to be quite appropriate for treating the problem that you have posed.

I have also carried out some calculations, but without being able to arrive at a decision as to whether the solution that you have found occupies a privileged position relative to all the other possible solutions. Another possible solution, which contains two cosmogonic constants, consists in taking the 10 former equations $R_{ij} = 0$, which still retain an invariant character here, and adding to them 12 equations of the form[4]

$$\Lambda^{\mu}_{\alpha\beta\mu} = \varphi_{\alpha,\beta} - \varphi_{\beta,\alpha} = ChS^{\gamma\delta\mu}\varphi_{\mu} \quad (1)$$

$$(hS^{\alpha\beta\mu})_{,\mu} = C'hS^{\alpha\beta\mu}\varphi_{\mu}; \quad (2)$$

in the equations (1), the indices $\alpha\beta\gamma\delta$ are fixed and represent a pairwise permutation of the indices 1, 2, [3], 4. C and C' are arbitrary numerical constants.

The corresponding physics would be *irreversible,* at least if $C \neq 0$; on changing the orientation of the tetrad—that is, the direction of time—the laws of physics would cease to be valid, but that would not be manifest to the first approximation. In your system, the 12 equations (1) and (2) would also play a role, with the constant C having a value of 0, and the constant C' the value 1. If $C = 0$, the equations (1) and (2) can be integrated by introducing two functions, ψ and χ:

$$\varphi_{\alpha} = \frac{1}{C'\psi}\frac{\partial\psi}{\partial x_{\alpha}},$$

$$hS^{\alpha\beta\gamma} = \frac{1}{\psi}\frac{\partial\chi}{\partial x_{\delta}};$$

The existence of these two universal functions defined up to two constants (ψ can be replaced by $a\psi$ and χ by $a\chi + b$) is rather curious. In the old physics, there were indeed two functions of this type: the gravitational potential and the time; like ψ, the gravitational potential is reversible, and like χ, the time is irreversible (it changes its sign on changing the orientation of spacetime). If $C \neq 0$, two more universal functions are introduced, defined by the transformations of a group with two parameters.

I have not yet been able to completely resolve the problem of finding out whether other systems of 22 equations exist besides yours and the one which I am about to describe below.[5] I still need to identify to the zero point a certain cubic form of the $\Lambda^{\gamma}_{\alpha\beta}$ which depends upon 16 parameters! That part of the equations which contains the derivatives $\Lambda^{\gamma}_{\alpha\beta;\mu}$ is certainly determined, but these are complementary terms which, even if one supposes them to be simply quadratic, will introduce all the complications—and I am continually surprised that you were able to find your 22 equations!

There are some other possibilities which furnish even richer geometric schemes while remaining deterministic. One could, for a start, take a system of 15 equations: six of them express the fact that the vector

$$b \neq 0 \qquad a\varphi_1 + bhS^{234},\ a\varphi_2 + bhS^{143},\ a\varphi_3 + bhS^{124},\ a\varphi_4 + bhS^{132}$$

[(]where a and b are numerical constants) is the gradient of a universal function; the other nine have the form

$$\begin{aligned} &\Lambda^{\beta}_{\underline{\alpha}\underline{\mu}\mu} + \Lambda^{\alpha}_{\underline{\beta}\underline{\mu}\mu} + c(\varphi_{\underline{\alpha};\beta} + \varphi_{\underline{\beta};\alpha}) + d(hS^{\gamma\delta\mu})_{;\mu} \\ &\qquad -\tfrac{1}{2}[\Lambda^{\rho}_{\underline{\rho}\underline{\mu}\mu} + c\varphi_{\underline{\rho};\rho}] = A^{\alpha\beta} - \tfrac{1}{4}A^{\mu\mu}, \end{aligned} \qquad (3)$$

c and d being arbitrary numerical constants, $A^{\alpha\beta}$ are functions of $\Lambda^{\delta}_{\mu\nu}$ *subjected to the unique condition of forming a symmetric tensor.* If one takes quadratic forms for $A^{\alpha\beta}$, such a tensor would be linearly dependent on ten arbitrary numerical constants. The corresponding physics is reversible if the constant a ⟨and b⟩ is zero and if the constant d is also zero; it is also necessary in this case to limit the tensor $A^{\alpha\beta}$, which must now depend upon only eight arbitrary constants.

Finally, *perhaps* there also exist solutions which are formed from sixteen equations; but the study of that case leads to calculations which are just as complicated as those of the 22 equations, and I have not yet had the good fortune to discover a possible system. The general form of the equations of such a system would be obtained by first taking 9 equations of the form (3) indicated above, then an equation of the form

$$\varphi_{\underline{\mu};\mu} = C,$$

C being a scalar tensor (with 5 possible constants); and, finally, 6 equations such that

$$(hS^{34\mu})_{;\mu} + a(\varphi_{1;2} + \varphi_{2;1}) + bh(\varphi_{\underline{3};4} + \varphi_{\underline{4};3}) = B_{12} \qquad (4)$$

where $B_{ij} = -B_{ji}$ is an antisymmetric tensor. If a or d is nonzero, the physics would be irreversible, and its irreversibility would be apparent in the first approximation.

I said above that the solutions with 15 equations and those (if they exist) with 16 equations have a richer content than those of the 22 equations; the general solution of the systems of 22 equations in fact depends on 12 arbitrary functions of 3 variables, while that of the systems of 16 equations depends on 16 arbitrary functions and that of the systems of 15 equations on 18 arbitrary functions. To see the precise meaning of these statements, you will have to refer to the enclosed technical note. It seems to me, after having thought about it, that the degree of generality of the geometric scheme corresponding to your 22 equations is somewhat weak, while the earlier classical theories of gravitation and of electromagnetism give the associated physics a considerably greater degree of generality. I would be grateful to hear your opinion about this.[6]

Finally, you will see from the note that the number of identities which must exist between the derivatives of the first members of a possible system of n equations in order for that system to be *not only compatible*, but also to be *in involution*, is *at least equal to* $n - 12$, but in certain cases it must be even *greater*: this is what also happens for your system of 22 equations; the derivatives of the first members are in reality connected by $12 = n - 12 + 2$ identical linear relations. Only by modifying the system through the introduction of an auxiliary function ψ were you able to find your a priori number of $n - 12$ identities. Of course, if one wished only to guarantee that the system were compatible, there would be no a priori necessity of such a number of identities, as one can see from the examples.

I hope that you had a good journey home and that your health did not suffer too much from your trip to Paris. In any case, we were all very happy that you introduced us to your more recent research. I thank you again for the confidence that you have shown in my abilities as a mathematician, a confidence which I hope to justify, at least *a posteriori*.

Please accept, dear and illustrious master, my most kind regards,

E. Cartan

A few days ago, I received the proofs of my note for the *Math. Annalen*;[7] I sent them back after correction without any modifications to the text, though certain remarks[8] that I made are now no longer necessary.

After having written my little note, I realized that it is indeed rather long; may it also at least be clear!

134. From Selig Brodetsky

77, Great Russell Street, London W. C. 1, London, 4 December 1929

[See documentary edition for English text.]

135. Foreword to Newton's *Opticks*[1]

[Berlin, 5 December 1929][2]

[See Doc. 136 in documentary edition for published English version.]

136. Foreword to Newton's *Opticks*

[*Einstein 1931a*]

DATED 5 December 1930[1]
RECEIVED before 12 December 1929[2]
PUBLISHED 1931

IN: *Newton 1931*, pp. vii–viii.

[See documentary edition for English text.]

137. On the Hilfsverein der Deutschen Juden[1]

"Der Hilfsverein der Deutschen Juden"
[*Einstein 1929xx*]

DATED between 5 and 13 December 1929[2]
PUBLISHED 13 December 1929

IN: *Jüdische Rundschau,* 13 December 1929, p. 662.[3]

We Jews are continuously being threatened because we live *everywhere as a minority*[4] that distinguishes itself from its environment through its complex traditions. Because of this, the individual is threatened both economically and morally: economically by being excluded from professions,[5] morally through isolation, which then leads mostly to pitiful egotism. Only Jewish solidarity can prevail against these damages.[6] Even if this solidarity must be expressed primarily through personal behavior, from person to person, there is nevertheless a need for organizations which work continuously and systematically[7] to remedy the most glaring distresses. Among these organizations, the Hilfsverein der Deutschen Juden occupies a high place, partly because of its great achievements in the past, partly because of its nonpartisanship within the Jewish community. In this time of economic depression,[8] many tend to save in the wrong places—namely, by shirking their social duties rather than limiting their private needs.[9] Think about it yourself: I don't have it in me to be a preacher in the desert.

138. To Eduard Einstein

[Berlin,] 5 December 1929

Dear Tete,

That's how it went for me too, in school.[1] You feel as passive as a force-fed goose, but it doesn't make you fat. With regard to superficial force-feeding, medical study is the worst, especially for someone without a strong memory. But there is one thing: observing the organic arouses a feeling of awe before the incomprehensible, an astonishment that is gradually extended to everything, connected with the feeling of a childlike, intellectual swoon. You talk about degenerate characteristics associated with unnatural living conditions. Ultimately, however, it is not yet established that all this has to be interpreted as degeneration. There can be a kind of lack of practice that is not bequeathed but rather conditioned only by the individual's history. For that, there might be something else more developed. I heard something regarding Russia that is relevant to this topic.[2] Many sons of proletarians are chosen to become students. They're very hard-working and capable, but less persevering than the sons of intellectuals, and mental work frequently causes them to go kaput after a few years. Should we say, then, that the sons of intellectuals are degenerate if they have a weaker physique? All this must not be judged so casually. For now, I'm pleased by your more intellectual attitude, because it ⟨makes⟩ you freer and more playful than obsessed with achievement. It doesn't seem to me to be degeneration—that's just my feeling, without reasons. I am glad that you've found such good laboratory comrades;[3] that's worth a great deal. But true talent is revealed only later—mainly by its absence. Moreover, I believe that talent is anchored less in the understanding than in a certain passionate aptitude that is connected with the superiority of *men*.

I am glad that my general field theory is completed. Its sublime character is confirmed by contemporaries' lack of interest in it. Veneration is to be had more cheaply than understanding.

Wouldn't you like to come for Christmas? Or rather for Easter? It's dumb that these holidays come in the season of short days and bad weather. But apart from that, there's something attractive in the big city. I recently stayed overnight at Albert's place[4] and found the daughter-in-law quite nice. Albert is happy with her; that's easy to see. He hasn't inherited his father's gypsy blood.[5]

Best wishes from your

Papa

139. To Élie Cartan

[Berlin/Caputh,] 8 December 1929

Dear Mr. Cartan,

I am both touched and delighted that you have taken so many pains over the problem. For the present, I am responding only to your letter; an answer to your note[1] will follow later.

You set up the equations

$$R_{ik} = 0 \qquad (1)$$

$$\Lambda^{\mu}_{\alpha\beta;\mu} = \varphi_{\alpha,\beta} - \varphi_{\beta,\alpha} = ChS^{\nu\delta\mu}\varphi_{\mu} \quad (2)$$

$$(hS^{\alpha\beta\mu})_{,\mu} = C'hS^{\alpha\beta\mu}\varphi_{\mu} \quad (3)$$

The compatibility of these equations is to be shown by the integrating assumption

$$\varphi_{\alpha} = \frac{1}{C'}\frac{1}{\psi}\frac{\partial\psi}{\partial x_{\alpha}} \ldots \quad (4)$$

$$hS^{\alpha\beta\gamma} = \frac{1}{\psi}\frac{\partial\chi}{\partial x^{\delta}} \ldots \quad (5)$$

But from (4), it follows that $\varphi_{\alpha,\beta} - \varphi_{\beta,\alpha}$ vanishes and thus from (2), $S^{\cdots}$ vanishes. So (3) no longer has any content. Assumption (4) is thus not allowable, and the compatibility of (1), (2), and (3) remains unproven.

A general remark: When equations are generally integrable, as e.g., $\Lambda^{\alpha}_{\mu\nu;\alpha} = 0$ (through $\varphi_{\mu} = \frac{\partial \lg\psi}{\partial x^{\mu}}$), then the *integrated* equations should be included in the count. It seems correct. to me. to say that these are not 6 but rather 4 equations.

Generalizing your case, one can write the following field equations:

$$R_{ik} = 0$$

$$\varphi_{\alpha} = A(\psi,\chi)\frac{\partial\psi}{\partial x_{\alpha}} + B(\psi,\chi)\frac{\partial\chi}{\partial x_{\alpha}}$$

$$hS^{\sigma\tau\rho}\delta_{\sigma\tau\rho\alpha} = C(\psi,\chi)\frac{\partial\psi}{\partial x_{\alpha}} + D(\psi,\chi)\frac{\partial\chi}{\partial x_{\alpha}}$$

where A, B, C, D are given functions of two variables. Then there are $10+4+4$ equations with 4 identities for $16+2$ variables, which is allowed. But I think this is too weak an overdetermination. In my system, there are 7 identities instead of only 4. This is where I see the essential point.

If one is willing to be satisfied with 4 identities, then one can use Hamilton's principle

$$\delta\left\{\int Hh\,d\tau\right\} = 0$$

where H is a scalar quadratic in Λ. There are very many possibilities for H, but I am convinced that one must demand a stronger determination.

Your system:

$$R_{ik} = 0 \qquad \alpha\varphi_1 + bS_1 = \frac{\partial\psi}{\partial x^1} \qquad \Lambda^{\beta}_{\underline{\alpha\mu};\mu} + . — . = A^{\alpha\beta} - \frac{1}{4}A^{\mu\mu}(g^{\alpha\beta})$$

also appears unacceptable to me. There are $10 + 4 + 9$ equations with 4 identities for $16 + 1 + 9$ field variables. ⟨Or is it meant differently?⟩ Thus, there are two equations too many, unless one can show the existence of two more identities.

Finally, the system of $9 + 1 + 6$ equations, whose scalar equation you have written as $\varphi_{\underline{\mu};\mu} = C$. Here, no identity is obvious at all, so I cannot see the compatibility. But even if 4 identities were to be discovered, the degree of determination would be weaker than in my system. *And the degree of determination is precisely the heuristic principle*; (in my opinion, a theory is the more valuable the more strongly it restricts possibilities without coming into conflict with reality).[2] It is like a wanted poster which is supposed to characterize a criminal; *the more precisely* it points him out, the better. It seems to me that the degree of determination is to be expressed by the number of (independent) identities (where it is assumed that this is equal to $N - 12$).

You propose the question of the interpretation of the case when the number of independent identities is larger than $N - 12$, where N is the number of equations. I have not been able to understand what you mean when you say that, in the case of my original equations ($\Lambda^{\alpha}_{\mu\nu;\alpha} = 0$), the number of identities is only two greater than the number $N - 12$. I would remark that, with respect to the identities,

$$\frac{\partial F_{\mu\nu}}{\partial x^{\sigma}} + \frac{\partial F_{\nu\sigma}}{\partial x^{\mu}} + \frac{\partial F_{\sigma\mu}}{\partial x^{\nu}} \equiv 0$$

and

$$\frac{\partial[\psi h K^{\mu\nu}]}{\partial x^{\nu}} \equiv 0$$

(where $K^{\mu\nu}$ is an abbreviation for an expression in which, in addition to $\underline{G}^{\mu\nu}$ and $F^{\mu\nu}$, F_{μ} also appears) only 6 count, and not all 8. For taking $F_{\mu\nu\sigma}$ for the left-hand side of the first system and $\mathfrak{K}^{\mu}$ for the left side of the second, then $(F_{\mu\nu\sigma}\delta^{\mu\nu\sigma\tau})_{,\tau}$ and $\mathfrak{K}^{\mu}{}_{,\mu}$ vanish identically, i.e., independently of how the quantities $F_{\mu\nu\sigma}$ and $K^{\mu\nu}$ are expressed in terms of $F^{\mu\nu}$ and $G^{\mu\nu}$. However, the second identity (even in the form $(hK^{\mu\nu})_{,\nu} + hK^{\mu\nu}\varphi_{\nu} \equiv 0$) is not a pure identical relation between the left-hand sides of the original field equations since the F_μ enter in. But I do not know how one could guarantee the compatibility of the equations without introducing the ψ. What the case $N-12$ < the number of identities means is not entirely clear to me. I suspect however that, in this case, the system is underdetermined.

More after a study of your detailed exposition.

Kind regards. Yours,

A. Einstein

P. S. I have no clear idea which of the solution manifolds exists in the case that there are less than $N-12$ identities (e.g., $\Lambda^{\alpha}_{\underline{\mu\nu};\nu} = 0$). There is no deterministic form, in this case.

I am eagerly studying your manuscript *and am very thankful for your interest in the problem.*

Translators' note: Translation based on *Debever 1979*, pp. 56–63.

140. From Paul Ehrenfest

Wirbalen-> Riga [8 December 1929][1]

[Not selected for translation.]

141. Compatibility of Field Equations in the Unified Field Theory

"Die Kompatibilität der Feldgleichungen in der einheitlichen Feldtheorie"
[Einstein 1930j]

PRESENTED 12 December1929
PUBLISHED 6 February 1930

IN: *Preussische Akademie der Wissenschaften* (Berlin). *Physikalisch-mathematische Klasse. Sitzungsberichte* (1930): 18–23.

Several months ago, I described the mathematical basis of the unified field theory in a summary article published in the Mathematische Annalen.[1] In the present treatise, I want to briefly summarize the essentials and, at the same time, to show the points in my earlier works on this topic (these Reports, "On the Unified Field Theory", 1929 I,[2] and "The Unified Field Theory and Hamilton's Principle," 1929 X)[3] which were in need of improvement. The proof of compatibility has been somewhat simplified compared to the version in the Mathematische Annalen, based on correspondence with, and thanks to, Mr. Cartan (see §3, [16]).

§1. Critical Remarks Concerning My Earlier Works.

The divergence operation in relation to the tensor identity, which I introduced in §1 of the first of the above-cited articles,[4] is not expedient. It is better to retain the divergence concept, which is defined as the contraction of the extension of a tensor, since only with this latter definition does the divergence of the fundamental tensor vanish identically.

The identity (3a) or (3b), *loc. cit.*, then takes on the form

$$\Lambda^{\alpha}_{\kappa l;\alpha} - (\phi_{\kappa,l} - \phi_{l,\kappa}) \equiv 0, \text{[5]} \tag{1}$$

where we have set

$$\phi_{\kappa} = \Lambda^{\sigma}_{\kappa\sigma}. \tag{1a}$$

As already explained earlier,[6] the compatibility proof for the field equations in that work was based on the incorrect assumption that there are four identities which interrelate the equations (10).

The second of the above-cited works[7] contains a disastrous error. It is, namely, not correct that the $G^{*\mu\alpha}$ depend homogeneously and quadratically on the $S^{\alpha}_{\underline{\mu}\underline{\nu}}$. This invalidates the derivation of equation (21) there, which was interpreted as a field equation for the electromagnetic field.

§2. Overview of the Mathematical Apparatus for the Theory.

The spatial structure, or the field, is described by the Gaussian components $h_s^{\ \nu}$ of the local orthogonal tetrad (the ν-th component of the s-th axis of the tetrad). The transformation law, under changes of the Gaussian coordinate system maintaining the same rotation of all the local tetrads, is given by

$$h_s^{\ \nu'} = \alpha_{st}\frac{\partial x^{\nu'}}{\partial x^{\sigma}}h_t^{\ \sigma}, \qquad (2)$$

where the constants α_{st} form an orthogonal system.

The normalized subdeterminants $h_{s\nu}$ of the $h_s^{\ \nu}$ obey a transformation law:

$$h'_{s\nu} = \alpha_{st}\frac{\partial x^{\sigma}}{\partial x^{\nu'}}h_{t\sigma}. \qquad (3)$$

Systems of quantities which differ in their transformation properties from the h only in terms of the number of their indices are called tensors. The quantities $h_{s\nu}$ and $h_s^{\ \nu}$ comprise the fundamental tensor.

Addition, subtraction, and multiplication are as in the usual tensor theory. Contraction is possible with respect to two local axes (Roman indices) or with respect to two coordinate axes (Greek indices) of differing character.

Changing the index character of a tensor by means of the fundamental tensor through multiplication and contraction is always possible, for example:

$$A_s = h_{s\nu}A^{\nu}.$$

Since A_sA_s is taken to be the magnitude of the vector (A_s), it follows that the coefficients of the Riemannian metric $g_{\mu\nu}$ are given by the quadratic form

$$g_{\mu\nu} = h_{s\mu}h_{s\nu}. \qquad (4)$$

From the restriction that the local tetrads should all be parallel, the law of (integrable) elementary parallel displacements follows,

$$\left.\begin{aligned} \delta A^{\mu} &= -\Delta^{\mu}_{\alpha\beta}A^{\alpha}\delta x^{\beta} \\ \Delta^{\mu}_{\alpha\beta} &= h_s^{\ \mu}h_{s\alpha,\,\beta} \end{aligned}\right\}, \qquad (5)$$

where the comma denotes ordinary differentiation. From this, the law of (absolute) differentiation follows:

$$A^{\mu}{}_{;\sigma} = A^{\mu}{}_{,\sigma} + A^{\alpha}\Delta^{\mu}_{\alpha\sigma} \tag{6}$$

$$A_{\mu;\sigma} = A_{\mu,\sigma} - A_{\sigma}\Delta^{\alpha}_{\mu\sigma}. \tag{7}$$

For tensors with several Greek and Roman indices, a corresponding term occurs for each Greek index.

From the tensor $\Phi_{;\sigma;\tau}$ obtained through twofold differentiation of the scalar Φ or from the tensor character of $(\Phi_{;\sigma;\tau} - \Phi_{;\tau;\sigma})$, the tensor character of

$$\Lambda^{\alpha}_{\mu\nu} = \Delta^{\alpha}_{\mu\nu} - \Delta^{\alpha}_{\nu\mu}. \tag{8}$$

can readily be verified. The vanishing of all the $\Lambda^{\alpha}_{\mu\nu}$ provides the condition for the continuum to have Euclidean character.

The tensor (Λ), owing to the fact that it can be expressed in terms of the *h* quantities, or also due to the integrability of the Δ parallel displacement law, obeys the identity

$$(\Lambda^{\iota}_{\kappa\lambda;\mu} + \Lambda^{\iota}_{\lambda\mu;\kappa} + \Lambda^{\iota}_{\mu\kappa;\lambda}) + (\Lambda^{\iota}_{\kappa\alpha}\Lambda^{\alpha}_{\lambda\mu} + \Lambda^{\iota}_{\lambda\alpha}\Lambda^{\alpha}_{\mu\kappa} + \Lambda^{\iota}_{\mu\alpha}\Lambda^{\alpha}_{\kappa\lambda}) \equiv 0, \text{[8]} \tag{9}$$

from which the identity (1) follows by contraction.

The product rule holds for absolute differentiation. The absolute derivatives of the *h* vanish identically; likewise those of the $g_{\nu\mu}$ (or the $g^{\mu\nu}$). The fundamental tensor as a factor thus commutes with the symbol for differentiation (;).

For the twofold absolute differentiation of an arbitrary tensor T: (the dots indicate arbitrary indices), the commutation law for differentiation applies:

$$T^{:}_{;\sigma;\tau} - T^{:}_{;\tau;\sigma} \equiv -T^{:}_{;\alpha}\Lambda^{\alpha}_{\sigma\tau}. \tag{10}$$

If T has no Greek indices (scalar character), the proof of the above is easy to show; the proof for arbitrary tensors is found from the fact that multiplication of the tensor by parallel vectors (vectors whose absolute derivative vanishes everywhere) converts them into tensors with scalar character.[9]

If the tensor T: under consideration has two contravariant indices, it can be contracted with respect to those and σ or τ; one then obtains from (10) a commutation rule for the divergence.

The special properties of the physical four-dimensional continuum are taken into account by the following stipulations: the coordinate x^4 is purely imaginary (as is the fourth local coordinate), while the others are real. Tensor components are purely imaginary if they have an odd number of 4th indices, otherwise real.

Finally, a formal point of notation:[10] Changing the position of a Greek index ("raising" or "lowering") can be denoted by underlining the corresponding index.

§3. The Field Equations and Their Compatibility.

The field equations must, of course, be covariant, and one can also assume that they are equations of second order and linear in the second derivatives of the field variables with respect to the coordinates. Although these requirements suffice in the original version of general relativity to determine (at least) the field equations for gravitation, this is no longer the case in the present theory. Owing to the tensor character of the Λ, one now has a much larger manifold of tensors than in the Riemannian scheme.

General covariance entails the requirement that 4 field variables must remain arbitrary. The 16 quantities h may be subjected to only 12 mutually independent conditions. Thus, if the number N of field equations is greater than 12, then there must be at least $N - 12$ identities which relate them.

There is no simple way to establish a covariant system of only 12 equations. Therefore, we must establish equations which are related by identities. The larger the number of such equations (and thus the number of identities which relate them), the more determined statements, going beyond the requirements of mere determinism, the theory makes. And such a theory is all the more valuable, to the extent that it makes predictions which are in agreement with empirical facts.(1)[11] The requirement of the existence of an "overdetermined" system of equations with the requisite number of identities gives us a means of finding the field equations.

As field equations, I suggest the following two systems of equations:[12]

$$G^{\mu\alpha} = \Lambda^{\alpha}_{\mu\nu;\nu} - \Lambda^{\tau}_{\mu\tau}\Lambda^{\alpha}_{\sigma\tau} = 0 \tag{11}$$

$$F_{\mu\alpha} = \Lambda^{\sigma}_{\mu\alpha;\sigma} = 0; \tag{12}$$

These comprise 16 + 6 equations for the 16 field variables $h_{s\nu}$. I arrived at them by applying the commutation rule for the divergence to the tensor $\Lambda^{\alpha}_{\mu\nu}$. In fact, we have

$$\Lambda^{\alpha}_{\underline{\mu\nu};\nu;\alpha} - \Lambda^{\alpha}_{\underline{\mu\nu};\alpha;\nu} \equiv -\Lambda^{\alpha}_{\underline{\mu\nu};\sigma}\Lambda^{\sigma}_{\nu\alpha}.$$

We can rewrite the right-hand side as

$$-(\Lambda^{\alpha}_{\underline{\mu\nu}}\Lambda^{\sigma}_{\nu\alpha}) + \Lambda^{\alpha}_{\underline{\mu\nu}}\Lambda^{\alpha}_{\nu\alpha;\sigma}.$$

Taking this into account, by a suitable designation of the summation indices, we can put the above identity into the form

$$G^{\mu\alpha}{}_{;\alpha} - F^{\mu\alpha}{}_{;\alpha} + \Lambda^{\sigma}_{\underline{\mu\tau}}F_{\sigma\tau} \equiv 0. \text{ [13]} \tag{13}$$

(1) In the existing theory of gravitation, there are, e.g., 10 equations for the 10 field variables, and they are related by 4 identities.

These are 4 identities relating the equations (11) and (12), which provided the motivation for setting them as field equations.

Equations (12), together with identity (1), furthermore lead immediately to the identity

$$F_{\mu\nu,\rho} + F_{\nu\rho,\mu} + F_{\rho\mu,\nu} \equiv 0. \tag{14}$$

We note that equations (12) can also be replaced by

$$F_{\mu\nu} = \phi_{\mu,\alpha} - \phi_{\alpha,\mu} = 0, \tag{12a}$$

or by

$$F_{\mu} = \phi_{\mu} - \frac{\partial \log\psi}{\partial x^{\mu}} = 0, \tag{12b}$$

where ψ is a scalar. Furthermore, $F_{\mu\nu}$ can be expressed in terms of F_{μ} by means of the relation

$$F_{\mu\nu} \equiv F_{\mu,\nu} - F_{\nu,\mu}. \tag{15}$$

A third system of identities can be obtained by defining $G^{\mu\alpha}{}_{;\mu}$. It is initially found from (11):

$$G^{\mu\alpha}{}_{;\mu} \equiv \Lambda^{\alpha}_{\underline{\mu}\underline{\nu};\nu;\mu} - \Lambda^{\tau}_{\underline{\sigma}\underline{\mu};\mu}\Lambda^{\alpha}_{\sigma\tau} - \Lambda^{\sigma}_{\underline{\mu}\underline{\tau}}\Lambda^{\alpha}_{\sigma\tau;\mu}.$$

If one applies the commutation relation for the divergence to $\Lambda^{\alpha}_{\underline{\mu}\underline{\nu}}$ with respect to the indices ν and μ, the result is

$$\Lambda^{\alpha}{}_{\underline{\mu}\underline{\nu};\nu;\mu} \equiv -\tfrac{1}{2}\Lambda^{\alpha}_{\underline{\mu}\underline{\nu};\sigma}\Lambda^{\sigma}_{\nu\mu}.$$

If one uses this latter relation to replace the first term on the right-hand side of the above identity, then the first and third terms can be replaced together by

$$-\Lambda^{\sigma}_{\underline{\mu}\underline{\tau}}(\Lambda^{\alpha}_{\sigma\tau;\mu} + \tfrac{1}{2}\Lambda^{\alpha}_{\tau\mu;\sigma}),$$

or by

$$-\tfrac{1}{2}\Lambda^{\sigma}_{\underline{\mu}\underline{\tau}}(\Lambda^{\alpha}_{\sigma\tau;\mu} + \Lambda^{\alpha}_{\tau\mu;\sigma} + \Lambda^{\alpha}_{\mu\sigma;\tau}).$$

The expression in parentheses can, however, taking account of (9), be expressed in terms of Λ itself, so that one obtains

$$+\tfrac{1}{2}\Lambda^{\sigma}_{\underline{\mu}\underline{\tau}}(\Lambda^{\alpha}_{\sigma\lambda}\Lambda^{\lambda}_{\tau\mu} + \Lambda^{\alpha}_{\tau\lambda}\Lambda^{\lambda}_{\mu\sigma} + \Lambda^{\alpha}_{\mu\lambda}\Lambda^{\lambda}_{\sigma\tau}),$$

or, since the first term in parentheses cancels and the two others may be combined,

$$\Lambda^{\sigma}_{\underline{\mu}\underline{\tau}}\Lambda^{\lambda}_{\sigma\tau}\Lambda^{\alpha}_{\mu\lambda}.$$

From this, we obtain

$$G^{\mu\alpha}{}_{;\mu} \equiv -\Lambda^{\alpha}_{\sigma\tau}(\Lambda^{\tau}_{\sigma\mu;\mu} - \Lambda^{\rho}_{\sigma\lambda}\Lambda^{\tau}_{\rho\lambda})$$

or, finally,

$$G^{\mu\alpha}{}_{;\mu} + \Lambda^{\alpha}_{\sigma\tau}G^{\sigma\tau} \equiv 0 . \qquad (16)$$ [14]

(13), (14), and (16) are the identities which relate the field equations (11) and (12).

That these identities indeed imply the compatibility of equations (11) and (12) can be elucidated by the following consideration. It will be possible to fulfill all the equations (11) and (12) on a section $x^4 = a$. Likewise, it will be possible to fulfill all those equations throughout all of space which are characterized by setting the following quantities equal to zero:

$$\begin{matrix} G^{11} & G^{12} & G^{13} \\ G^{21} & G^{22} & G^{23} \\ G^{31} & G^{32} & G^{33} \\ F_{14} & F_{24} & F_{34} \end{matrix}$$

Furthermore, one can choose the latter solution in such a way that it is a uniform continuation of the solution given for the section $x^4 = a$. We then assert that this solution also fulfills everywhere those equations which are characterized by setting the following equal to zero:

$$G^{14} G^{24} G^{34} G^{41} G^{42} G^{43} G^{44} F_{23} F_{31} F_{12} .$$

First of all, it follows from this that F_{14}, F_{24} and F_{34} vanish everywhere and, taking (14) into account, that also $\frac{\partial F_{23}}{\partial x^4}, \frac{\partial F_{31}}{\partial x^4}$ and $\frac{\partial F_{12}}{\partial x^4}$ are zero everywhere. Since, however, F_{23}, F_{31} and F_{12} vanish on the section $x^4 = a$, they then also vanish everywhere. Furthermore, it follows from (13) and (16) that, on the section $x^4 = a$, the derivatives of $G^{14}, G^{41}, \ldots G^{44}$ with respect to x^4 vanish; these quantities, and thus all the $G^{\mu\alpha}$, also vanish on the infinitesimally neighboring sections $x^4 = a + da$. By repeated application of this chain of reasoning, it finally follows that all the $G^{\mu\alpha}$ vanish everywhere. This completes the compatibility proof for the field equations (11) and (12).

The first approximation. We investigate those fields which differ only infinitesimally from the Euclidean special case:

$$h_{s\nu} = \delta_{s\nu} + \bar{h}_{s\nu}. \tag{17}$$

$\delta_{s\nu}$ is equal to 1 or 0, depending on whether $s = \nu$ or $s \neq \nu$, respectively; the $\bar{h}_{s\nu}$ are infinitesimal as compared to 1. If we neglect terms quadratic in $\bar{h}$ (second-order terms), then we can replace the above field equations by

$$\bar{h}_{\alpha\mu,\,\nu,\,\nu} - \bar{h}_{\alpha\nu,\,\nu,\,\mu} = 0 \tag{11a}$$

$$\bar{h}_{\alpha\mu,\,\alpha,\,\nu} - \bar{h}_{\alpha\nu,\,\alpha,\,\mu} = 0 \tag{12a}$$

The *ansatz* (17) still permits an infinitesimal transformation of the Gaussian coordinates. It can then be shown that, given the equations (12a), we can choose the coordinates in such a way that

$$\bar{h}_{\mu\alpha,\,\alpha} = \bar{h}_{\alpha\mu,\,\alpha} = 0 \tag{18}$$

is satisfied, whereby the only field equations left are

$$\bar{h}_{\alpha\mu,\,\nu,\,\nu} = 0 \tag{11b}$$

If we denote the doubly-symmetric part of $\bar{h}_{\alpha\mu}$ by $\bar{g}_{\alpha\mu}$, and the doubly-antisymmetric part by $a_{\alpha\mu}$, then we can split the field equations into the two systems

$$\left.\begin{aligned} \bar{g}_{\alpha\mu,\,\nu,\,\nu} &= 0 \\ \bar{g}_{\alpha\mu,\,\mu} &= 0 \end{aligned}\right\} \tag{19}$$

$$\left.\begin{aligned} a_{\alpha\mu,\,\nu,\,\nu} &= 0 \\ a_{\alpha\mu,\,\mu} &= 0 \end{aligned}\right\}. \tag{20}$$

In my opinion, (19) expresses the laws of the gravitational field, while (20) expresses those of the electromagnetic field, with the $a_{\alpha\mu}$ playing the role of the electromagnetic field strengths.[15] Strictly considered, splitting up the field into a gravitational field and an electromagnetic field is not possible. More details can be found in my article in the Mathematische Annalen.[16]

The most important aspect of the questions arising from the (strict) field equations is the existence of solutions which are free of singularities and which could represent the electrons and protons.

142. From Élie Cartan

Le Chesnay, 13 December 1929

Dear Sir,

I did indeed receive your letter[1] and have read and thought about it. What you have told me about the degree of indetermination was of interest to me, as I thought, for reasons that are explained in my note[2] but which are rather more reasons of feeling, that your system of 22 equations was perhaps too determined and that its solution does not have a sufficient degree of generality. But you are infinitely more competent than I in this regard. I nevertheless believe that my notion of the indicator of generality, which has, in fact, a precise meaning, can be of use. If, for example, your system depends only on 3 or 4 arbitrary functions of 3 variables in the sense that I indicated in my note, it would certainly not be sufficiently general. My indicator of generality evidently varies inversely to the number of identities, at least in the case that this number is $N - 12$.

You have indicated to me, relative to certain systems which I had mentioned to you, some objections which, I am convinced, are due only to the poor manner of expression in my letter. Thus, for the system of 22 equations, which I can write as

$$R_{ik} = 0 \tag{1}$$

$$F_{\alpha\beta} \equiv \varphi_{\alpha,\beta} - \varphi_{\beta,\alpha} - C(\varphi_\alpha S_\beta - \varphi_\beta S_\alpha) = 0 \tag{2}$$

$$G_{\alpha\beta} \equiv S_{\alpha,\beta} - S_{\beta,\alpha} - C'(\varphi_\alpha S_\beta - \varphi_\beta S_\alpha) = 0 \tag{3}$$

I have never had the idea of demonstrating their compatibility by integrating equations (2) and (3), in particular, not by integrating them in an incorrect manner. The compatibility results for me from the existence of $n - 12 + 2 = 22 - 12 + 2 = 12$ identities, which are, apart from the 4 identities of Bianchi, the following 8:

$$\frac{\partial F_{\beta\gamma}}{\partial x_\alpha} + \frac{\partial F_{\gamma\alpha}}{\partial x_\beta} + \frac{\partial F_{\alpha\beta}}{\partial x_\gamma} - C(S_\alpha F_{\beta\gamma} + \ldots - \varphi_\alpha G_{\beta\gamma} \ldots) \equiv 0,$$

$$\frac{\partial G_{\beta\gamma}}{\partial x_\alpha} + \frac{\partial G_{\gamma\alpha}}{\partial x_\beta} + \frac{\partial G_{\alpha\beta}}{\partial x_\gamma} - C'(S_\alpha F_{\beta\gamma} + \ldots - \varphi_\alpha G_{\beta\gamma} \ldots) \equiv 0.$$

I indeed believe that I called your attention to the fact that, in your system of 22 equations, the 12 equations (2) and (3) occur with a value of 0 for the constant C and the value -1 for the constant C', and added that, *in the special case of* $C = 0$, one can integrate not only the equations (2), but also the equations (3), by introducing the two new functions ψ and χ. The integration can furthermore be carried out when $C \neq 0$; it is only necessary to set

$$\Sigma_\alpha = S_\alpha - \frac{C'}{C}\varphi_\alpha$$

and to treat φ_α and Σ_α as unknown functions; equations (2) and (3) conserve their form, but the new constant C' is zero.

I would like to say something about the generalization that you proposed with arbitrary functions of ψ and χ, but it would not be worth the effort.

I now come to the system of 15 equations about which I spoke previously; it is made up of

First, the 6 equations

$$F_{\alpha\beta} \equiv \frac{\partial(aS_\alpha + b\varphi_\alpha)}{\partial x_\beta} - \frac{\partial(aS_\beta + b\varphi_\beta)}{\partial x_\alpha} = 0$$

Second, of 9 equations whose first terms are subjected to the sole condition that they form a symmetric tensor, with a contracted scale tensor of zero.

I have never thought of including the equations $R_{ik} = 0$ into this system, which would make it incompatible. This system of 15 equations is very interesting from the mathematical point of view (it is understood that it is not sufficiently determined for useful applications in physics). Let us write one of the 9 last equations, for example

$$G^{12} \equiv \Lambda^{2}_{\underline{1}\underline{\mu};\mu} + \Lambda^{1}_{\underline{2}\underline{\mu};\mu} + c(\varphi_{1,2} + \varphi_{2,1}) + d(S_{1,2} + S_{2,1}) - A_{12} = 0$$

A_{12} depends only upon the $\Lambda^{\gamma}_{\alpha\beta}$ and not on their derivatives.

In the general case, that is, if one has

$$a - bd \neq 0, \quad c \neq -\frac{1}{3},$$

the system is deterministic and in involution. But if $a - bd$ or $c + \frac{1}{3}$ is zero, the system ceases to be deterministic and also ceases to be in involution: One can no longer affirm that it is compatible, but it is also not even certain that it has any solution.

As it seems to me that we are perhaps not absolutely in agreement on the sense of the word *identities*, I will take the liberty of elaborating precisely to you just what the point of view is to which I am led by the theory of systems in involution.

Let us take N equations with linear partial derivatives of first order with p unknown functions

$$F_\alpha = 0 \quad (\alpha = 1, 2, \ldots, N),$$

and let us suppose that the first terms are independent linear functions of the derivatives of the p unknown functions with respect to any one of the variables. There then exists a *maximum* number[3]

$$N - p + r_1 + r_2$$

of independent linear combinations of the $F_{\alpha, \beta}$ playing on the property that, in each of these combinations, the second derivatives eliminate themselves.

Given this proposition, there are three possibles cases:

1. This maximum number is not attained.

2. The maximum number is attained, but the $N - p + r_1 + r_2$ combinations found (which consequently depend only on the unknown functions and their first derivatives) are not all zero if one takes account of the given equations $F_\alpha = 0$.

3. The maximum number is attained, but the $N - p + r_1 + r_2$ combinations found are all zero when the given equations are taken into account.

In this third case, I would claim that the first members of the equations are connected by $N - p + r_1 + r_2$ identities: This is the necessary and sufficient condition that the system be completely integrable.

In the second case, the system is incompatible, or at least all the solutions of the system necessarily satisfy new equations with partial derivatives of the first order which are not *algebraic* consequences of the given equations. Since this is an eventuality that we wish to exclude, we can say that the system is incompatible.

In the first case, the system is not completely integrable, but one cannot affirm without more elaborate examination whether it would be incompatible. One is obliged to consider the system obtained by joining the differentiated equations to the given equations (the first derivatives of the unknown functions being themselves new unknown functions), and then to investigate whether the conditions of involution are fulfilled by this new system, and continuing thus. In the end, one is certain to arrive at one of the cases third or second.

The system of 16 equations of which I spoke to you, and of which I wrote down only a few,

$$X_{12} \equiv S_{1,2} - S_{2,1} + a(\varphi_{1,2} - \varphi_{2,1}) + b(\varphi_{3,4} - \varphi_{4,3}) - A_{12} = 0,$$

$$Z \equiv \varphi_{\alpha,\alpha} - C = 0$$

$$\require{cancel}\cancel{Y_{12} \equiv \Lambda_{\underline{1}\underline{\alpha};\alpha} + 2(c\varphi_{1,1} + dS_{1,1}) - \{(}$$

$$Y^{12} \equiv \Lambda^{2}_{\underline{1}\underline{\alpha};\alpha} - \Lambda^{1}_{\underline{2}\alpha;\alpha} + c(\varphi_{\underline{1},\underline{2}} + \varphi_{\underline{2},\underline{1}}) + d(S_{\underline{1},\underline{2}} + S_{\underline{2},\underline{1}})$$

$$-g^{12}[\Lambda^{\mu}_{\mu\underline{\alpha};\alpha} + \dots\ ] = 0$$

recurs in the second case; there are 4 linear combinations of the derivatives of the first terms, whose second derivatives eliminate themselves; one of these combinations is

$$Y_{1\alpha;\alpha} - \frac{3c+1}{2}Z_{;1} - dX_{1\alpha;\alpha} - \frac{1+c-ad}{b}h(X_{23;4} + X_{34;2} + X_{42;3})$$

If the quantities A_{12}, C, etc. are arbitrary, the system is impossible, but I do not know that it is not possible to choose them in such a way that the system would be completely integrable.

What you told me about the subject of the number of *real* identities of your system comes down, perhaps, to a question of convention. When I say that there are 12 identities, I am presuming that I am making use only of the equations such as they are, without modifying them by the introduction of any auxiliary functions. My 12 identities, to wit:

$$G^{\alpha\mu}{}_{;\mu} + F^{\alpha\mu}{}_{;\mu} + \Lambda^{\rho}_{\alpha\mu}F_{\mu\rho} \equiv 0,$$

$$G^{\mu\alpha}{}_{;\mu} + \Lambda^{\alpha}_{\rho\sigma}G^{\rho\sigma} \equiv 0,$$

$$F_{\alpha\beta,\gamma} + F_{\beta\gamma,\alpha} + F_{\gamma\alpha,\beta} \equiv 0$$

are indeed quite independent. I am convinced that, in order to be able to decide with certainty in every case, it would be necessary to have a method which can be applied to the equations *such as they are given*, and I am certain that my method can be applied.

I apologize, dear Sir, for once again allowing myself to become a bit too wordy, and I ask you to accept my highest regards and sincere respect,

E. Cartan

143. To Max Born

[Berlin,] 14 December 1929

[Not selected for translation.]

144. To Selig Brodetsky

[Berlin,] 14 December 1929

Dear Mr. Brodetsky,

I am sorry that you were so greatly upset by my criticism.[1] What I object to in your talk is less what you said than what you failed to say. Namely, it lacks an analysis of the causes of the uproar against us in the Arab world, a purely causal perspective, without which the question cannot, in my opinion, be resolved. I consider it my duty to express my opinion, but I have neither the time nor the energy to take part in your polemics. I am glad we have no power. If nationalistic pigheadedness gains the upper hand in our ranks, we'll be at each others' throats, as we deserve.

Kind regards, your

145. To Swami Jnanananda[1]

[Berlin,] 14 December 1929

[Not selected for translation.]

146. To Joseph Sauter

Berlin, 14 December [1929][1]

Dear Mr. Sauter,

Your remarks were quite correct.[2] I have really left the notations ${}^{s}h_{\nu}$ or $h^{s}{}_{\nu}$ ⟨long since⟩[3] behind me, since the orthogonality makes ⟨the double index⟩[4] unfounded. I have now specified the notation as follows: local indices are in Roman type, and[5] lowered, coordinate indices are Greek. Since one[6] can easily ⟨convert⟩[8] Greek[7] into Roman indices by means of the *h*, an extension of the tensor concept is in order.

We indeed have[9]

$$A_{\sigma}{}^{\tau} = h_{\sigma s} A_{s}{}^{\tau}$$

$$A_{s}{}^{\tau} = h_{s\sigma} A^{\sigma\tau} \quad \text{etc.}$$

Transformation law for local rotations and coordinate transformation:

$$A_{s\alpha}{}^{\beta'} = \alpha_{st} \frac{\partial x^{\lambda}}{\partial x^{\alpha'}} \frac{\partial x^{\beta'}}{\partial x^{\mu}} A_{t\lambda}{}^{\mu}.$$

I have now succeeded in stating the natural field laws for such a manifold.

All of the earlier work suffered from forbidden overdetermination. The number of independent equations must namely be 12, and the total number of equations Z must be as large as possible. Among the Z equations, there must be I autonomous identities, so that

$$Z - I = 12.$$

The equations that I have presently found are very satisfying.

I have now departed from the earlier interpretation[10] of the electromagnetic field. When the article has appeared, I will send you a copy.[11]

I am very happy to see that you take such an interest in the matter. I am confident that I am on a good path, in spite of the absolute skepticism of my professional colleagues. Even Weyl[12] has expressed himself energetically against it.

Best wishes to you and Besso,[13] your

A. Einstein

Your comical letter in reply[14] was a lot of fun for me. But I will keep it![15]

147. On the Hedwig Wangel Foundation

[Berlin, 15 December 1929]

⟨A large part of the⟩ Many of the people who populate the state prisons are afflicted with a congenital mental defect. We must have compassion for them, protect society from them, into which we cannot incorporate them by normal means.[1] Many, however, drift into prison despite having a normal disposition. Circumstance, need, unhealthy influences, sometimes even noble motives often bring about conflict with the law. They are the victim of the inadequacy of our social institutions; in part also the shortcomings of existent laws.[2] Anyone who does not understand what I mean by that can consult Domela's Memoirs[3] for an example.

The most disastrous part in the execution of our penal system lies not in the punishment itself but rather in the subsequent fate of the "criminal." Brusque rejection, exploitation, scolding of every form awaits those released from prison and, all too often, urges them anew onto the path to crime, to their detriment and severely damaging to society. To intercede and assuage here is one of the most important and quite neglected social duties.[4]

Driven by a sense of social duty of a wholly rare magnitude and purity, Mrs. Hedwig Wangel has, in addition to a highly stressful artistic career, placed all her labor ⟨and her means⟩, her income, and her spiritual knowledge in the service of caring for women released from prison. Such a degree of devotion and self-denial for external goals, I have never seen. It is therefore disgraceful to see how little support has been bestowed upon her efforts thus far. She herself cannot be permitted to collapse from exhaustion, and her work from lack of means. Means and ways must be found so that the financial and administrative worries are taken from her, so that she herself can return to the very work for which she has been created: the care, education, and professional training of her protégés.

Anyone who is able and willing to contribute in some way so that this goal will be achieved, I would ask them to get in contact with Mrs. Wangel.[5]

Albert Einstein 1930.

148. From Hugo Bergmann

Jerusalem, 17 December 1929

Dear Professor,

I have received your letter of the 25th.[1] Basically, I am in complete agreement with your decision not to publish the correspondence, since my name now appears on the Zionists' blacklist, and almost all Zionist periodicals (and not only the revisionist ones, but also the official ones, such as Vienna's "Stimme,") brand me as a traitor and demand my resignation from the library's board of directors.[2] Therefore, it would make no sense to publish my views now, even if you were to give your consent to it. On the other hand, what I consider extremely important is that, in the next few weeks, you yourself, on an occasion that presents itself to you—for instance, in a letter to the Zionist Congress of Delegates[3] or the like—forthrightly express your own view again. The interview you gave in Paris[4] has already been very useful in the outside world, and in the Near East, for example, which is close to English official circles, it has been seen with approval as a sign that a readiness for peace is being documented in the Zionist camp.

Domestically, however, in the Zionist camp itself, the situation is as bad as it could be. For instance, in the press here, your interview[5] is either suppressed or presented in such a way that nothing essential remains of it. Our public refuses, at any price, to face the facts. That said, I consider the revisionists[6] relatively harmless. They have an overt program that can be overtly attacked. The pseudo-revisionists are far worse. For purely tactical reasons, they do not belong to the revisionist party, but instead support the program of the Jewish state, which can be realized only by oppressing the Arabs, and oppose every reasonable demand made by the latter. This species really includes all Zionists, from the "general Zionists"[7] to the workers' parties.

To give you an example, I am sending you the Palestine Bulletin, the Zionist daily published here, with an article written by a British journalist.[8] This article is being widely disseminated as we speak by the London government's press office and appeared at the same time in the *official* organ of the London government, the American Zionist Federation, and also in an English and in a Hebrew daily,[9] and probably in many other papers. In this article, it is said that the Balfour Declaration implies a Jewish state, that the retraction of the White Book written by Samuel and countersigned by Weizmann (the most reasonable political document Zionism has!) must be demanded, that Arabs must be given no representative institution because the Turks have regarded the Palestinian Arabs' every wish to express political views as a comical impertinence, and other cordial notions. And yet, the Arabs have had representatives in the Turkish parliament since 1908.[10]

With such views disseminated by government officials, what could the Arabs expect other than further pogroms? And our press here—and, once again, not only the revisionist press, but also the moderate *Ha'aretz*, for example—is already preparing the ideology for the next pogroms: Brit Shalom[11] is to blame, because it teaches the Arabs that something is always achieved by pogroms. For years, we have said: We have to come to a settlement with the Arabs. We were told: The time has not yet come. We've said it again after pogroms, and now we are told: We cannot yield under the pressure of events. You are entirely right, Professor. "If it goes on like this, a catastrophic development is very much to be feared."[12]

You will probably have occasion to see Weizmann in Berlin.[13] He can save the situation if he has the inner courage to publicly oppose the advocates of a Jewish state *of any kind*, to speak the full, unvarnished truth regarding the situation in Palestine, just as he once had, at the time of the high tide of immigration to Palestine,[14] the courage to express openly very unpopular truths about the economic danger of the immigration. We all want to lighten his heavy burden in any way we can, but he must—of this, he cannot be spared—draw a dividing line in a clear and unambiguous way and distance himself from those who endanger his great achievement in the most dreadful ways and are driving us toward the abyss.

With devoted regards for you and your wife,[15] your

Hugo Bergman

149. From Richard von Mises[1]

Berlin NW 97, 17 December 1929

Dear Colleague,

A short time ago, you told me that you would be happy to have a younger man who was interested in field theory and would be suited to do some work on it. I believe that I can recommend Dr. Walther Mayer,[2] *Privatdozent* in mathematics at the University of Vienna, for that position. He is very capable and certainly not without talent, and he recently received the *Habilitation* under Wirtinger[3] in Vienna, in spite of considerable political[4] opposition. His areas of specialization have been in the main the theory of invariants and Riemannian geometry, and he has familiarized himself thoroughly with general relativity theory. A textbook on differential geometry by Duschek and Mayer[5] is soon to be published by Teubner, whose second volume, the differential geometry of the *n*-dimensional Riemannian space, was written by Mayer.

Dr. Mayer is not entirely without means, and he could possibly live for a while in Berlin, if he were to be given some financial support.[6] He would then be on leave from the University of Vienna. If you are at all inclined to consider hiring him, I would suggest to Dr. Mayer that he should perhaps come to Berlin to discuss the matter with you during the Christmas vacation, without any obligation on your part.[7] I would be grateful for a kind and timely reply from you.

With best wishes, sincerely yours,

Mises

150. To Élie Cartan

Berlin W, 18 December 1929

Dear Mr. Cartan,

I am very fortunate that I have acquired you as a coworker. For you have exactly that which I lack: an enviable facility in mathematics. I have not yet fully understood your explanation of the *indice de généralité*,[1] at least not the proofs. I beg you to send me those of your papers from which I can properly study the theory.

But I am very grateful to you for the identities

$$G^{\mu\alpha}{}_{;\mu} + \Lambda^{\alpha}_{\rho\sigma} G^{\rho\sigma} \equiv 0$$

which, remarkably, had escaped me.[2] It must be for this reason that I had to take the long way round concerning the function ψ. I have made use of these identities in a new article in the *Sitzungsberichten*, in which I took the liberty of citing you as the source.[3]

Now, to the system

$$(1) \qquad R_{ik} = 0$$

$$(2) \qquad F_{\alpha\beta} = \varphi_{\alpha,\beta} - \varphi_{\beta,\alpha} - C(\varphi_\alpha S_\beta - \varphi_\beta S_\alpha) = 0$$

$$(3) \qquad G_{\alpha\beta} = S_{\alpha,\beta} - S_{\beta,\alpha} - C'(\varphi_\alpha S_\beta - \varphi_\beta S_\alpha) = 0$$

I should like to tell you why I do not think it likely that the laws of nature are expressed by these equations. (1) contains only the g_{ik}. Thus, it would provide a *determined* set of laws for the pure gravitational field alone, without allowing the parallel structure to react upon it. It would give a causal connection of the form

metric ⟶ parallel structure,

but no reverse causal connection. This is much too peculiar. And besides, from my point of view, this system is less determined than the one offered by me. For I can satisfy (2) and (3) with the particular system

$$\varphi_\alpha = \frac{\partial \varphi}{\partial x^\alpha} \qquad (2')$$

$$S_\alpha = \frac{\partial S}{\partial x^\alpha}, \tag{3'}$$

where φ and S are two scalars. But (1), (2′), and (3′) are 18 equations for 18 unknown functions $h_{s\nu}$, φ, and S. On the other hand, if one writes my system in the corresponding form, one has 20 equations for 13 unknowns, and thus—it appears to me—a stronger determination. Or is this not the case?

After reading your detailed definition, it seems to me that I have the same understanding as you with regard to identities.

When I asserted, for example, that

$$F_{\alpha\beta,\,\gamma} + F_{\beta\gamma,\,\alpha} + F_{\gamma\alpha,\,\beta} \equiv 0 \ (\equiv F_{\alpha\beta\gamma})$$

are only 3 independent identities, I meant that the identity $0 \equiv F_{\alpha\beta\gamma,\,\sigma}\delta^{\alpha\beta\gamma\sigma}$ (= 1 or -1, according to the character of the permutation $\alpha\beta\gamma\sigma$) is always satisfied, whatever one takes for $F_{\alpha\beta}$. So this last identity says nothing about how the $F_{\alpha\beta}$ is expressed in terms of the h.[4] The same thing holds for my third identity.

As for the application of your identities, I have proven their compatibility as follows: Let us label the identities as follows

$$G^{\alpha\mu}{}_{;\mu} + F^{\alpha\mu}{}_{;\mu} + \Lambda^{\rho}_{\underline{\alpha}\underline{\mu}} F_{\mu\rho} \equiv 0 \tag{1}$$

$$G^{\alpha\mu}{}_{;\alpha} + \Lambda^{\mu}_{\rho\sigma} G^{\rho\sigma} \equiv 0 \tag{2}$$

$$F_{\alpha\beta,\,\gamma} + \cdot + \cdot \equiv 0 \tag{3}$$

Let a solution of all the equations have been found on a cross-section $x^4 = a$. This can certainly be extended in so continuous a manner that the 12 equations arising from setting

$$\begin{matrix} G^{11} & G^{12} & G^{13} \\ G^{21} & G^{22} & G^{23} \\ G^{31} & G^{32} & g^{33} \\ F_{14} & F_{24} & F_{34} \end{matrix}$$

equal to zero are all satisfied. From the last three together with (3), it follows that $\frac{\partial F_{23}}{\partial x^4}$, $\frac{\partial F_{31}}{\partial x^4}$, and $\frac{\partial F_{12}}{\partial x^4}$ also vanish everywhere. Thus, the left-hand sides, F_{23}, etc., vanish everywhere (since they do on $x^4 = a$).

Now, it follows from (1) and (2) that, on $x^4 = a$, the derivatives of $G^{14}, G^{41} \ldots G^{44}$ with respect to x^4 vanish; hence, these quantities vanish on

$x^4 = a + da$. That these equations are satisfied in the whole domain follows from continuation. Your general methods probably make such a demonstration superfluous; but this is the way that is helpful for me.

One more remark. The equations

$$\begin{cases} (\Lambda^{\alpha}_{\mu\nu;\sigma} + \cdot + \cdot\,) + (\Lambda^{\alpha}_{\underline{\mu}\rho}\Lambda^{\rho}_{\nu\sigma} + \cdot + \cdot\,) = 0 \\ \Lambda^{\alpha}_{\underline{\mu}\underline{\nu};\nu} + \Lambda^{\sigma}_{\mu\tau}\Lambda^{\alpha}_{\sigma\tau} = 0 \\ \Lambda^{\alpha}_{\mu\nu;\alpha} : = 0 \end{cases}$$

have the unpleasant property that they contain Δ in their quadratic terms (in the carrying out of;). Surely it is strange that there should be no system in which *only* the Λ appear. Hence, the system is not "canonical." Indeed, one can add on the h as a field variable and set

$$h_{s,\mu,\nu} = h_{s\alpha}\Delta^{\alpha}_{\mu\nu}$$

If one now considers the h and Δ as field variables, then the derivatives arise only linearly. But this is unpleasant, because the Δ are not tensors.

Kind regards, your

A. Einstein.

The most important thing for me is the question: from the point of view of the degree of their determination, do you consider my system of equations to be especially privileged? For only if one may correctly assert this does the whole theory acquire any cogency.

Translators' note: Translation based on *Debever 1979*, pp. 72–79.

151. To Semen L. Frank

[Berlin W,] 18 December 1929

Dear Prof. Frank,

I thank you for your friendly letter and the article.[1] The cancerous evil in Russia lies precisely in the absolutist methods, which necessarily lead to every abuse on the part of those who have power. Also, it is clear that, under such circumstances, morally inferior people will strive for positions of power with particular intensity. These general statements appear to me much more important than the question of how many injustices have occurred in a particular case.[2]

Kind regards, your

152. To Leo Kohn

[Berlin,] 18 December 1929

Dear Mr. Kohn,

It pleases me greatly that I'll see you and Dr. Weizmann soon.[1] I am also not of the opinion that official negotiations with Arab career politicians who live off of the discord can have much meaning.[2] But I do, however, believe that dynamic relations with the larger Arab masses must be pursued in order to prevent such considerable and dangerous tensions from developing, as in recent times. The question is not English or Arab; if anything, relations with England would ease, if a respectable *modus vivendi* with the Arabs were to be found.

Kind regards, your

P.S. I have received your private letter to Caputh.[3]

153. From Myron Mathisson[1]

Warsaw, 18 December 1929

Sir,

Your note published in the *Sitzungsberichte der Preuss. Akademie* (meeting of 8. December 1927), titled "Allgemeine Relativitätstheorie und Bewegungsgesetz," deals with a problem to which I have found a more complete solution one and a half years ago.[2]

Here, first of all, are some observations concerning your note:

You limit yourself to a quasi-stationary moment. As you ascertain that, for the field within an electron, the deviations from spherical symmetry are of the same order as the external field, you thus dispense with radiation—that is to say, to the modifications of spherical symmetry by the derivatives of the velocity of the electron, those derivatives not necessarily being zero (at least a priori)—when the external field vanishes.

Apparently quite involuntarily, your neglect of radiation was forced upon you by the insufficiency of your mathematical method. In fact, your calculations are inseparably tied to the condition that a well-determined function whose surface integral you evaluate should be of the order of $\frac{1}{r^2}$ around a pole. Since, otherwise, the result of the integration for a spherical surface whose radius r is infinitely small could become infinitely large. And, nevertheless, the quadratic terms in relation to the internal field of the energy tensor T_{ik} responsible for the radiation are of the order of $\frac{1}{r^4}$. Thus, you suffer the same defect as that inherent in Weyl's deduction of the dynamic equations.[3]

The other cunning trick that Weyl invented and which you adopted was those virtual fields which fill the interiors of the "singularity tubes" after having mentally emptied them of the real fields (containing a singularity). In this manner, it is true that one can establish the uniqueness of the solution of the equations which determine the gravitational-tensor potential. But the procedure becomes illusory (in general) if T_{ik} contains terms of the order of $\frac{1}{r^3}$, $\frac{1}{r^4}$, which one does not neglect. Furthermore, the "virtual fields" are only a subterfuge, imagined ad hoc and lacking in any theoretical basis.

Now we find ourselves facing a grave difficulty, as pointed out in your note: having once admitted functions with singularities, the ordinary conditions (at infinity) no longer suffice to define a unique solution to our equations. I have found the key to the puzzle after a year of stubborn reflection and calculations. In addition, my calculations have proven nicely that one must further pursue further approximations: they will never lead to supernumerary equations which might render the *quantum* phenomena conceivable—an eventuality which you declare to be possible toward the end of your note.

Before entering into a rapid summary of my work concerning the problem of the motion of an electron, I wish to assure you, dear sir, that the reproaches which I have just addressed to you, rather petty and rather ridiculous toward you, have caused me considerable sorrow to write. Please do not hold them against me. I sheepishly appeal to your generosity.

The evaluation of the reaction force of the radiation for an electron-singularity seems quite interesting to me. One might expect an expression different from that of the old theory, even a zero reaction force. (I note in passing that the so-called classical expression was deduced so far as possible, but its uniqueness was never established.) I have thus succeeded at the task, which seemed rather rough to me. The calculations, at the same time delicate and complicated, reflect an irreproachable elegance.

Those obstacles which I have just mentioned appeared to me to be quite banal in comparison to the difficulties of a higher order, both mathematical and physical, which are heaped around a most disturbing question: the apparent polyform nature of the potentials. After a bitter struggle, I have been able to eliminate this duality. The results of my labor were of an unexpected scope. The existence of material particles exhibits itself as subject to certain numerical conditions between their masses and their electric charges, those masses and charges assuming discrete values, the same for all particles, the charges having necessarily two signs but absolutely equal magnitudes. It is quite remarkable that the concept of particle-singularity has shown itself to be essential for the existence of the properties of matter, as I have just indicated. And what protects the masses and the charges of the particles during their wanderings through powerful fields, what prevents them from diminishing or increasing, is precisely the same mathematical connection which guarantees the uniqueness of the potentials. All of a sudden, the supernumerary equations have been eliminated, free motion maintained, and the true correlative aspect of the constancy of the numerical characteristics has been established.

Since the initial approximations were so favorable, a higher approximation led me to zero constants for the electrons. The gravitational equations including their so-called cosmological extensions have shown themselves to be insufficient.

But, however, I attained my original goal. I was in possession of a criterion which permits distinguishing the true, unique potential, and I could continue with my calculations.

As for your admirable works on the unified field theory, are they even able to assist me in my researches of a numerical connection between the constants of the electron? I have no idea. That question would require some calculations, which I will postpone for a later time. To tell you the truth, my heights and my depths have shown me the inanity of a physics of pure reason, which seeks to guess the Hamiltonian function that rules the universe and errs among the innumerable possibilities. The close connection that I have detected between the equations of physics and the existence of particles will perhaps furnish some concrete indications of the corrections that must be undertaken.

I feel myself to be attached for life and death to the general theory of relativity; I am riveted to it. Its decadence would also be mine.

Ille dies utramque ducet ruinam.[4]

(I often think of Heinrich von Kleist, whose drama was revealed by Cassirer).[5]

Your general theory of relativity is marvelously solid; I know that. It is like a rock in the turbulent waters. It has contributed powerfully to the consolidation of all of physics. But its role in that physics which is soon to be born is rather modest. Many experimental physicists who are not closely familiar with your theory have taken up the idea that newer results on the general theory of relativity are relevant only to minor effects in astronomy and astrophysics. What a humiliation for a *Physical Theory*, fashioned to the measure of gigantic conquests! The fault lies with the relativists who have become apostles of your doctrine instead of cultivating the vineyards of the Lord. There is also a more profound cause. It is necessary that, between a great discovery and its first worthy application, a considerable amount of time should elapse. Your wine has not yet aged to ripeness in their minds. The good wine! Praised be the vines which have produced it! Whoever drinks of it will be raving. Even those who drink nothing will become dizzy simply by inhaling the fragrance of the vintage in passing.

The new micromechanics gets along without general relativity. (It also dispenses with a number of other things: with intuitive and realistic evidence, even with logic.) And yet, the natural predestination of general relativity which is worthy of it is to fill the abyss between the two disparate domains of physics, between the two notions of light. I know how to make your theory play that role. I have advanced quite far with my calculations. I have a net notion of the real sense of Schrödinger's equation[6] (it was going to be relativistic ever since its birth, without having Mr. Dirac[7] interfere with it), of the "spin" of the electron,[8] even if the electron may be, in my opinion, a singularity. Interference, diffraction are maintained, so to speak, by themselves.

I would not know how to work without having before my eyes a large plan, authorized by the calculations, embedding, in principle, all of physics. The recent optical theories, cowardly and opportunistic, apply themselves to developing, in earnest, concepts which are only bits of ideas, "working hypotheses": one would say that they were children, putting cuff links into their scribblings of little men. Is it possible in this way to regain the fundamentals, to rejoin the origins of the curve by marking it as tangent to a little element placed at its end? I believe not. It is very significant that you, sir, are a spectator to this bowling match without engaging yourself in the detailed development of the diverse hypotheses, even though you were among the first of those who enunciated them.

All the attempts to explain the atom by means of general relativity have been only defensive. One has argued thus (or roughly so): given the nearly inextricable complexity of the equations of general relativity, one can expect in their solutions some complications in favor of quanta. One was counting on kindly complications. As for the solutions, one did not venture there, and one restricted oneself, as distinguished mathematicians, to differential equations "whose manipulation is just as convenient as it is elegant." One titles such daydreams as "the electrodynamics of the atom" (Eyraud).[9] And this stirring-up ends by compromising the tool which had been employed by unskilled workers, that is to say, general relativity.

As for my calculations, they establish, at the least, a solution of the equations of general relativity by successive approximations. Since the approximation that has already been reached is quite far along, this result in itself exhibits its permanent value.

The truth that you have revealed to the world has been proclaimed, by those pretending to be wise, to be eroded and sterile. But the day will come. Then the dessicated plant, the rose of Jericho, will suddenly blossom, grow, and fill the air with its fragrance.

I can supply some copies of my first work on dynamics, written in German and printed.[10] I will need a week for that. But in order to go to work seriously, I would need assured honoraries, in case my work were to be accepted. If you find it poor, I will accept my failure without complaints. Do not accuse me of greed, sir. Giving Hebrew lessons, poorly paid (and difficult to arrange), are my means of living. And, at the end of the day, one feels quite stupid, empty. Inspiration does not flow to beasts of burden. Publicity would be precious to me, if it would permit me to catch my breath in preparation for the final struggle. Otherwise, I will necessarily give up.

How sad life is here, for men of my race!

Here the mud is made with our tears![11]

If only one could escape from this prison! Large industrial companies are searching for clever theoreticians, so one hears. Mr. Thirring, who began by rotating enormous masses in order to thus generate *den euklidischen Hintergrund des Führungsfeldes*, ended up rotating grammophone discs.[12] As for me, my head is full of technical ideas. I have an idea about how to effect the rapid vertical take-off of an aircraft, very different from that of a helicopter.

All this makes you laugh, perhaps? Myself, I feel like crying. At times, one is tempted to let go of everything, to pack all the calculations into a big carton and give it into the hands of a friend, after having inscribed upon it the words of the King of Sweden, dying in battle:

"*I have enough: save thyself, brother.*"[13]

I would be happy to be favored with your reply, even if it be a simple "no."

I am not writing to you in Hebrew (that would be quite comfortable for me) because I do not know if you would be pleased with that.

Please excuse, sir, any disturbance that I may have occasioned. Yours sincerely, in deep devotion and admiration,

Myron Mathisson

154. From Max Born

Göttingen, 19 December 1929

[Not selected for translation.]

155. From Wolfgang Pauli

Zurich, 19 December 1929

Dear Mr. Einstein,

I thank you heartily for sending me the proofs of your latest work in the Mathematischen Annalen,[1] which gives such a convenient and beautiful overview of the mathematical properties of a continuum with a Riemannian metric and teleparallelism. I would like to add a few remarks to my report, and from a major portion of the younger generation of physicists, concerning the physical aspects of the matter.[2]

Contrary to what I said to you last spring,[2] from the point of view of quantum theory, there is now no longer any argument which could be advanced in favor of teleparallelism. For as Weyl and Fock have shown, integrating the Dirac theory of the electron into the relativistic theory of gravitation is possible, indeed, when the tetrad quantities h_s^{ν} are introduced, but only if the equations remain invariant when the tetrads at spatially distant points are *arbitrarily* rotated relative to one another.

You will answer this by saying that, for the moment, you do not want to hear any more about the quantum theory. I know that, but I find it very regrettable. I must furthermore say to you that the derivation of your field equations (29), (30)[4] does not seem at all compelling to me, and even the simplest conclusions that you can draw from them hardly resemble the physical facts as observed empirically. First of all, it must be censured that, already in the first approximation, the one system of the Maxwell equations emerges only in differential form ($a_{\alpha\mu,\,\sigma,\sigma} = 0$ instead of $a_{\alpha\beta,\,\gamma} + a_{\beta\gamma,\,\alpha} + a_{\gamma\alpha,\,\beta} = 0$.) Second, there is no integral for the total energy and the total momentum of the field, and it does not seem possible to construct an energy-momentum tensor from it. And furthermore, where is the interpretation of the rotation of the perihelion of Mercury's orbit and the deflection of light by the Sun? They would appear to have gotten lost in your extensive deconstruction of the general relativity theory. I, however, continue to support that beautiful theory, even if it is betrayed by you![5]

With your remark that you are still far from being able to assert the physical validity of the derived equations, you have, so to speak, cut off in mid-sentence what critical physicists might say. The only thing left for them to do is to congratulate you (or should I rather say: to offer their condolences?) on having gone over to join the pure mathematicians. I am also not so naive that I would expect that you would change your opinion due to any sort of criticism from others. But I would wager any amount with you that, within a year at the latest, you will have given up on the whole teleparallelism, just as you earlier gave up on the affine theory. And I do not want to further provoke you to defensiveness by continuing this letter, in order not to delay the approach of that natural end to the teleparallelism theory.

With a certain expectation that I will win ⟨my⟩ wager, I wish you a Happy Christmas and greet you warmly. Yours respectfully,

W. Pauli

156. From Alexandre Proca[1]

Paris, 19 December 1929

[Not selected for translation.]

157. From Heinrich Zangger

[Zurich,] 19 December 1929

[Not selected for translation.]

158. From Carl Zimmermann[1]

Bern, 19 December 1929

Dear Professor,

Please forgive me for bothering you with my problem.

During the time when you were living in Bern, Switzerland, I was able to plan and construct a complicated apparatus for you, which took a considerable amount of time, and with a quite modest payment for my work.[2] You remarked at the time that, later, when you would be in a better position, you would pay me some more.

In the meantime, through your high intelligence and your never-tiring energy, you have attained one of the highest positions in science, on which I congratulate you very heartily.

My family and I myself have not been so fortunate. Illness and bad luck have visited us; a son and a daughter were taken from us in the bloom of youth by the influenza. I have been retired with a small pension for the past six years, due to my advanced age, and at present, I am in a precarious position.

May I ask you, esteemed Professor, for the supplementary payment, whose amount I leave up to you. Please accept in advance my sincere thanks.

With respectful greetings, yours truly,

Carl Zimmermann, Machinist

159. Einstein to the Arab World

"Einstein an die arabische Welt"
[*Einstein 1930h*]

DATED [before 20 December 1929]

PUBLISHED 31 January 1930

IN: *Jüdische Rundschau*, 31 January 1930, p. 37.

[See Doc. 207 in documentary edition for published English version.]

160. "The Problem of Space, Field, and Ether in Physics"

"Das Raum-, Feld- und Äther-Problem in der Physik"
[*Einstein 1930o*]

DATED [after 20 December 1929][1]
PUBLISHED end of February1930

IN: *Die Koralle* (end of February 1930): 486–487

Concepts and systems of concepts always serve to bring order and "meaning" to our experiences. If we want to have a clear idea of the role and significance of concepts, it is by no means sufficient simply to demonstrate their mutual logical relations; the experiences must also be identified to which the concepts refer.

For concepts which are near to the sphere of sensory experiences, this goes without saying; but for those which belong to the so-called abstract concepts, which include the concept of space, it is not so obvious. Such concepts often appear to be purely mental, independent of the native soil of our sensory experiences. Such an approach is, in my opinion, always wrong.[2] Considered from the viewpoint sketched above, the concept of space would seem to precede that of material bodies. If this concept has been established, then such complexes of experiences that we call "the positioning of material objects" would seem to become particularly simple.[3] It is clear that the positional relations of objects are real in the same sense as the objects themselves.

The ancient Greek Euclidean geometry is none other than the attempt to apprehend those positional relations within a logical-deductive system. Instead of the various, manifold shapes of objects, it deals with elements consisting of idealized forms, denoted by the terms point, straight line, plane, etc.; their positional properties are determined by the so-called axioms, and all other forms and their positional relations are derived from these in a purely logical manner.[4] All positional relations can be referred back to contact or touching of the objects.

If the above is a discussion of spatial relations, then it is not a great leap to the concept of "space," which, strictly speaking, does not occur at all in Euclidean geometry. Instead of investigating the contact relations of the objects among themselves, one can investigate the contact relations of all the objects to an imaginary universal object, which is indeed what we call space.[5] Space is to a geometer what the semirigid surface of the Earth is for the geometric treatment of daily life, or the drawing sheet for those who wish to illustrate the relations of plane figures to one another.[6]

Only with analytic geometry, founded by Descartes, does (three-dimensional) space become a fundamental concept. His introduction of it as a coordinate space made it possible to arrive at an extraordinary simplification of the logical system of geometry.[7] As a basis for that system, the theorem that the metrically determined distance between two infinitesimally neighboring points, according to Pythagoras' law (square root of the sum of the squares), can be determined from the coordinate differences of the two points themselves; that is, one need only define a "Euclidean metric," from which, then, all the concepts and theorems of geometry can be derived.[8]

According to what we have said thus far, the spatial relations of objects indeed possess a physical reality, but not space itself. The latter, however, gains a physical reality in Newtonian mechanics. There, the fundamental concept for the law of motion is the acceleration.[9] Acceleration is thereby a state of motion relative to space which cannot be reduced simply to the relative positions of the objects alone.[10]

The metric and inertia are thus the most essential properties of space, according to Newtonian mechanics. In this classical theory, space is absolute.[11] This means that the positional relations of the objects (ideally rigid bodies), as well as their inertial behavior, are not affected by any sort of physical causes which may be acting in their environment.

In addition to the concepts of space, time, and matter, Faraday and Maxwell introduced a new concept into physics—that of a field—which soon exceeded the scope of the mechanical description of nature.[12] Fields are continuous formations which can occupy otherwise empty space. We distinguish between the electromagnetic and the gravitational fields; meanwhile, the field constituting light was recognized as electromagnetic. Initially, an effort was made to conceive the field as a mechanical state of an omnipresent material, termed the ether. When that effort did not lead to satisfactory results, the ether as a particular sort of matter whose state was supposed to produce the field was retained, but the mechanical interpretation of those states was dropped.[13] Near the end of the last century, H. A. Lorentz[14] demonstrated that the ether must not be supposed to have any sort of translational motion with respect to space if one wished to give a quantitatively correct description of the electromagnetic phenomena. At that point, space and the ether could have already been identified as the same, if one had not held fast to the prejudice that space must be absolute, i.e., that it could not be subject to any sort of changes.

That prejudice was eliminated only with the advent of the general theory of relativity, after the propagation law of light had forced to combine the three-dimensional continuum of space and the one-dimensional continuum of time into a four-dimensional space (continuum).[15] That general theory of relativity, in fact, showed that the empirical law of the equivalence of inertial and gravitational mass could be justified in a natural manner only by assuming the following: The metric of space in the absence of a gravitational field is non-Euclidean. The deviation of the metric from Euclidean behavior on the one hand, and the gravitational field on the other, are only different appearances of the (metric) structure of space.

With this, space had lost its absolute character. It had become subject to variable (lawful) states and processes, so that it could even take over the functions of the ether, and—in what concerns the gravitational field—it did take them over in reality. The only remaining unclear topic was the formal interpretation of the electromagnetic field, which could not be clarified by simply referring to the metric[16] structure of space. But after the formulation of the general theory of relativity, it could no longer be seriously doubted that the gravitational field and the electromagnetic field were to be interpreted as a unified structure of (four-dimensional) space.

The "unified field theory" which has recently been proposed is an attempt in this direction. The structure of space which underlies it is such that any two given line elements can be compared meaningfully not only in terms of their lengths, but also in terms of their directions (orientation). In this sense, this fundamental spatial structure is more similar to that of Euclidean space than to the non-Euclidean geometries which have thus far been studied.[17]

The search for laws of nature leads, as a mathematical problem, to the following procedure: One seeks the logically simplest laws which can be conceived within a (spatial) structure of the type being considered.[18]

We can see that theoretical research can be guided by a belief in the harmony or the simplicity of nature. But this belief is not blind; for a theory is always subject to the experience as its highest judge.

161. To Count Richard Nikolaus von Coudenhove-Kalergi[1]

[Berlin,] 21 December 1929

Dear Count Coudenhove-Kalergi,

I have read your statements about antisemitism[2] which you said in advance of your father's work. The certainty in the assessment of the causal connections that is expressed in these lines is really exceptional, no less the love of justice.

But you have also very aptly characterized the peculiar dilemma in which Zionism ensnares the cosmopolitan Jew. Namely, on the one hand, we see the great merits of Zionism in our struggle for self-respect and dignity; on the other, we see the unsightly growth of nationalism spreading in our own ranks, which the better among us abhor.[3] If we could attain the security necessary for existence without Palestine, it would be far better for us. As it is, we must hold on to this tangible matter and console ourselves that external circumstances will prevent nationalism from assuming dangerous forms within our ranks or even leading to questionable actions. Meanwhile, antisemitism will powerfully contribute to sustaining the Jewish community, which, according to my conviction based on life experience, holds meaningful traditional values.[4]

In this, I express to you my joy that European matters are represented by a man of such energy and such subtle refinement.[5] With friendly greetings and wishes, I am your

162. To Richard von Mises

Berlin W, 21 December 1929

Dear Colleague,

Thank you very much for your friendly letter.[1] Your proposal seems really good to me. I have the feeling that the collaboration with this man could be fruitful. I would love it if something could come out of it. It is fine with me if you suggest to him that he come here sometime. I am sending you a letter to him that I hope you will enclose with your own.

Best wishes and thanks, your

A. Einstein

163. From Cornel Lanczos

Frankfurt am Main, Robert Mayerstr. 2, 21 December 1929

Dear Professor,

It is with an anxious heart that I write these lines to you; for it is not easy to speak out in one's own behalf, but that is nevertheless sometimes unavoidable when much is at stake.

Although in the meantime, many weeks have passed since its submission—I have as yet heard nothing more about the fate of Prof. Freundlich's application; but I assume that he would have informed me immediately if it had been approved successfully.[1]

When I consider that a single word from you was sufficient to establish the framework for my activities in Berlin, then I will not be wrong in assuming that your influence in the matter of the application that is now in suspense would also be very great.[2] And if I again appeal to you at this point, it is in the firm belief that you may not be aware just how serious my interest in the success of Freundlich's project is, how much such a position would appeal to my whole scientific and human being. I could hardly imagine anything which would be more fitting to my disposition, though the suggestion of this plan came by no means from me, but from Prof. Freundlich or from his co-workers.

You know me sufficiently well to be aware that I do not overestimate my own abilities, that I attribute, at best, only a moderate talent to myself, which is, however, of a very specific kind and, if given a chance, can be quite effective in a positive way. And a pure research institute would be much more suited to my disposition than the usual academic environment, which is burdened with all sorts of trivial tasks.

Even if you are not entirely convinced of the necessity of the objective justification given, I would ask you to consider supporting me for humanitarian reasons. You well know how important subjective factors—in particular, mutual trust and personal sympathy—can be for the success of scientific work. And then we have the favorable circumstance that I was able to establish a close contact to all of Freundlich's circle right from the beginning, which is by no means a matter of course; and when it comes about spontaneously, this can be considered to be a very positive factor.

I wanted to present this request to you, even though I am aware that you are overloaded with many other tasks. But precisely for that reason, I want to summarize for you just what this project means to me, so that you can see clearly and can arrive at your judgment correspondingly.

Please be so kind as to give my warmest wishes to your wife and to Miss Margot.[3]

With hearty greetings, yours truly,

Lanczos

164. To Eduard Einstein

[Berlin,] 22 December 1929

Dear Tetel!

I hear from Zangger that you are feeling miserable.[1] However it may be, *come to me and express yourself.* We will soon find the solution. Perhaps the variety of medicinal fare is just as little suited for *your* stomach as it would have been for mine. In any case, don't take it and yourself too seriously, as one easily does so long as one is young. Send me news quickly; a warm hug from your

Papa.

I write so little because I hope to see you soon. Friendly greetings to your mama.[2]

165. From Élie Cartan

Le Chesnay (Seine et Oise), 27 avenue de Montespan, 22 December 1929

Dear Sir,

I wanted to avoid the necessity of your reading the papers in which I presented my theory of systems in involution;[1] they are rather long. Furthermore, they treat differential systems in the form of equations with total differentials, while it is the form of equations with partial derivatives which represent the systems that are of interest to you. I believe I can give you a summary which will be sufficiently precise, with demonstrations, within a limited number of pages (the note that I sent you does not contain any demonstrations).[2] I expect to send you that new note within a few days.

Let us now take up the points raised in your letter.[3] I did not understand very well what you wrote on the subject of the reasons that make the system which includes $R_{ik} = 0$ more realistic. In your sentence: "Es gäbe also für das reine Gravitationsfeld *allein* eine *determinierte* Gesetzlichkeit, ohne dass die Parallelstruktur darauf zurückwirkte," does the word *allein* refer to "*Gravitationsfeld*" or to "*eine determinierte* etc…"? My German is not strong enough to allow me to decide. I also do not understand very well what is meant by this "Kausale Verbindung Metrik → Parallelstruktur." It is certain that the 10 equations $R_{ik} = 0$ only cause the metric to intervene. But could one not also say that the 6 equations $\Lambda^{\mu}_{\alpha\beta;\mu} = 0$ in your system only make parallelism intervene? For the rest, I indicated this system to you only as a possible solution to the *mathematical* problem posed.

I am not in agreement with you on the greater or lesser degree of determination of the two systems of 22 equations; the one containing the $R_{ik} = 0$ and yours. They certainly have the same degree of determination. Your system could be put into a form quite analogous to the other, to wit

$$\varphi_\alpha = \frac{1}{\psi}\frac{\partial\psi}{\partial x^\alpha}, \tag{1}$$

$$S_\alpha = -\frac{1}{\psi}\frac{\partial\chi}{\partial x^\alpha}, \tag{2}$$

$$G^{\alpha\beta} + G^{\beta\alpha} = 0, \tag{3}$$

with two scalars, ψ and χ; this then still makes 18 equations for 18 unknown functions, $h_{s\nu}$, ψ and χ. Equations (1) and (2) result from the integration of the equations

$$F^{\alpha\beta} = 0, \quad G^{\alpha\beta} - G^{\beta\alpha} = 0\,;$$

in fact, one has, for example,

$$h(G^{34} - G^{43} - F^{34}) \equiv \frac{\partial S_1}{\partial x^2} - \frac{\partial S_2}{\partial x^1} + \varphi_1 S_2 - \varphi_2 S_1\,.$$

The two systems thus have structures which are indeed quite analogous, and their indices of generality are the same.

One thus cannot say that your system would be privileged with respect to its degree of determination; nevertheless, I believe that I can confirm, after having considered all of the possible identities and reviewed all of the different systems which might be suitable, that *there exists no system* <u>*more*</u> *determined than yours.*

I have now understood perfectly what you meant by saying that the 4 identities $F_{\alpha\beta\gamma} = 0$ are not independent. It just is that we do not use the word *independent* in the same sense. For me, it is a question of *algebraic* independence between the first members of the 4 identities considered as linear combinations of the 24 partial derivatives of the quantities $F_{\alpha\beta}$ with respect to x^1, x^2, x^3, x^4. Our points of view were thus completely different.

For a moment, I had thought that, in every case when there were $N - 12 + r_2 > N - 12$ identities, there would exist among the identities r_2 identical relations in your sense. But that is not the case, and I am now convinced that the non-independence (in your sense) of the identities plays no role at all in the question of compatibility. And that is quite fortunate, since otherwise you would have to demonstrate that the $N - 12$ necessary identities, once you have found them, are independent *in your sense*, which would not always be easy to do.

I will not speak to you today of your new method for presenting a demonstration of compatibility; it is, furthermore, quite analogous to the manner in which I demonstrate the existence of solutions to my systems in involution; but I believe that it is not sufficient to demonstrate the existence of one solution throughout all of space by starting with one solution in the section $x^4 = a$ of the partial system

$$G^{11} \quad G^{12} \quad G^{13}.$$

It is still necessary to demonstrate the compatibility of that partial system. That brings us back to my point of view, that is the consideration of the solutions in one, two, and three dimensions.

I will also not speak of the system of $16 + 22$ equations which you find "ugly." I further believe that I could present to you in an elementary manner—though not absolutely satisfying, in my sense—the problem of the degree of generality of Riemannian spaces with absolute parallelism considered in terms of what are their *essentials*, that is to say, independently of the choice of variables.

May I now ask you what you think of the considerations that I presented at the beginning of my note on *integral* determinism (in the sense of old mechanics) and of *local* determinism. That question is connected to the following:

Does every solution of the Maxwell equations define only what is *physically* admissible within a small region of spacetime? Put differently, suppose that we have some functions which define the electromagnetic field within this domain, for

$$|x - x_0| < a\,, \quad |y - y_0| < b\,, \quad |z - z_0| < c\,, \quad |t - t_0| < h$$

and which satisfy, *within this limited domain*, the Maxwell equations; do you think that such a field could occur in reality? The local state is certainly firmly related to whatever can occur elsewhere, outside the domain under consideration, but can one make use of the possibilities, infinite in number, which exist in the interior of the domain in order to render possible, in its interior, any given *local* solution of the Maxwell equations?

This question seems to me to be important, because if it were to be answered in the affirmative, that would give a *physical* sense, and not only a mathematical sense, to the indicator of the generality of the system of equations of partial derivatives which determines the field.

But this is surely enough for today. I ask you, dear and honored Sir, to accept my admiration and cordial regards,

E. Cartan

166. From Heinrich Zangger

[Zurich,] 22 December 1929

[Not selected for translation.]

167. To Heinrich Zangger

[Berlin,] 23 December 1929

Dear Zangger,

Thank you for your letter.[1] I knew the boy has weak nerves; likewise that he is not suited to processing a great deal of material in his head. This is also as abhorrent to him as it would have been to me. Perhaps his condition is only just a psychological reaction, like someone who has eaten a rotten meal pukes. All of this does not have to mean much and could be corrected by a change in the choice of profession.

It is also be conceivable that it could be a progressive process, inherited from the mother's family. You already know that, at just over twenty, Mileva's sister[2] developed dementia praecox and (after a good disposition in her youth) fell victim to a progressive idiocy. The boy has never been completely normal (predisposition to hypochondria, great irritability, tremendous fluctuations in self-esteem, depression, also often a very peculiar, objectively unfounded mistrust with no objective basis.)[3] It is also striking that Mileva has not written to me—as if she wanted to hide something. This would be psychologically understandable because, from her youth onwards, she was trained to see any sort of family defects as a scandal to be concealed.

Hopefully, this is not the case but above all the unvarnished truth, as far as one can attain it! I would very much like to have the boy there, if only for the reason that he expresses himself the most easily with me and lets himself go, perhaps more than with the very withdrawn and difficult mother. Perhaps you can advise him to visit me, if you would find it at all advisable.

Hopefully, you are in good health. I am, myself, quite content, particularly since I do not have the compulsion to do great things anymore since I got, at least so far, with my problem that it must awaken the interest of a similarly organized head. The colleagues (aside from Langevin and the famous Parisian mathematician, Cartan)[4] consider it all as a half-learned, half-senile whim. But with such matters, it is like with a girl: it is enough if *someone* really likes her.

The boy is my favorite in my entire family, probably due to the inner similarity and relatively complete empathy. Nothing about that would be changed if he were to become kind of plant from a kind of animal. What do our powers and goals ultimately mean?

Write me soon and candidly as to how you find things.

Best regards and wishes for 1930! Your

A. Einstein

168. From Eduard Einstein

Zurich, 23 December [1929][1]

Dear Papa,

Last week, as has already been reported to you, I had unfortunate conditions—conditions which undoubtedly would have been of interest to me as a professional, had they not been aimed at me so personally.[2] I was already on the edge the whole quarter, and now I am at the mercy of the neurologist for a while. I comfort myself with the thought: earlier people had their god, had prayers and confession; sometimes, we need a substitute. I must now rest and limit and adjust myself. Zangger (who was very helpful) wants to suggest that you should also come to a quiet place halfway up here.[3] Would that work? I feel completely well now, will share more about it another time. For today, friendly greetings to everyone for the holidays, your

Teddy

P.S. In case one of you should be poorly at some point, I recommend as reading material *The Pickwick Papers* by Dickens.[4]

169. From Hans Wolfgang Maier[1]

Zurich 8, 23 December 1929 [2]

[Not selected for translation.]

170. To Peter L. Gellings[1]

Berlin, 24 December 1929

[Not selected for translation.]

171. To Wolfgang Pauli

[Berlin,] 24 December 1929

Dear Mr. Pauli,

Your letter is quite amusing,[1] but your report seems rather superficial to me. Only someone who is certain of viewing the unity of the natural forces from the correct point of view would permit himself to write in this manner. I by no means assert that the path which I have embarked upon must necessarily be the correct one. But I do assert that it is conceptually the most natural path that I have thus far encountered. Before the mathematical consequences have been properly thought through, it is by no means justified to make negative judgments about it. It is not correct that I repudiate the progress that has been made in quantum theory. I simply believe that one cannot penetrate to the depths of the problem with these semiempirical methods, and being satisfied with only statistical laws will later be rejected. If you were to study it more deeply, you would certainly realize that the system of equations which I have established stands in a necessary relation to the fundamental space structure which underlies it, especially since the proof of compatibility of the equations has, in the meantime, been able to be simplified.[2] Forget what you have said, and plunge into the problem with such an attitude as if you had just arrived from the Moon and had to form a fresh opinion from the start. And then say something about it only after at least a quarter of a year has passed.

With kind regards, your

172. To Max von Laue

[Berlin, 25 December 1929][1]

Dear Laue,

I have already received a second letter from Lanczos, who quite seriously wants to come here. I will send the second letter to you.[2] I have some reservations about supporting his request, quite apart from the difficult question of where we should find the funds to support him.

Lanzcos is an intelligent, imaginative, and decent fellow, no question about that. But his mind is, in fact, not flexible when it is a question of evaluating theories for an empiricist; at least, that is what I believe. Even with me, only a limited degree of working together was possible due to his fixed attitude. What do you think about this?

Another matter. I spoke to Bottlinger,[3] who also shares the opinion of other astronomers.[4]

He complains that quite worthless people have been hired in Potsdam, whereby family relations also enter into the picture in an aggravating fashion for the administrators.[5] Müller[6] is a case in point, who is now supposed to become an observer; also Hassenstein,[7] who was recently hired. Becker,[8] in contrast, is much better. Bruggenkate[9] is held in high esteem by everyone. But he could also be offered a professorship; in any case, he is very competent as an observer and could easily become stunted. B. is no friend of Freundlich's[10] and asserts that the people in Göttingen are unreliable.

It is now certain in my view that we were in the wrong to give in to Ludendo[rff][11] and that the ministry was justified in its mistrust. We could suggest to them to request professional opinions in the personnel question from Chapley[12] and from de Sitter[13] (Holland). If you are in agreement, we can both go on a private visit to the ministry. If that would be unpleasant for you, I will go alone.

I feel that it is my duty, given the mistake that was made.[14]

Send an answer soon to your

[A. Einstein]

173. "To a Young Scholar"

"An einen jungen Gelehrten"
[*Einstein 1929yy*]

Published 25 December 1929

In: *Berliner Tageblatt*, 25 December 1929, 1. Beiblatt, No. 607

Dear Mr. X,

You have reservations about starting a course of study in mathematics because, in this economically difficult time, the choice of profession would have to be completely adapted to external necessities. And yet, I know from our previous conversations that you are imbued with a passionate urge to understand all things, but especially the simplest, logically ascertainable ones.

I understand your qualms, but advise you not to allow yourself to be bewildered by such anxious considerations. Only he who remains true to himself becomes a complete fellow, and only such a fellow can truly be something to others. But that is ultimately what everyone must strive for.

But what should become of me, once I have studied mathematics or physics? So I hear you ask. "Do the teaching and research professions offer sufficient prospects?"

My answer to this will seem somewhat odd to you: It is totally unnecessary for you to become a teacher or a researcher. This depends not only on your suitability, but at least as much on the demand. If you spend your money, you demand for it something that you currently need. So it goes perforce with society as well: it is ready to give you your food if you do for it something that is needed.

If you now think: "Why, then, do I need to study mathematics," then it is actually better if you do not study. However, if you think: "In his opinion, I can still study mathematics without being neglectful of my duty to myself and my family," then you should.

For this optimistic point of view, I can give you two reasons that you must not scorn. In particular, I studied mathematics and physics in Switzerland and then could not find any employment as a teacher.[1] Without any technical knowledge, I was employed as a preliminary examiner at the Swiss patent office;[2] it quickly became apparent that the general overview of physical connections was often more valuable than expert knowledge and routine. That which is acquired from pure joy in knowledge is a versatile, useful instrument in the hand of a living person.

But there is still another thing to consider here. There is still very much the question of whether the researcher by profession is in a more advantageous position with regard to that research than one who earns his bread through a trade. In the first case, how easily one sees oneself forced into superficial, prolific writing and busyness in order to "get ahead!"

Perhaps academia belongs to those delicate women who are better as mistresses than as housewives. Now, I dare not make any decision over that.

Good luck!

Your

Albert Einstein

174. From Élie Cartan

Le Chesnay, 25 December 1929

Dear Sir,

I am sending you the second note, where you will find the complete theory of systems in involution, with demonstrations;[1] I have written it from the point of view of systems of equations with partial derivatives, and not, as in my articles, from the point of view of systems of equations with total differentials. I have provided several examples.

If you can get through to the end without too much difficulty, I would like to ask what you think of the last paragraph on page 17.[2]

I will make use of the opportunity, dear and honored Sir, to once again address my best wishes to you for the new year, for your health, and for a rich harvest in your new theory of fields.

Yours truly,

E. Cartan

175. From Max von Laue

Zehlendorf, 27 December 1929

Dear Einstein,

Thank you very much for your friendly letter from the 25th of this month.[1] I am also not in favor of the project to bring Mr. Lanczos[2] here, specifically, for not paying him out of funds from the Kaiser Wilhelm Institute. Apart from the fact that the KWI currently lacks the means of this sort, I do not want to take any steps that would bind me in future staffing decisions, now that the expansion to an actual institute finally appears to be approaching in the foreseeable future.[3] Furthermore, I also think that a position like the one Lanczos now seeks is not the best for him.

Ever since I heard him speak at the Physicists' Society about his continuation of Dirac's theory, I am of the opinion that one should make use of this magnificent, truly quite uncommon teaching talent at a university right away. In contrast, I do not think much of his research. If he could exchange talents with Nernst[4] to such an extent that Nernst transfers to him some sense of reality in regard to mathematical insight, perhaps he would become a great physicist. But now, there is unfortunately something lacking in him.

I incidentally want to travel to Potsdam shortly and discuss the matter with Freundlich.[5]

Now, as far as the observatory position in Potsdam is concerned, I would not like to attach myself to your intended step with the ministry. First, for formal reasons. Admittedly, no statute exists for the board of trustees that (as with the faculty) stipulates that dissenting votes of individual members must be announced in the respective session. But I think it would be good if the board were to follow that custom as well. Ultimately, such a matter must also come to an end. But then I am also governed by factual motives. It is not as if Ludendorff[6] would stand alone in his attitude toward questions of appointment. Guthnick[7] and, above al,l Strömgren,[8] who, in my opinion, truly thinks objectively, have both declared themselves warmly for Müller.[9] Whether they or the opposing party are in the right, only the future can really decide. And under these circumstances, I do not think it would be good if an observator came to Potsdam who was not acceptable to the director. That would hardly facilitate the work to be done there.

If one accuses the board of trustees of handling the matter too quickly in the Hassenstein[10] case, they may be correct. At that time, I did not yet know the circumstances in the astronomy department. But in the case of Müller, we certainly have not made the matter easy for ourselves.

Under these circumstances, I consider the matter as *res judicata*. Apart from that, I wish you and yours a good new year from the heart. With best wishes, your

M. Laue

176. To Élie Cartan

[Berlin, 28 December 1929][1]

Dear Colleague,

Immediately after I sent off my letter,[2] it became clear to me that, from the standpoint of the degree of determination, your system, $R_{ik} = 0$ etc. is fully equivalent to mine.[3] I had forgotten that my system, too, can be given an analogous form (with ψ and χ), such that, now, four identities appear.

On the other hand, I think that my other objection is sound. The equations $R_{ik} = 0$ alone fully determines the g_{ik}-field, and in this, the parallel structure plays no role. If the g and their first derivatives are given on a space-like cross-section, $x^4 =$ const., then the analytic continuation of the g is fully determined (up to the arbitrariness corresponding to the freedom in the choice of coordinates). That is, the g_{ik} alone have an autonomous causality, independent of the parallel structure. This can be expressed as follows: the electromagnetic field has no effect upon the gravitational field.[4] This completely contradicts physical expectations.[5]

The same thing does not hold true for the equations $\Lambda^{\alpha}_{\mu\nu;\alpha} = 0$ of my system with respect to the parallel structure. The parallel structure can, by no means, be separated out from the whole collection of variables, i.e., there does not exist a total of 6 quantities which can be algebraically expressed in terms of the h so as to produce an analogy to the quantities

$$g_{\mu\nu} = h_{s\mu}h_{s\nu}$$

In the equations $\Lambda^{\alpha}_{\mu\nu;\alpha} = 0$, no uniquely determined combination of the h appears whose causality would be fixed by these equations *alone*.

Of course, the discovery of your system awakens the suspicion that perhaps yet another compatible system of equations can be given having the same degree of determination as mine but which do *not* have the aesthetic drawback of the equations $R_{ik} = 0$. In that case, I could no longer consider my equations to be privileged. What do you think about this?

It would be a good thing if you were to publish an article on the degree of determination of partial differential equations. This could also be of interest to mathematicians. I have not been entirely enlightened by your written explanation. I especially have difficulties with the fact that the equations—when one takes Λ and h as field variables—contain h in its differential form in the quadratic terms; something—it seems to me—that your general theory does not deal with.

I am delighted that, according to your investigations, more strongly determined, compatible systems of equations than our current ones do not exist. It's wonderful that you can prove such a thing.

You are, of course, right when you say that I have not proven, but only assumed, that all the equations can be satisfied at $x^4 = a$. But if we physicists committed no worse mathematical sins than this, we would be comparatively honest folk! I am lucky that I succeeded in interesting you in such matters, and I knew that you would bring much clarity to them.

Your question concerning the continuous extension of the solutions of differential equations *from the physical standpoint* is not totally clear to me. But I shall try to answer as best I can.

It seems to me to be necessary to look at what one demands of a solution from the physical point of view.[6]

For example, if we look at Maxwell's equations for empty space, then, because of their hyperbolic character, continuation of a solution in spacelike directions is not determined (waves coming in from infinity!). But I leave this aside because it does not touch upon the essence of your question. Therefore, I restrict myself to, say, static problems.

(Now, is every solution meaningfully continuable in a finite region? This depends on *what other properties of the field we believe we must require on physical grounds* (*the admissibility of singularities*, or finite or infinite boundary conditions of a certain kind).)[7] In the case of Maxwell's equations, for example, there are no singularity-free solutions which tend to zero at infinity. If you have a solution in your neighborhood, then, by continuation, you would arrive at a singularity or at growth of the field at infinity. The admissibility of your solutions in a finite region then depends on what kind of "extravagances" of the solution one believes to be permissible from the physical point of view.

As I've said, one does not get away without singularities in the case of Maxwell's equations. But no reasonable person believes that Maxwell's equations can hold rigorously. They are, in the best case scenario, first approximations for weak fields.

(It is now my belief that, for a serious and rigorous field theory, one must insist that *the field be free of singularities everywhere*.)[8] This will probably restrict the free choice of solution in a region in a very far-reaching way—over and above the restrictions which correspond to your degree of determination. Nevertheless, I think the latter theory is very important and clarifying. A detailed article on this from your pen would certainly be useful to everyone who is interested in these as fundamental issues.

Again, many thanks for all the pains you have taken with the research—and with me. With kind regards, gratefully yours,

A. Einstein

Translator's note: Amended translation from *Debever 1979*, pp. 88–93.

177. To Cornel Lanczos

[Berlin,] 28 December 1929

Dear Mr. Lanczos,

I have negotiated with Mr. Laue in your case.[1] We reached a consensus that no funds are available to offer you a position at one of the institutes. But we want to make an effort to permit you to receive a teaching appointment at the university in the near future. We both believe that you possess exceptional clarity of thought and a talent for teaching.

The proof of compatibility for the field equations has now been improved by the fact that, through a remark of Prof. Cartan's in Paris, the detour via the introduction of a new field variable has proven to be superfluous.[2] It can now hardly be doubted that the equations are the only natural ones in such a manifold.[3] Everything now depends upon the existence of solutions without singularities.

Kind regards, your

178. To Élie Cartan

[Berlin, before 29 December 1929?][1]

Dear Mr. Cartan,

I have read your manuscript—*enthusiastically*.[2] Now, everything is clear to me. You should publish this theory in detail; I believe it is of fundamental importance. Previously, my assistant, Prof. Müntz, and I had sought something similar—but we were unsuccessful. Just a few remarks.

First, your question concerning the nonexistence of a "substance." Substance, in your sense, means the existence of timelike lines of a special kind. This is the transference of the concept of particle to the case of a continuum. Such a transference into the continuum, or the necessity of the possibility of transference, seems totally unreasonable as a theoretical requirement. To realize the essential point of the idea of atomistics on the level of continuum theory, it is sufficient to have a field of high intensity in a spatially small region which, with respect to its "timelike" evolution, satisfies certain integral conservation laws; i.e., the *whole* complex must have a kind of individual evolution, but not its individual *points* or pieces.

Your wonderful explanations leave me only the following questions:

(1) Are there systems with nontrivial solutions whose *indice de généralité* = 0 (apart from the consequences of coordinate freedom)?[3] These would be systems which would be largely analogous to those of classical mechanics, in which the initial conditions would be fixed not by an arbitrary choice of given *functions*, but by a choice of parameters (numbers).

For some time, I was convinced that the true laws of nature would have to be of such a kind.

(2) But it might also be possible that such higher degree of constraint (which I have no doubt is realized in the true laws of nature) is based on something else. The indicated small measure of arbitrariness (in nature) could also be grounded in the requirement that singularities be excluded from all space!

My equations, then, would only have a possibility of validity if they were to possess *singularity-free* solutions appropriate for the representation of material charges.

(Can there possibly be general methods by which we may learn something about the existence and properties of such solutions?)[4]

(3) In my opinion, the explanation of *one* point would still be desirable: if we have a system that is *not* in involution, in which the necessary identities

$$\text{e.g., } \Lambda^{\alpha}_{\underline{\mu\nu};\nu} = 0,$$

do not hold, then what can one say about the solutions of such a system?

In this case, your equations

$$\frac{\partial F_i}{\partial y} - \sum_j A_{ij}\frac{\partial G_j}{\partial x} = 0 \tag{4}$$

are new conditions for *continuation* (or for the cross-section $x^4 = a$). Can one always further augment these additional conditions for continuation by differentiating and, accordingly, eliminating given derivatives? How can one get some kind of idea about the degree of generality of the solutions of such a system?

At first, I myself thought such equations possible as expressions of laws of nature. But I changed my mind by a study of the first approximation.

Let $G = 0$ (1) be the field equations.

We break them up into

$$\overline{G} + \overline{\overline{G}} = 0, \tag{1a}$$

where $\overline{G}$ represents all the terms which are *linear* in $h_{s\nu} - \delta_{s\nu} = \overline{h}_{s\nu}$, while $\overline{\overline{G}}$ is of higher order in $\overline{h}_{s\nu}$. Let $L(\overline{G}) \equiv 0$ be certain linear identities (i.e., L are linear differential expressions). Then it follows from the field equations that

$$L(\overline{\overline{G}}) = 0. \tag{2}$$

If the linear identities $L(\overline{G}) \equiv 0$ include no rigorous identities, then one can show the following remarkable fact.

One can generate a rigorous solution of the equations this way:

$$h_{s\nu} = \delta_{s\nu} + \overline{h}_{s\nu} + \overline{\overline{h}}_{s\nu} \ldots,$$

which stands for an expansion in orders of magnitude. In the first approximation h can be found by solving the equations

$$\overline{G}(\overline{h}_{s\nu}) = 0 \ldots \tag{3}$$

Substituting the series solution above in (2), one obtains, in the first approximation, the equations

$$L(\overline{\overline{G}}(\overline{h})) = 0 \ldots \tag{4}$$

These are quadratic equations for $\overline{h}$ which, in general, (where there are no rigorous identities) are *not* conclusions from (3). Equations (4) now restrict the solutions of (3) so severely that it is impossible to believe that nature would have hit upon such a drastic restriction. I have convinced myself of this by examples!

But it would still be desirable if you were to draw up a general clarifying study.

One more remark on this subject. Were it to be arranged that the above mentioned rigorous identities do exist, then the system would be in involution.

The equations (2) hold rigorously. Since, in the case of my system of equations, the $L(\;)$ are simple divergences, one can apply Gauss's Theorem and obtain a surface integral relation $\int \overline{\overline{G}}_n df = 0$.

If now, *on the surface* (but not inside it), $h_{s\nu} - \delta_{s\nu}$ are small, then one can replace h by $\overline{h}$ and obtain, in the first approximation, a nontrivial integral relation. (This supplies the equations of motion, in the case of the Riemannian relativity theory.) *I have not yet* looked at what follows from my new theory.

With warm wishes for 1930, gratefully yours,

A. Einstein

Translators' note: Amended translation from *Debever 1979*, pp. 94–99.

179. To K. K. Suntoke[1]

[Berlin, after 30 December 1929][2]

I believe in an unrestricted determinism, although I am also convinced that such a belief cannot be justified logically at all, and empirically only approximately.[3]

Respectfully,

A. E.

180. Poem for Ethel Michanowsky-Charol

[1930]

[Not selected for translation.]

181. Aphorism

1930

The art of living is to accept fate patiently, thereby preserving life-giving activities.

A. Einstein

182. To Walther Mayer

[Berlin,] 1 January 1930

Dear Mr. Mayer,

I thank you sincerely for your most appropriate comment. The footnote can no longer be included, since, according to what Mr. Blumenthal mentioned, the article is already in press.[1] But it is, however, such a small error that this is not very important.

I am glad to note that you have given the matter so much attention. It is indeed interesting that their compatibility, together with the requirement of a certain agreement, seems to practically determine the equations. Nearly all of our colleagues react to the theory with considerable reserve because it puts the successes of the earlier general theory of relativity in doubt again.[2] I will be here again from mid-January on,[3] and I await your visit, so that we can discuss the current problems. Then, we need to be clear about the conditions of a collaboration, if we want to undertake one. It would be good if I were informed about your financial needs, in case you would like to come here for some longer time.[4] I know from experience that such matters always take a certain amount of time and effort to be arranged. I am especially curious to know whether you consider the project described in the article to be sufficiently promising to invest a great deal of effort and time in it. In this connection, I expect you to be completely honest, for a marriage without love is worthless; indeed, it is a calamity.

In the meantime, friendly greetings from yours truly,

A. Einstein.

PS. Prof. Cartan in Paris has now applied the compatibility proof to a very nice general theory.[5] I will show this to you when you are here.

183. To Josef Pohl[1]

[Berlin,] 1 January 1930

[Not selected for translation.]

184. To Robert Weltsch

Berlin W., 1 January 1930

Dear Mr. Weltsch,

I am very pleased that reason has filtered through, at least to some extent. Weizmann's speech[1] was, apart from its brilliant formulation, also better in terms of content than I had dared to hope.

I will not allow the Arab to come.[2] But consider providing me the opportunity to speak with Arabs from less murky origins, even if it is with people who themselves have no political influence. (For example, Arab students, if there are any.)[3] Please send me the letter[4] to the Arabic newspaper, as well as the address of the latter.

Kind regards, your

A. Einstein.

185. To Heinrich York-Steiner

[Berlin], 1 January 1930

Dear Mr. York-Steiner,

I too, of course, am not in favor of a general parliament.[1] It would legalize our mutilation. But on the other hand, I think it is a mistake to underestimate the Arabs. A people who can accomplish such a boycott are not to be disregarded politically.[2] The people have, in that regard, every reason to feel the discontent they must, insofar as they have no legal means through which to oppose their complete displacement from Palestine.

In my opinion, it is just as unwise, as it is dishonorable, for us to rely exclusively on British military power against the Arabs.[3] We must seek to promote direct co-operation with the Arabs in small and large matters, and work to ensure that they receive legalized representation capable of negotiating that would be consistent with the Zionist organization and would be recognized by us and the English government as representing Arab interests.[4] I get the impression that some in our ranks place too much value on prestige and act with too little understanding of others.

Kind regards, your

186. From Élie Cartan

Le Chesnay, 3 January 1930

Dear Sir,

I have indeed received your two letters; I am quite happy that my manuscript was of interest to you and that you judge my theory to be of possible use.[1] It is relatively unknown, probably because I published it in the form of a report on systems of equations with total differentials;[2] some few mathematicians know that I have derived important results from it, for example, the theory of the structure of continuous infinite groups. But in the form which I described it to you in my note, it would evidently attract the attention of a much wider audience, and I intend to take up the matter of publishing its essential points.

I thank you for your answers to the questions that I took the liberty of asking; first of all, on the physical existence of local solutions of the differential systems, and then on the problem of substance.[3] I had rather wanted to ask you, as well, how you explain the fact that, in the first approximation in your system, you find only the equations for the field in the *vacuum*; it is no doubt because the matter and electricity are due to condensations of the field which are not treated, by its very nature, in the first approximation?

I shall now attempt to respond to the questions which you have asked of me.

(1) I do not believe that there exist systems in involution which would be more determined than yours. A priori, it could happen that any solution in four dimensions could be completely determined by a solution with two dimensions, or even with one dimension, or even with zero; but it seems to me—always with the reservation that I have carried out only a summary study which is not definitive—that these various cases cannot occur if the system is of the analytic form we have assumed from the beginning: linear equations with respect to the first derivatives of the $\Lambda^{\gamma}_{\alpha\beta}$ (and a system in involution).

(2) Concerning the solutions *with no singularities*, this question is extremely difficult, it seems to me. In reality, there are two questions that are apparently independent but, in fact, are intimately related. First, what is, from the point of view of *analysis situs*, the space, or the continuum, in which we wish to localize the phenomena? Second, having chosen this continuum, what are the solutions with no singularities within that continuum? It is quite possible that the existence of solutions with no singularities imposes purely topological conditions on the continuum, requiring, for example, that the continuum be closed, like a spherical space in four dimensions.

One can get an idea of these difficulties by considering the system

$$\Lambda^{\gamma}_{\alpha\beta;\mu} = 0, \qquad (1)$$

which generates spaces representing the transformations of a finite and continuous group. The solution of this system first leads to the study of a system of constants Λ^{k}_{ij} that satisfy the relations

$$\Lambda^{m}_{ij}\Lambda^{h}_{km} + \Lambda^{m}_{jk}\Lambda^{h}_{im} + \Lambda^{m}_{ki}\Lambda^{h}_{jm} = 0 \qquad (2)$$

(equations of the structure given by S. Lie). Let us suppose that we have found a system of constants that satisfies these relations (2); the constants represent the components of the torsion *with respect to a given n-Bein coordinate system*. It is not a priori evident that a space with *n* dimensions even exists in which one could find functions $h_{s\nu}$ which would be everywhere regular and would lead to the torsion in question; this theorem is, however, exact, and I give a summary demonstration of it in the next issue of *Mémorial des Sciences mathématiques*; but that demonstration rests on some theorems which involve the whole theory of the structure of groups. The space in which a group exists thus depends, from *the topological point of view*, on the constants Λ^{k}_{ij}, and each choice of these constants gives rise to a space (or to a family of spaces) which is topologically defined. Thus, in a definitive sense, *each solution of the system without singularities (1) creates, from the topological point of view, the continuum within which it exists*, and these solutions with no singularities correspond to the algebraic solutions of the relations (2).

I have dwelt upon this particular case because it probably gives a foretaste of the difficulties of the general problem. In all the cases, I know of no method, nor even the embryo of a method, which would permit us to overcome this problem.

Concerning your system, the theory of groups permits us to find a certain number of solutions with no singularities, but those solutions are *isolated*;[5] they exist in an open space, like Euclidian space. One obtains them by searching for the systems of constants Λ_{ij}^{k} which satisfy the relations (2) and the relations

$$\Lambda_{i\underline{k}}^{m}\Lambda_{km}^{j} = 0\,;$$

I could indicate some of them to you, if you are interested; but I doubt that you could take any part of them and find a physical interpretation. Each one of these solutions corresponds to a group with 4 parameters; unfortunately, none of them correspond to the most interesting group connected to a spherical space of three dimensions and an indefinite time.

(3) I now take up the question related to those systems which are not in involution. The example that you gave me,

$$G_{\mu}^{\alpha} \equiv \Lambda_{\mu\underline{\nu};\nu}^{\alpha} = 0\;,$$

is unfortunately not well chosen, since these 16 equations form a system in involution due to the 4 identities,

$$G^{\mu\alpha}{}_{;\mu} + \Lambda_{\rho\sigma}^{\alpha}G^{\rho\sigma} \equiv 0;$$

its index of generality is 16.[6]

If we choose a system which is not in involution, there is a fundamental theorem, the following: If one extends the system a certain number of times by adjunction, with new unknown functions of the derivatives of the given functions up to a certain order, *one finally arrives at*

either an incompatible system

or a system in involution.

One thus knows, *in every case,* the degree of generality of the solution of the system.

For the rest, it may happen that a system has *singular* solutions; that does not mean that the solutions contain analytical singularities, but it means that all the sections (in three dimensions, for example) are characteristic. Your system does not allow for singular solutions of this nature; but, in geometry of similar singular solutions, it occurs frequently, and it can happen that they are more important than the general solutions.

The unclear question remains of studying all the systems in involution which have the same degree of determination as yours.[7] Perhaps there is a continuous series of systems of this type, a series which would contain your system and the system with the R_{ik}?! I hope to be able to return to this problem, which unfortunately requires much patience and also the complete absence of calculational errors! After extensive ca1calculations, one need only to identify, to 0, a cubic form of the $\Lambda^{\gamma}_{\alpha\beta}$ which does not take more than one page! But there is perhaps a method for avoiding all those calculations; in any case, I am not aware of it.

Dear Sir, please accept my most sincere regards,

E. Cartan

187. To Elsa Einstein

Fürstenberg, 4 January 1930, Saturday afternoon

Dear Else,

You cannot imagine how happy I am here.[1] The area is completely wild and uninhabited virgin forest, heather, marsh, and lake. Every day, I take two walks and discover how wonderful simple existence in silent surroundings is. The book by Levisohn[2] is hauntingly beautiful; I am reading it very slowly, so that nothing escapes me. Ask Alice[3] for his address; I want to write to him. I have not come up with much yet; I'm too comfortable for that to happen. Berlin doesn't exist; I still have not even read Zangger's letter[4] I am so lost in thought. It is a shame that I am lodged in such good rooms, because you now think that is the reason why I feel so happy.

Margot[5] would be so delighted with this wilderness. But in the summer, supposedly, the day-trippers get lost.

Best wishes from your

Albert

P.S. If the man from Warsaw comes, then take care of him. You'll get everything back.[6]

188. To Heinrich Zangger

[Fürstenberg,] 4 January 1930

Dear Zangger,

I must now seem ungrateful to you. But if I am not supposed to say everything openly to Zangger, then to whom? I am unhappy, namely because medical authority always keeps my son from me.[1] I am currently, for example, in solitude in Mecklenburg, totally alone in a wild, barren, but highly picturesque region for two weeks.[2] Why does not anyone send me my son when he is in need of rest? I know that he feels happy and relaxed with me—I believe more than anywhere in the world. Ask him yourself! Mileva happily sends him to me; she just does not like surrendering him to the second wife.[3] I ask you, therefore, to arrange it so that he comes to me at the semester's end.[4] I'll then go straightaway to an isolated place with him and hike. So do not do anything against it again!

And another thing. I am not in favor of Tetl completely suspending his studies. He will develop an inferiority complex, because he is quite delicate as regards his balance. He must have the feeling of achieving something normal; otherwise he will suffer a blow.

And still a third thing. I believe you must cut back. It is not good to be running around until the last breath.[5] These later prime years in which we find ourselves bring with them the right to an element of contemplation (see for example, Plato's *Republic.*) *This way*, you could perhaps be there more for the others than if you completely ruined your heart. At our age, one must primarily care for the preservation of his creative tradition, I think. Because this is the only sensible way of continued existence.

Pauli has an edited version of my work.[6] Let them give it to you, but do not ask him (or Wenzel)[7] yourself. Pauli has already scolded me for it.[8] Possible that the matter will prove a failure, but the fellows cannot know that. They only pass judgment due to a fashion in which they are enthralled.

Best wishes from your

Einstein

189. To Élie Cartan

[Fürstenberg], 7 January 1930

Dear Mr. Cartan,

I am delighted that you intend to publish your theory of systems in involution.[1] But please do it in the same detail as you did for me. Learning has become not at *all* easy for me anymore. And this is really a theory of fundamental importance.

In your letter, you have already answered the question about the nonlinearity of the equations with respect to electrical and mass densities just as I would have.[2] Argument: because of the appearance of mass densities, the superposition principle does not hold; hence, nonlinearity of the corresponding equations.

In addition, I must further apologize for the insistence of my questioning. There's a lovely old saying: A fool can ask more questions than a wise man can answer.

I am happy that you feel confidence in your answer that there exists no involutive system of less freedom. I myself have tried in vain to find one. Naturally, I restricted myself to systems which are linear in their second derivatives.

Concerning the connectivity properties of space, I can say nothing, but probably it is unavoidable to demand that the solutions be free of singularities. Apart from the fact that this is a highly desirable requirement in itself, it is also a necessity of a more special kind.

My equations are rigorously satisfied by an $h_{s\nu}$ of the form[3]

$$\begin{matrix} 1 & 0 & 0 & 0 \\ 0 & 1 & 0 & 0 \\ 0 & 0 & 1 & 0 \\ 0 & 0 & 0 & h_{44} \end{matrix} \qquad h_{44} = \frac{1}{1+\sum \frac{\alpha}{r}}$$

The sum ranges over an arbitrary discrete number of singular points; r is the distance of the origin from one of these singular points.

Were one to allow such singularities $(h_{44} = 0)$, these would represent (1) uncharged point masses (2) in equilibrium with respect to each other, both of which contradict experience. The theory could then only be maintained if these two contradictions to experience were to vanish with the exclusion of singularities. Unfortunately, in nature, corpuscles do not appear to be spherically symmetric (magnetic moment), so to decide the question of whether such singularity-free solutions exist is mathematically quite difficult.

And further, it is clear that, *prior* to the solution of this problem, the behavior of a corpuscle in a field (the lawof motion) cannot be tackled. And without the solution of *this* problem, the correctness of the theory cannot be tested!

The worst is that our theoretical physicists do not wish to collaborate, but rather insult me because they have no feeling for the naturalness of this approach[4] (except for Langevin!).[5]

When I read your remark that the system

$$G_{\mu}{}^{\alpha} \equiv \Lambda^{\alpha}_{\mu\underline{\nu};\nu} = 0 \qquad \text{(I)}$$

is in involution, I experienced a great shock. For the nonexistence of a four-identity in this case formed one of the starting points from which I set out. But you must be mistaken here, for the identity[6]

$$G^{\mu\alpha}{}_{;\mu} + \Lambda^{\alpha}_{\rho\sigma} G^{\rho\sigma} \equiv 0$$

holds for the quantities

$$G_{\mu}{}^{\alpha} \equiv \Lambda^{\alpha}_{\mu\underline{\nu};\nu} - \Lambda^{\sigma}_{\mu\underline{\tau}}\Lambda^{\alpha}_{\sigma\tau}, \qquad \text{(II)}$$

(as you yourself have pointed out) and not for the quantities (I). In fact, in the case of (I), the commutation rule for differentiation yields

$$G^{\mu\alpha}{}_{;\mu} \equiv -\frac{1}{2}\Lambda^{\alpha}_{\mu\nu;\sigma}\Lambda^{\sigma}_{\mu\nu},$$

and the factor G can in no way be taken into the term on the right-hand side, for which purpose, only the cyclic fundamental identity is at our disposal.

The question whether (I) possesses only completely trivial solutions thus appears justified. The crude way in which I answered it,—just for my own needs—I have already communicated to you. Did you understand my arguments about this?

Your assertion that noninvolutive systems can be reduced either to involutive or to contradictory systems corresponds to my conjecture which my friend and I tried in vain to prove. It would be very good if you were to add this proof to your article, for this, too, is a crucially important point. I am truly thankful to you for the clarity you have given me.

On the question of the existence of another system of equations with the same degree of determination as the two known ones, I believe—and I hope—the answer to be "no." As for my questions, I refer back to the old maxim mentioned above.

Kind regards, your

A. Einstein

Translators' note: Translation from *Debever 1979*, pp. 108–113.

190. To Élie Cartan

[Berlin?], 8 January 1930

Dear Mr. Cartan,

I am ashamed to bother you again so soon,[1] but I will be short, and I hope the question raised here will electrify you as much as it has done me. It concerns a paradox, behind which there seems to lie something important touching your Index de généralité (I).

It can be shown that I is bigger for the approximate equations than for the full equations, both in the case of the old theory of gravitation and the new field theory,[2]

(a) Old theory

full equations $R_{ik} = 0$ $\quad I = 8$ (with fixed choice of coordinates)

$$\text{approximate equations} \quad \left.\begin{aligned} \gamma_{ik,\,\alpha\alpha} &= 0 \\ \gamma_{i\alpha,\,\alpha} &= 0 \end{aligned}\right\} I = 12$$

$$\left(\gamma_{ik} = \bar{g}_{ik} - \frac{1}{2}\delta_{ik}\bar{g}_{\alpha\alpha}\,; \quad \bar{g}_{ik} = g_{ik} - \delta_{ik} \quad \infty \text{ small}\right)$$

(b) New theory

$$\text{full equations} \quad \left.\begin{aligned} G^{ik} &= 0 \\ F^{ik} &= 0 \end{aligned}\right\} I = 10 \text{ (with fixed choice of coordinates)}$$

$$\text{approximate equations} \quad \left.\begin{aligned} \bar{h}_{ik,\,\alpha\alpha} &= 0 \\ \bar{h}_{i\alpha,\,\alpha} &= 0 \\ \bar{h}_{\alpha k,\,\alpha} &= 0 \end{aligned}\right\} I = 17$$

Thus, clearly, it is not true that, to every solution of the approximate equations, there belongs a rigorous solution. Hence, supplementary conditions must be added to the approximate equations to insure that the solution has a neighboring rigorous solution.[3] It is clear that the formulation of these additional conditions would enormously promote the understanding of the subject. We must hold tight to this thread!

Kind regards, your,

A. Einstein

Translators' note: Translation from Debever 1979, pp. 115–117.

191. To Elsa Einstein

[Fürstenberg,] Tuesday evening [8 January 1930]

Dear Elsa,

So, I'll come during the Friday so as not to miss Lina.[1] It is wonderful here, and I see that the winter is no less appealing in nature than summer is. I can't possibly subscribe to Nossig's preaching. I have written an address myself that better suits the occasion through its simplicity.[2] I will then stay in Berlin, because my new coworkers[3] are coming, and I don't want to keep myself away from academic life any longer. But life in the city is wrong for me, and, in the future, I want to be in Caputh permanently, even if Berlin remains the official place of residence. What more we need as a result, I must just try to earn. Have you given up on Uncle Caesar?[4] I will also have to officially cancel. Give Miss Dukas[5] the letters.

Warm regards to you and Margot[6] from your

Albert

192. Poem

[Berlin, after 8 January 1930][1]

[Not selected for translation.]

193. Poem for Siegfried and Josephine Bieber[1]

[Fürstenberg,] 9 January 1930

[Not selected for translation.]

194. To Walther Mayer

[Berlin,] 10 January 1930

Dear Colleague,

I am now back in Berlin and look forward to your visit.[1] Also, I look forward to our collaboration, and I am very confident that the path on which we have started will finally lead to success.

Kind regards, your

A. E.

195. To Élie Cartan

[Berlin,] 10 January 1930

Dear Mr. Cartan,

I believe I ought to communicate two points to you where—it seems to me—we are of different minds.

(1) The earlier theory of relativity, taking into account the electromagnetic field, which can be written,

$$R_{ik} - \frac{1}{2} g_{ik} R = -T_{ik}$$

$$\frac{\partial f_{ik}}{\partial x^l} + \frac{\partial f_{kl}}{\partial x^i} + . = 0$$

$$\frac{\partial f^{il}\sqrt{-g}}{\partial x^{\alpha}} = 0$$

likewise has $I = 12$. The equation $R_{ik} = 0$ has $I = 8$. In addition to that, due to the two divergence equations for the electric and the magnetic field, there is a specification of $6-2$ components (e.g., $\mathfrak{e}_1$, $\mathfrak{e}_2$, $\mathfrak{h}_1$, $\mathfrak{h}_2$) on the cross-section $x^4 = \text{const}$. Then the continuation is fully determined.

I would like to tell you briefly the reasons why, on formal grounds, the new equations please me more. First, the field interpretation is not unified in that g and φ have logically nothing to do with each other. Second, the left- and right-hand sides of the gravitational field equations have no logical connection. Third, this theory can never grasp charged masses otherwise than as singularities,—that is, they cannot grasp them at all—and this because of the last equation of the third set, whose left-hand side has the form of a pure divergence. All these drawbacks are overcome in the new theory.

(2) The equations which you have in mind;

$$R_{ik} = 0$$

$$\varphi_{\alpha} = \frac{\partial \lg \psi}{\partial x_{\alpha}}$$

$$S_{\alpha} = \frac{1}{\psi} \frac{\partial \chi}{\partial x_{\alpha}}$$

do *not* have $I = 12$, but rather $I = 16$. The equations R_{ik} alone have $I = 8$, if one views the g_{ik} alone as field variables. Now, one can express the $h_{s\nu}$ from the g_{ik} and 6 further variables $\lambda_{(\mu)}$. Then the second and third field equations will be linear in the first derivatives of $\lambda_{(\mu)}$, ψ, and χ with respect to x_4. Then everything is determined when, besides the 8 functions $g_{\mu\nu}$, resp. $g_{\mu\nu,\alpha}$—which also must be given in the pure gravitational case—on the chosen section $x^4 = \text{const}$, there are also given another 8 functions, $\lambda_{(\mu)}$, ψ, χ. Hence, $I = 8+8$ and not 12.

Thus, up to now, no second system of equations has presented itself with an *ind. d. généralité* $I = 12$.

Do not take my perfunctory letter of the other day badly, but for the future be kind enough to be patient with your

A. Einstein

P.S. It is remarkable that the *Mathematische Annalen* has such terrible constipation that, after so many months, it has not been able to excrete what it has ingested.[1]

Translators' note: Translation from *Debever 1979*, pp. 118–120.

196. To Élie Cartan

[Berlin, 10 or 11 January 1930][1]

Pater peccavi![2]

Even the most beautiful theory can be applied wrongheadedly! Forgive me.

Kind regards,

A. E.

(One time, I fixed the coordinate system by means of differential equations, the other time by fixing four variables.)[3]

197. To Walther Mayer

[Berlin, between 10 and 27 January 1930][1]

Dear Mr. Mayer,

It is not good that you are struggling with the equation when the result has become quite clear to me. The general transformation rule for the $h_s{}^\nu$ is

$$h_s{}^{\nu\prime} = a_{st}\frac{\partial x^{\nu\prime}}{\partial x^\alpha}h_t{}^\alpha.$$

If I take the same rotation for the coordinate transformation as for the n-Beine,

$$\frac{\partial x^{\nu'}}{\partial x^{\alpha}} = a_{\nu\alpha},$$

then the transformation formula becomes

$$h_s^{\ \nu'} = a_{st} a_{\nu\alpha} h_t^{\ \alpha}.$$

This is, however, the tensor transformation in a Euclidean space. $h_s^{\ \nu}$ thus behaves as a tensor. I separate it into a symmetric $(\bar{\bar{h}}_s^{\ \nu})$ and an antisymmetric part $(\bar{h}_s^{\ \nu})$. Then the central symmetry requires—as we already know—that

$$\bar{\bar{h}}_s^{\ \nu} = \chi R_1 + \mu R_2 x_s x_\mu$$

$$\bar{\bar{h}}_s^{\ \nu} = R_3 \delta_{s\mu\sigma} x_\sigma,$$

where R_1, R_2, R_3 are functions of the radius alone.

When no magnetic field is present, R_3 must vanish (at least in the first approximation). Furthermore, by means of an *r*-transformation, R_2 can be made equal to zero. In the first approximation, then, the approximate field equations also require that $R_1 = \text{const}$.

Thus, for our problem, the ansatz

$$\begin{matrix} 1 & 0 & 0 & 0 \\ 0 & 1 & 0 & 0 \\ 0 & 0 & 1 & 0 \\ j\frac{\partial R}{\partial x} & j\frac{\partial R}{\partial y} & j\frac{\partial R}{\partial 2} & \varphi \end{matrix}$$

is probably the correct one. We want to try it out, first of all, with this approach (R and φ come from r).

198. To Walther Mayer

[Berlin, between 10 and 27 January 1930][1]

Dear Mr. Mayer,

I thank you for your lovely and precise resumé. I myself had already related the matter to what we previously knew in the following way:

The transformation formula

$$h_s^{\nu'} = a_{st}\frac{\partial x^{\nu'}}{\partial x^{\alpha}}h_t^{\alpha},$$

for the special case of a "rerotation" of the local systems with a rotational transformation

$$x^{\nu'} = a_{\nu\alpha}x^{\alpha},$$

becomes

$$h_s^{\nu'} = a_{st}a_{\nu\alpha}h_t^{\alpha}.$$

This is, however, the transformation formula for a tensor with respect to linear orthogonal transformations. The symmetry of the h must therefore be the same as that of a tensor in the case of central symmetry.

Thus, the symmetric part has the structure

$$\lambda\delta_{\mu\nu} + \sigma x^{\mu}x^{\nu},$$

and the antisymmetric part has the structure

$$\kappa\delta_{\mu\nu\alpha}x^{\alpha}.$$

This consideration is interesting already because it can be applied also to the rotationally symmetrical case.

Now, one can, by means of a simple r-transformation of the spatial coordinates, ensure that σ vanishes. Furthermore, through a transformation

$$x^{4'} = x^4 + f(r),$$

one can, in the centrally symmetric case, make the h_s^{4} also vanish so that, in the centrally-symmetric case, the h_s^{ν} have a structure like

$$(j = \sqrt{-1}) \qquad \begin{matrix} \lambda & 0 & 0 & 0 \\ 0 & \lambda & 0 & 0 \\ 0 & 0 & \lambda & 0 \\ j\sigma x_1 & j\sigma x_2 & j\sigma x_3 & \mu \end{matrix} \qquad \begin{matrix} \lambda\sigma \text{ \& } \mu \\ \text{are funct. of } r \text{ alone.} \end{matrix}$$

The non-mirror-symmetric field which corresponds to the antisymmetric part of h_s^{ν} ([---] 123) must, naturally, likewise be set to zero, because this term corresponds to a centrally symmetric magnetic field, whose existence we had wished to exclude from the outset.

My wife has already looked around.[2] Full board is less favorable than a room without board. I eagerly look forward to our further collaboration.

Best wishes, your

A. Einstein

199. From Heinrich Zangger

[Pontresina, 11 January 1930][1]

[Not selected for translation.]

200. From Arnold Berliner

Berlin W9, 11 January 1930

Dear Mr. Einstein,

Here you can see what a calamity you've caused. What turned the poor young man's head; was it the concept of relativity, or the debate about Euclidean geometry? On one occasion, the experiment with the circular disk was attributed to Abbe,[1] and it appears even in the biography of Jung Stilling.[2] In any case, I wanted to let you know that you have got the experiment with the circular disk up and running again.

Also, are you aware that you've recently aroused another great uproar in New York? The American Museum of Natural History had announced a lecture on the theory of relativity, and it has said that there were 4,000 people there, and an iron gate at the entrance was stormed by the crowd and broken open.[3] Soon, I'll have to commission reports about the same kind of goings-on in the natural sciences.

By the way, will you visit me again next week in the clinic? I have to go there once more.

Best regards, your

A. Berliner

201. From Élie Cartan

Le Chesnay (Se. et O.), 27 avenue de Montespan, 11 January 1930

Dear Sir,

I, in fact, received your two letters at the same time, and then your card.[1] I was just thinking of writing an answer when the latter arrived. Am I really the person to whom you should confess your sins? I can only say for my own part: *Pater, peccavi!*[2] as I was certainly distracted by your question about the system $\Lambda^{\beta}_{\underline{\alpha\mu};\mu} = 0$. When I read your letter, I told myself, "There is a question that needs to be examined." But then, when I thought about it the following day, I had totally forgotten that the quadratic terms must not enter into the first members of the equations,[3] and, to tell the truth, I did not even think to look back at the terms given in your letter! You will certainly have to excuse me for this confusion.

Concerning your argumentation on the subject of the impossibility of the system in question, I believe that I have understood that it, at bottom, rests on the existence, for the approximated equations, of solutions which cannot be the first approximations to rigorous solutions. If that is indeed your idea, it recalls the observation in your last letter. This was that, in fact, it is easy to prove that *if a system is in involution, every solution to the equations in first approximation can be considered to be the first approximation of an infinity of rigorous solutions*. One can show this in various ways; the easiest way to understand it is perhaps the following:

Let us suppose that we can begin with a particular solution, which we can always take to be $f_\alpha = 0$. The equations in the first approximation, which give solutions that are infinitely close to each other, are obtained by keeping only those terms that are linear in f_α. Let us thus replace f_α in the equations by

$$f_\alpha = tf_\alpha^{(1)} + t^2 f_\alpha^{(2)} + t^3 f_\alpha^{(3)} + \dots,$$

t being a variable parameter, and let us consider the systems obtained successively taking the terms in t, the terms in t^2, etc., and the systems being

$$E^{(1)} = 0, \qquad (1)$$

$$E^{(2)} = 0. \qquad (2)$$

Equations $E^{(1)} = 0$ contain only the unknown functions $f_\alpha^{(1)}$; these are the equations of the first approximation (*equations with variations* of Poincaré and Darboux). Equations $E^{(2)} = 0$ contain the unknown functions $f_\alpha^{(2)}$ (as well as the $f_\alpha^{(1)}$), and so on.

If the existence of the identities that guarantee the primitive system to be in involution is given, equations (1) will be in involution; equations (2) will also be in involution *if one replaces the* $f_\alpha^{(1)}$ *there by a solution of* (1). Likewise, even equations (3), if one replaces the $f_\alpha^{(1)}$ and the $f_\alpha^{(2)}$ by a solution of (1) and (2), and so forth. Every solution of (1) can thus be regarded in an infinite number of ways as the first term of a series expansion in powers of t of a rigorous solution of the given equations.

This demonstration has the inconvenience of leaving the questions of convergence in suspense, but one can present it in a form which renders that objection inoperative.

I had already noted that your approximated equations include an index of generality that is too large, and I was on the point of writing to you about that in my last letter, but, since it was already rather long, I postponed that question for later. The way in which you choose the normal coordinates allows us a considerable degree of arbitrariness, since you could replace x_i by $x_i + t\xi_i$, where the functions ξ_i satisfy the equations

$$\xi_{i,\,kk} = 0 \qquad \xi_{k,\,k} = 0. \tag{3}$$

This results in only two gravitational fields, for which the components $g_{\alpha\beta}$ differ by functions of the form $\frac{\partial \xi_\alpha}{\partial x_\beta} + \frac{\partial \xi_\beta}{\partial x_\alpha}$, where the ξ nevertheless satisfy equations (3) and still correspond in the *same* Riemannian space with absolute parallelism; that is to say, they must not be physically distinguishable from each other. It then holds that, if these two gravitational fields cancel at infinity, the functions ξ_α are necessarily zero. There is nevertheless something a bit disconcerting: this is that it appears to be impossible to find, in the first approximation, a system of normal coordinates unambiguously determined in an invariant manner, i.e., under invariant conditions by an arbitrary displacement in Euclidean space. All that is still rather obscure to me.

I have again taken up my calculations for determining the systems of 22 equations, but without reaching definite conclusions thus far. I however still regard it as very probable that there must not be any other system besides those that we know of, or at least there could be only some *isolated* systems.

Dear and honored Sir, please accept my most sincere regards,

E. Cartan

202. To Ludwig Lewisohn[1]

[Berlin, 12 January 1930][2]

Dear Mr. Lewisohn,

I read your book *The Legacy in the Blood*[3] with great interest and true enjoyment, and indeed slowly and completely. I liked it so much because, on the one hand, it is masterful in style and, on the other hand, it seems destined to have an educational effect in the most beneficial way. Now, the bias! Attaining love for our ancestors—no longer seeing ourselves through the distorting lens of others, through which the sad centrifugal tendency arises to seek company among people like ourselves, except for in special cases—appears externally dignified and with self-esteem as Jews—I support all of that.[4] I also admit that our cultural tradition distinguishes us; altogether morally superior, intellectually one-sided but powerful, artistically and ideologically poor—on the whole, of an exquisitely preserving nature but of withering character. It was a stunted growth of a valuable race sifted by merciless "history," comparable with the stunted growth of the pines in frosty and stormy alpine mountains. Should we drift back to that? No, better to quit—quite apart from the fact that it is not possible. Also, the religious form can no longer be brought to life. The ⟨power of⟩ traditional religions have lost their power, without a fight, under the influence of education toward conscious thought in the educated classes. The moral and ideological contents have remained and been deepened, but are no longer the domain of a particular community of tradition.[5]

In my eyes, we Jews are a kind of moral nobility—even if partly degenerated by external influences. We must strive for solidarity and self-assurance without nationalistic arrogance, as well as preserving our political cosmopolitanism.[6] However, we cannot isolate ourselves on these grounds, because we mostly belong to the class of intellectual workers (in the broadest sense) and are dependent on prolific economic interaction.[7] This compels us to a certain adaptation in form to those with whom we must co-exist. However, this must remain restricted to form and should not erode the content.

This is all easily said and perhaps contradictory. But life does not allow itself to be ensnared in concepts.

If you come to Berlin one day, do not neglect to visit me, and I will attempt to do the same if I come to Paris.

Warm regards, your

A. Einstein

203. To Maxim Bing[1]

[Berlin,] 12 January 1930

Dear Sir,

My result, to which you are referring here, is the following:[2]

If one writes the equations of the general theory of relativity in such a form that a non-zero average density s in a quasi-static world is even possible, then one finds from those equations a relation between the radius of the world R in centimeters and that density, i.e.,[3]

$$R^2—1, 08 \bullet 10^{27}$$

Apart from the question of the exactness of the validity of the theory, the difficulty lies in our very imprecise knowledge of the average density of the matter in space. For that reason, I have dispensed with giving numerical values. It should be certain that s is not greater than 10^{-22}.[4] From that, one would find ten million light years as the lower bound on the value for R.[5]

Yours very respectfully,

204. From Élie Cartan

Le Chesnay (Se. et O.) 27 avenue de Montespan, 12 January 1930

Dear Sir,

Just a few words to let you know that I am nearly certain of having the result that we have searched for so long. *There is no other deterministic system in involution except the two known ones with 22 equations* (and those with 15 and 16 equations which you reject as not sufficiently determined). Finally, we can present the demonstration without requiring overly formidable calculations.

I have not yet had the time to consider your last letter of January 10.[1] But I believe that your conclusion relating to the index of generality of the system $R_{ik} = 0$ is not exact; I had calculated that index some time ago, and I have just repeated that calculation; I find $I = 4$ and not 8. As for your system in which the tensor T_{ik} enters, I will consider it at leisure and will write the result of my thoughts to you.

I have to give a talk the day after tomorrow in Langevin's laboratory on the geometric introduction to your new theory; I will speak mainly on the geometric notion of absolute parallelism and of torsion.[2] It will be followed by another talk in which I will perhaps speak about the results of my research in connection with our collaboration and under your direction, if you will permit that.

Dear and honored Sir, please accept my most sincere regards,

E. Cartan

205. From Hans Albert Einstein and Frieda Einstein-Knecht

Dortmund, 12 January 1930

Dear Papa,

Many thanks again for the beautiful Christmas greetings. We were very happy with them. But now, something comes from us, admittedly only a message, but a beautiful one: next summer, our family will increase by one. What do you think? I hear you already in spirit: I knew it straightaway; now the disaster is here…[1]

Dear Papa, please do me a favor: let worry about the future be our business. We considered it ⟨the matter⟩ at length and decided for it only after several medical assessments and will also fully bear the responsibility.

I hope that you will now also be happy about it; you do not speak in vain of a joyful event.

We are now, by necessity, very healthy; we practically don't drive anymore, because that is forbidden. We therefore go for walks, and the radio plays a large role. In regards to the radio, I also figured out the disturbance that caused the distortions. It was due to the low frequency transformer that was not well dimensioned; its resistance was too great. By parallel connecting a high frequency inductor, the evil was eliminated.

How are you, the family, the house, the boat, etc.? What do you think of Theddy?[2] Please write to us whether it would be all right with you if we came to Caputh in early summer. I will have to arrange my vacation for then.

Best regards from

Adn and Friedi

206. To August Kopff[1]

[Berlin,] 13 January 1930

Dear Colleague,

Probably, what I am about to tell you here is already well known. I ask you only as a cautionary measure.[2]

A young boy from Posen[3] asked me in a letter whether it was true that the Moon would crash into the Earth after some time. This myth must be based on the assumption that the tidal motions[4] cause a slowing down of the orbital velocity of the Moon. It strikes me that the effect must be the opposite, that the orbital velocity of the Moon must be accelerated, since the angular velocities of the lunar motion and Earth's rotation strive to become equal over time. The Moon must thus gradually be moving further away from the Earth, at the cost of the energy of the Earth's rotation.[5] An analogous cause and effect must be present in other binary star systems.[6]

Kind regards, your

207. To the Editor of *Falastin*[1]

[*Einstein 1930i*]

DATED [before 14 January 1930][2]
RECEIVED after 16 January 1930
PUBLISHED 1 February 1930

IN: *Falastin*, 1 February 1930, p. 1.

[See documentary edition for English text.]

208. To the *Jüdische Welt*[1]

[Berlin, after 14 January 1930][2]

In my view, it is self-evident that a German Jew regards and treats the Eastern Jew as a brother, especially since the Eastern Jew often distinguishes himself through competence and intellectual originality.[3] However, a few words must be added. ⟨External experience⟩ In misfortune, everyone is more picturesque and bearable than in success; this is also the case for the Eastern Jews. Thus, also take care that the eastern Jewish upstart no longer constitutes the eye-catching character of the Kurfürstendamm. An anonymous Vehmic court, whose function would consist merely in the sending of a message to the sinner, could render a wonderful service there.

209. On Disarmament[1]

[Berlin,] 16 January 1930

[See Doc. 210 in documentary edition for published English version.]

210. Einstein Sees Arms Injuring Nations

[*Einstein 1930f*]

DATED 16 January 1930

PUBLISHED 21 January 1930

IN: *The New York Times,* 21 January 1930, p. 5:1

[See documentary edition for English text.]

211. To the Prussian Academy of Sciences

Berlin, 16 January 1930

[Not selected for translation.]

212. From August Kopff

Berlin-Dahlem, 16 January 1930

Dear Prof. Einstein,

The notion that the Moon, making use of the energy of the Earth's rotation, is moving further and further away from the Earth[1] can be found in the form that you gave already in Poincaré.[2] One can also arrive at this same result by a different route. The crest of the tidal wave that is on the same side as the Moon is ahead of the lunar orbital motion as a result of the frictional forces associated with the Earth's rotation (see figure).[3] There is thus an acceleration component in the direction of the Moon's orbital motion.[4] How various people have come to the conclusion that the Moon is approaching the Earth is not clear to me. That is perhaps a vestige from an earlier time when it was believed that an acceleration of the Moon could be determined from the observations.[5] Since Earth's rotation is subject to a secular slowing down, that acceleration was only simulated by the assumption of a constant length for the day.[6] In reality, a retardation of the Moon is observable.[7]

Most respectfully, yours truly,

A. Kopff

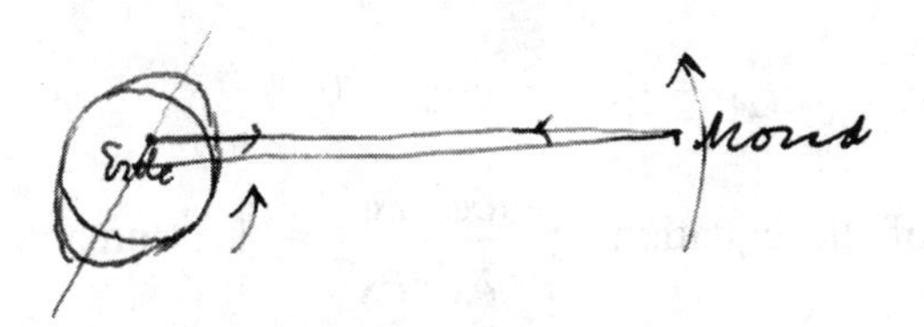

213. From Willy Scherrer[1]

Bern, Manuelstr. 76, 18 January 1930

[Not selected for translation.]

214. From Élie Cartan

Le Chesnay (Seine et Oise), 27 avenue de Montespan, 19 January 1930

Dear Sir,

I have been very busy this week, and that is the reason why I have not written to you sooner on the subject of the index of generality of the system $R_{ik} = 0$.

Let us first try to determine it by an elementary method which, unfortunately, will not suffice. The system considered is a system of equations in the second order partial derivatives with respect to the functions $g_{\alpha\beta}$; one can relate them to a system of first order by making use of the functions

$$g_{\alpha\beta\gamma} = \frac{\partial g_{\alpha\beta}}{\partial x^{\gamma}}.$$

The equations can be resolved with respect to $\frac{\partial g_{\alpha\beta 1}}{\partial x^4}, \frac{\partial g_{\alpha\beta 2}}{\partial x^4}, \frac{\partial g_{\alpha\beta 3}}{\partial x^4}$, since they contain

$$\frac{\partial g_{\alpha\beta 1}}{\partial x^4} - \frac{\partial g_{\alpha\beta 4}}{\partial x^1} = 0, \ldots$$

As for the derivatives $\frac{\partial g_{\alpha\beta 4}}{\partial x^4}$, which are 10 in number, there are only 6 equations to resolve them by means of the derivatives with respect to $x^1, x^2,$ and x^3. That corresponds to the fact that, the choice of variables being arbitrary, 4 of the $g_{\alpha\beta}$ are also arbitrary. In order to clarify the picture, one can indeed always arrange them so that one obtains

$$g_{44} = 1, \qquad g_{14} = 0, \qquad g_{24} = 0, \qquad g_{34} = 0$$

(the variable x^4 is thus a solution of the equation $g^{\rho\sigma}\frac{\partial\varphi}{\partial x^\rho}\frac{\partial\varphi}{\partial x^\sigma} = 1$, and the curves $x^1 = c^{\text{th}}, x^2 = c^{\text{th}}, x^3 = c^{\text{th}}$ are the orthogonal trajectories of the variety $x^4 = c^{\text{th}}$).

Up to this point, everything works well. Let us now look at the equations of the system which does not contain the derivatives $\frac{\partial g_{\alpha\beta\gamma}}{\partial x^4}$. It will be necessary to see how many derivatives $\frac{\partial g_{\alpha\beta\gamma}}{\partial x^3}$ there are which can be obtained by means of the derivatives taken with respect to x^1 and x^2. There are already the $\frac{\partial g_{\alpha\beta 1}}{\partial x^3} = \frac{\partial g_{\alpha\beta 3}}{\partial x^1}$ and the $\frac{\partial g_{\alpha\beta 2}}{\partial x^3} = \frac{\partial g_{\alpha\beta 3}}{\partial x^2}$. There still remain the 20 derivatives $\frac{\partial g_{\alpha\beta 3}}{\partial x^3}$ and $\frac{\partial g_{\alpha\beta 4}}{\partial x^3}$. We can already eliminate the 8 derivatives $\frac{\partial g_{\alpha 43}}{\partial x^3}$ and $\frac{\partial g_{\alpha 44}}{\partial x^3}$, since $g_{\alpha 4} = 0$ or 1. There thus remain 12 derivatives. Now, in order to obtain them, we have only 4 equations; then $12 - 4 = 8$ among them can be set arbitrarily; thus, 8 of the functions $g_{\alpha\beta\gamma}$ can be set arbitrarily in the section $x^4 = c^{\text{th}}$. It would thus seem that $I = 8$.

But we must take into account the fact that we can, thanks to the arbitrary choice of the independent variables, still particularize our coefficients $g_{\alpha\beta}$. In particular, for $x^4 = 0$, one can reduce the form

$$g_{11}(dx^1)^2 + 2g_{12}dx^1dx^2 + \ldots + g_{33}(dx^3)^2$$

to

$$(dx^3)^2 + g_{11}(dx^1)^2 + \ldots + g_{22}(dx^2)^2,$$

i.e., *one can suppose, for* $x^4 = 0$, that

$$g_{33} = 1, \; g_{13} = 0, \; g_{23} = 0;$$

this suppresses the 3 functions $g_{133}, g_{233}, g_{333}$ which are zero for $x^4 = 0$. That thus reduces the index of generality by three units; it will be equal to $\underline{\underline{5}}$ and not to $\underline{8}$ in the following.

But I do not know whether or not, *if I were more clever*, I would not be able to find an even greater reduction.

To see the matter in a way that is more independent of my cleverness, I note that what completely characterizes the Riemannian space is the series of its differential invariants, of which the first are the $R^{\;j}_{i\;kh}$, then their covariant derivatives, and so forth. Let us then take a section $x^4 = 0$, or some other arbitrary section in three dimensions. We then attach to each point in space a rectilinear tetrad, such that its fourth axis is oriented normal to the section considered. The quantities R_{ijkh} have their derivatives subjected to the following relations (Bianchi's identities):

$$R_{ij;khl} \equiv R_{ijkh;l} + R_{ijhl;k} + R_{ijlk;h} = 0 . \qquad (1)$$

If we suppose the space considered to be one of those which satisfy the relations $R_{ik} = 0$, then the relations (1) are linear with respect to the derivatives of the *10* quantities R_{ijkh} remaining. They reduce under these conditions to *16*; there are in fact $6 \times 4 = 24$ of them, but one has identically

$$R_{i1,\,234} + R_{i2,\,314} + R_{i3,\,124} + R_{i4,\,132} \equiv 0 \qquad (i = 1, 2, 3, 4)$$

$$R_{23,\,231} + R_{24,\,241} + R_{34,\,341} \equiv 0 ,$$

$$R_{13,\,132} + R_{14,\,142} + R_{34,\,342} \equiv 0 ,$$

$$R_{12,\,123} + R_{14,\,143} + R_{24,\,243} \equiv 0 ,$$

$$R_{12,\,124} + R_{13,\,134} + R_{23,\,234} \equiv 0 .$$

Of the 16 independent equations (1), 10 can be resolved with respect to derivatives (taken with respect to 4) of the 10 independent R_{ijkh}. There remain 6, not counting the derivatives with respect to 4; these are manifestly the 6 relations

$$R_{ij;123} = 0 ;$$

they are resolvable with respect to the derivatives, taken with respect to 3, of 6 of the quantities R_{ijkh}, to wit, $R_{12,\,12}, R_{13,\,13}, R_{1214}, R_{1223}, R_{1224}, R_{1234}$, and these 6 quantities are effectively independent. *As a consequence, only 4 of the* R_{ijkh} *can be set arbitrarily in the section* $x^4 = 0$. If one thus only pays attention to the *independent properties of the choice of variables*, the index of generality is effectively 4, and not 5, nor 8.

What I call the derivative with respect to 3 is the invariant derivative in the direction of the third axis of the tetrad.

The same reasoning would show that, in the case of *n* dimensions, the index of generality of the spaces which satisfy $R_{ik} = 0$ is $n(n-3)$, while the elementary method mentioned at the outset would yield only $I \leq n^2 - 3n + 1$.

If we now turn to the system

$$R_{ik} - \frac{1}{2}g_{ik}R = -T_{ik}$$

$$\frac{\partial f_{ik}}{\partial x^k} = 0$$

$$\partial f_2^{ik}\sqrt{-g} = \text{etc.},$$

we find $I = 4 + 4 = 8$. It is, however, necessary to note that this system only gives the field in the vacuum, since you exclude singularities a priori. In any case, the reasons which you have indicated to me for why you prefer the new system to the old one appear, to me, to be indeed natural, and they are just as valid in regard to the system of 22 equations with the $R_{ik} = 0$, a system whose index of generality is, in fact, 12.

Francis Perrin, the son of Jean Perrin, made an interesting remark to me the other day.[1] By modifying your system in the following manner:

$$\Lambda^{\mu}_{\alpha\beta;\mu} = 0\,,$$

$$G^{\beta}_{\underline{\alpha}} = kg^{\alpha\beta}\,,$$

k being an arbitrary *constant*, one obtains still another system in involution with the same degree of generality. This new system allows a non-singular solution in a spherical space (with indefinite time). One can, for the rest, suppose that $k = 1$ without loss of generality (that only effects a change in the units of length). This led me to note the following: the hypothesis that the equations sought contain the $\Lambda^{\gamma}_{\alpha\beta}$ to the second degree in a homogeneous manner recalls the hypothesis that *the equations sought are invariant with respect to a change of units of length.* (Do you think that there would be an a priori reason for believing that a privileged unit of length exists in nature?)[2] (or rather a privileged unit of *intervals*)? The idea of F. Perrin would thus, perhaps, be worth keeping in mind.

Dear Sir, please accept my most sincere regards,

E. Cartan

215. To the Conference of Jewish National Fund

[*Einstein 1930g*]

DATED before 20 January 1930
PUBLISHED 21 January 1930

IN: *Jewish Daily Bulletin*, 21 January 1930, p. 4

[See documentary edition for English text.]

216. To August Kopff

[Berlin,] 20 January 1930

Dear Colleague,

I thank you very much for your kind answer.[1] I would like to ask an additional question, whether the same principle has also been applied to the planets with respect to the Sun. I indeed believe that the large diameter of the solar system is due to the same cause.[2] Now, no doubt, the tidal friction on the Sun is unknown. The order of magnitude of the phenomenon would have to be estimated from an example. In the case of Venus, as far as I know, there is still an unexplained secular motion.[3] Could it perhaps be explained by a tidal-frictional moment whose axis would have to be practically the same as the rotational axis of the Sun? If that were the case, then one could make a rest estimate of the ages of the various planets.[4]

Kind regards, your

217. From Willy Scherrer

Bern, 20 January 1930

[Not selected for translation.]

218. To Élie Cartan

[Berlin,] 21 January 1930

Dear Mr. Cartan,

I feel it presumptuous of me to oppose you on a mathematical issue,[1] but it cannot be helped. I *cannot* understand your two arguments in favor of $I = 4$ for $R_{ik} = 0$.

In your first proof, you ask for I in the case of that system of equations arising from invariants through the specialization

$$g_{14} = g_{24} = g_{34} = 0 \qquad g_{44} = 0.$$

I understand this, since it corresponds to a possible, generally feasible, special choice of coordinates.

Then, we ought simply to ask: How many functions may we still freely prescribe on the cross-section x^4 = const. ? But we may not ask: How many of these functions can we still freely choose *for a fixed choice of coordinates on this surface*? For the answer to this question does not provide the value of I which interests us.

To me, it seems logical to say: the system of equations reads

$$R_{ik} = 0$$

$$g_{14} = 0 \qquad g_{24} = 0 \qquad g_{34} = 0 \qquad g_{44} = 1$$

What is the corresponding I? The analogous question applies in the case of my system of equations.

I do not understand the other proof for the following reason. We must decide whether we view

$$R^i_{jlk} \text{ or } R_{ijkl}$$

as the field variables. In the first case, their number is

$$6 \cdot 16 - 1$$

and, in the second case, only $\frac{6 \cdot 7}{2} - 1$

because of additional symmetries.

Now, I can form the R_{ik} by contraction from the R^i_{jkl} but not from the R_{iklm}, because then the g_{ik} enter in, which is precisely what was to be avoided.

Thus, please have *still more* patience and compassion for a poor physicist whose destiny has led him to these troubled pastures.

Kind regards, your,

A. Einstein

Translators' note: Translation from *Debever 1979*, pp. 140–143.

219. From Eduard Einstein

Zurich, 22 January [1930][1]

Dear Papa,

Sorry that you hear about me almost solely third hand:[2] one writes so reluctantly when at odds with himself. I am somewhat settled again, through a consistent routine. First of all, there are some contentious matters to address. Your astonishment that I did not come to Berlin surprised me.[3] I had to travel three to five hours three times this holiday and arrived completely exhausted each time. Concerning the future, it seems to me that you, rather than Zangger, are a little biased toward local patriotism: your lake in the spring, with the traditional twenty-four hours of rain daily, probably a rustic, improvised meal preferable in the south. Why are you so panicked about every changed landscape? Perhaps you will still change your mind?

Now to happier things. I heard with pleasure that you are active journalistically and on the radio. Such diversions do a person good. I attend as few lectures as possible, missing out on the first hour in the morning (besides chemistry), don't dissect often. (I allowed the co-preparer to underline his name on the note which states we had both completed the prepared specimen.) Also, I have resumed literary habits, reading a thick book about Stefan George.[4] I saw a terrible opera: *Machinist Hopkins*.[5] Mr. Besso was here the other day; we met Mr. Weyl;[6] they spoke (incomprehensible to me) about two numbers—number 135 and 16^{32}—and did so as if the composition of the world depended upon it. I hope not.

Altogether, kind regards, your

Teddy

Just now, it occurred to me that, with the radio broadcasting activities, it might be a case of confusion with Alfred Einstein, or Carl Einstein,[7] or some other Stein. How about that?

220. From Mileva Einstein-Marić

[Zurich, 22 January 1930][1]

Dear Albert,

Teddi was unfortunately quite sick; he must also now still take it very easy and nurse himself with food, with sleep, and so forth. Because of that, don't take some concerns about a trip to Berlin as reluctance to see you.[2] The desire is actually very great. A short stay in an area where one has *sun* and also good *food*, which he is very much in need of, would be best for him. All of that would also do you much good! Could you two not go somewhere together? I believe that would be the greatest joy for Teddi. Couldn't you do it? Please do it, if at all possible!

Kind regards,

Mileva

221. On Apolitical Pacifism

[*Einstein 1930bb*]

DATED 23 January 1930
PUBLISHED 1 June 1930

IN *Die Wahrheit*, 1 June 1930, p. 4

Dr. Albert Einstein:[1]

An apolitical union of all pacifistic groups is just as important as it is difficult to achieve. Such an initiative could only succeed if it came from the most significant existing organizations. Strict confinement to the purely pacifist goal would already be necessary for the attainment of unanimity. However, in the area limited in this way, a radical position (obligation to conscientious objection) must be sought, for he who is not willing to take any risks is worth nothing as an ally.[2]

222. From Willy Scherrer

Bern, 23 January 1930

[Not selected for translation.]

223. To Eduard Einstein

[Berlin,] 24 January 1930

Dear Tetel,

Zangger and Maier[1] have portrayed your condition to me more rosily than you yourself, probably to spare me. But they should not do that, because, in the long run, unreserved candor is better. One must be reconciled in one's powerlessness to fatalism, so as not to repeatedly run one's head into the wall.

I understand now that you find the journey too difficult, or at least that you perceive it as such.[2] But I cannot leave, because the Academy got a good, paid mathematician from Vienna[3] to work with me for half a year. He arrives here on 1 February, and I have to make the most of the time. It involves the calculation of the consequences of my new theory, and so a matter of burning interest to me. The juicy part is once again that, as yet, no one believes in it other than myself.

I realize, even from your letter, that you are not as laid-back as before. You will prepare yourself for a placid life and must feel out how much work is healthy for you. If you find the study of medicine too burdensome, then just let it go. What one does not finish with the feeling of a certain ease and pleasure, one should leave alone. If it was allotted to you by fate not ever to practice any actual profession, then you must reconcile yourself with that. But you should not lapse into the error of many people who somehow suffer: taking yourself too seriously and thinking everything will naturally be there for you; rather, always consider that everything that you need for yourself is a gift from others and a product of their work.

Warm wishes to you and your mother[4] from your

Papa

I gave something to Transradio about disarmament. Do you mean that, or the New Year's greeting for the radio?[5]

224. From Élie Cartan

Le Chesnay, 26 January 1930

Dear Sir,

I did indeed receive your letter of the 21st.[1] My second demonstration is, in fact, very deserving of criticism, and it takes a rather overly summary view of matters. I had hoped to be able to get along without a completely general demonstration, but that requires rather delicate explanations; I can see that I will be obliged to exit from this war machine, which is not very readily maneuverable!

Before speaking of this general demonstration, I will consider your criticisms of my first demonstration.

On that point, I am not yet quite in agreement with you. You allow the characterization

$$g_{14} = g_{24} = g_{34} = 0, \; g_{44} = 1,$$

and it is thus a question of seeing what happens in the section $x^4 = c^{th}$. But, in my opinion, one needs to see what is happening *independently of the choice of the variables in that section.*

Let us take things to a higher level. We are searching, if I am not mistaken (I think that this is also the question that you wish to resolve), for the degree of indetermination of the Riemannian spaces which satisfy the conditions $R_{ik} = 0$.[2] *We will thus consider two such spaces as identical if, by a suitable change of variables, the $g_{\alpha\beta}$ of the first space can be identified with the $g_{\alpha\beta}$ of the second.* Now, let us take some arbitrary Riemannian space; we can always carry out a change of variables such that we have, identically,

$$g_{14} = 0, \; g_{24} = 0, \; g_{34} = 0, \; g_{44} = 1;$$

we are in agreement on this point. But in this space, we can choose, among all the possible changes of variables that satisfy the preceding condition, an infinity of such a type that, *for* $x^4 = 0$, one has

$$g_{33} = 1, \; g_{13} = g_{23} = 0;$$

this is a matter which is also incontestable. I can thus be sure of obtaining *all* the Riemannian spaces where $R_{ik} = 0$ by taking the system of equations

$$R_{ik} = 0, \; g_{14} = 0, \; g_{24} = 0, \; g_{34} = 0, \; g_{44} = 1 \qquad (1)$$

and by searching *within the particular section* $x^4 = 0$ for those solutions in three dimensions which satisfy the supplementary conditions:

$$g_{13} = 0, \; g_{23} = 0, \; g_{33} = 1. \qquad (2)$$

This solution in three dimensions determines a complete solution to the system (1), and that solution will certainly furnish me with *all* the Riemannian spaces that satisfy $R_{ik} = 0$; or at least I can affirm that every Riemannian space that satisfies $R_{ik} = 0$ is *applicable* on one of the Riemannian spaces that I have found (and in reality on an infinity of them).[3] The degree of generality sought is thus not greater than the degree of generality of the solution in three dimensions ($x^4 = 0$) of systems (1) and (2). The only thing which I cannot affirm is that it is equal; it is *perhaps* smaller: $I \leq 5$.

Summing up, it seems to me that one can pose only two problems:

1. What is the index of generality of the system

$$R_{ik} = 0$$

considered as accepting the $g_{\alpha\beta}$ as unknown functions? The answer is: The general solution depends on 4 arbitrary functions of 4 variables.

2. What is the index of generality of the same system, without considering two solutions which can be deduced from each other by a change of variables as distinct? The answer is: The general solution depends upon $I \leq 5$ arbitrary functions of 3 variables.

I do not see very clearly what might be the intermediate problem of which you must be thinking, if your criticisms are, in fact, well founded.

I now take up the general method which permits one to find the index of indetermination with certainty, if one regards two solutions which are reduced to one another by a change of variables as being identical.[4] For the meantime, I will limit myself to the simple case of Riemannian spaces *with absolute parallelism.*

Let us choose a Riemannian space with absolute parallelism which satisfies, for example, your system.[5] We attach tetrads to the different points in this space which are all parallel to each other (one tetrad at each point), and denote by Λ_{ij}^{k} the components of the torsion with respect to these tetrads, and by $\Lambda_{ij;h}^{k}$ the (ordinary) derivatives of $\Lambda_{ij}^{\cdot k}$ taken in the direction of the hth axis.

I dare to make the following remark: We take four of the Λ_{ij}^{k} which are independent functions of x^1, x^2, x^3, x^4; for brevity, we denote them as $\Lambda^{(1)}$, $\Lambda^{(2)}$, $\Lambda^{(3)}$, and $\Lambda^{(4)}$. The quantities Λ_{ij}^{k} and $\Lambda_{ij;h}^{k}$ are, in my Riemannian space, well-determined functions of $\Lambda^{(1)}$, $\Lambda^{(2)}$, $\Lambda^{(3)}$, $\Lambda^{(4)}$:

$$\Lambda_{ij}^{k} = \varphi_{ij}^{k}(\Lambda^{(1)}, \Lambda^{(2)}, \Lambda^{(3)}, \Lambda^{(4)}),$$

$$\Lambda_{ij;h}^{k} = \varphi_{ijh}^{k}(\Lambda^{(1)}, \Lambda^{(2)}, \Lambda^{(3)}, \Lambda^{(4)}).$$

I now assert that *if, for two different spaces* (with absolute parallelism), *the functions* φ_{ij}^{k} and φ_{ijh}^{k} *are the same*, these two spaces are identical (that is, they differ only by the choice of variables).[6] Indeed, let x^{α} and $h_{s\alpha}(x)$ be the variables and the functions $h_{s\alpha}$ of the first space, and $\bar{x}^{\alpha}$ and $\bar{h}_{s\alpha}$ the corresponding quantities of the second. We can establish an exact correspondence between the two spaces, defined by the 4 relations

$$\Lambda^{(1)} = \bar{\Lambda}^{(1)},\ \Lambda^{(2)} = \bar{\Lambda}^{(2)},\ \Lambda^{(3)} = \bar{\Lambda}^{(3)},\ \Lambda^{(4)} = \bar{\Lambda}^{(4)}; \qquad (3)$$

we would then have, due to this exact correspondence,

$$\Lambda_{ij}^{k} = \bar{\Lambda}_{ij}^{k};\ \Lambda_{ij;h}^{k} = \bar{\Lambda}_{ij;h}^{k};$$

in particular, we would have

$$\Lambda^{(i)}{}_{;k} = \bar{\Lambda}^{(i)}{}_{;k} \quad (i, k = 1, 2, 3, 4) . \tag{4}$$

Now, one has

$$d\Lambda^{(i)} = \Lambda^{(i)}{}_{;s} h_{s\alpha} dx^{\alpha}$$

and

$$d\bar{\Lambda}^{(i)} = \bar{\Lambda}^{(i)}{}_{;s} \bar{h}_{s\alpha} d\bar{x}^{\alpha} .$$

As a result, the exact correspondence that we have established between the two spaces by means of the relations (3) leads to, taking account of (4),

$$h_{s\alpha} dx^{\alpha} \equiv \bar{h}_{s\alpha} d\bar{x}^{\alpha} ;$$

the two spaces are thus identical (in the sense that we give to that term).

What essentially characterizes a Riemannian space with absolute parallelism are thus the functions φ^{k}_{ij} and φ^{k}_{ijh} *of the four variables* $\Lambda^{(1)}, \Lambda^{(2)}, \Lambda^{(3)}, \Lambda^{(4)}$. The question that one must ask is then the following: How must we construct the differential system which defines these functions, and how do we find the index of generality of that system? Here is how we can accomplish that:

We begin for this with the field equations (having unknown functions $h_{s\alpha}$ and $\Lambda^{\gamma}_{\alpha\beta}$), but we do not write those equations which define the $\Lambda^{\gamma}_{\alpha\beta}$ by using the partial derivatives of the $h_{s\alpha}$. As for the others, we write them in a form which involves the components Λ^{k}_{ij} with respect to a system of parallel tetrads, as well as their derivatives $\Lambda^{k}_{ij;h}$ taken in the directions of the axes of the tetrads. We will obtain the 38 equations

$$\text{(I)} \quad \begin{cases} H^{l}_{ijk} \equiv \Lambda^{l}_{ij;k} + \Lambda^{l}_{jk;i} + \Lambda^{l}_{ki;j} + \Lambda^{m}_{ij}\Lambda^{l}_{km} + \Lambda^{m}_{jk}\Lambda^{l}_{jm} + \Lambda^{m}_{ki}\Lambda^{l}_{jm} = 0 \\ F_{ij} \equiv \Lambda^{k}_{ij;k} = 0 \\ G^{j}_{i} \equiv \Lambda^{j}_{ik;k} + \Lambda^{m}_{ik}\Lambda^{j}_{km} = 0 \end{cases}$$

Let us recall that, of these 38 equations, 2, of which we will denote the first members as A_1, A_2, contain only the $\Lambda^{k}_{ij;1}$ and $\Lambda^{k}_{ij;2}$; 12 others, which are independent, contain only the $\Lambda^{k}_{ij;1}$, $\Lambda^{k}_{ij;2}$ and $\Lambda^{k}_{ij;3}$; and finally, the 24 remaining are linearly independent with respect to $\Lambda^{k}_{ij;4}$. We thus denote the system (I) by

$$(\mathrm{I}') \quad \begin{cases} A_i = 0 & i = 1, 2 \\ B_j = 0 & j = 1, 2, \ldots, 12 \\ C_k = 0 & k = 1, 2, \ldots, 24 \end{cases}$$

Now, we are going to take the $\Lambda^k_{ij;h}$, which we denote by Λ^k_{ijh}, as new unknown functions. They satisfy the equations

$$(\mathrm{II}) \quad \begin{cases} L^l_{ijkh} \equiv \Lambda^l_{ijk;h} - \Lambda^l_{ijh;k} + \Lambda^m_{kh}\Lambda^l_{ijm} = 0 \\ H^l_{ijkh} \equiv \Lambda^l_{ijk;h} + \Lambda^l_{jki;h} + \Lambda^l_{kij;h} + \Lambda^m_{ij}\Lambda^j_{kmh} + \ldots = 0 \\ F_{ijh} \equiv \Lambda^k_{ijk;h} = 0 \\ G^j_{ih} \equiv \Lambda^k_{ijk;h} + \Lambda^m_{ij}\Lambda^j_{kmh} + \Lambda^m_{ijh}\Lambda^j_{km} = 0 \end{cases}$$

Equations (II) are, for the rest, not all linearly independent, since they are evidently connected by 16 identity relations (taking (I) and (II) into account).

We can arrange system (II) just as we have arranged system (I). We would then have a series of independent equations

$$(\mathrm{II}') \quad \begin{cases} 1. \quad A_{i1} = 0 \quad B_{j1} = 0 \quad C_{k1} = 0 & (38) \\ 2. \; A_{i2} = 0 \quad B_{j2} = 0 \quad C_{k2} = 0 \quad L^l_{ij12} = 0 & (62) \\ 3. \quad B_{j3} = 0 \quad C_{k3} = 0 \quad L^l_{ij13} = 0 \quad L^i_{ij23} = 0 & (84) \\ 4. \; C_{k4} = 0 \; L_{ij14} = 0 \quad L_{ij24} = 0 \quad L_{ij34} = 0 & (96) \end{cases}$$

Equations 1 contain only the derivatives $\Lambda^l_{ijk;1}$; equations 2 contain only the derivatives $\Lambda^l_{ijk;1}$ and $\Lambda^l_{ijk;2}$; equations 3 contain only the derivatives $\Lambda^l_{ijk;1}, \ldots, \Lambda^l_{ijk;3}$. Equations 4 are evidently independent with respect to 24×4 derivatives $\Lambda^k_{ijh;4}$. As for equations 3, they permit us to determine only $24 \times 4 - 12$ derivatives $\Lambda^k_{ijh;3}$ by means of the derivatives ;1 ;2, and the 12 other derivatives $\Lambda^k_{ijh;3}$. *This number 12 is the same as the number of derivatives* $\Lambda^k_{ij;3}$ *which cannot be determined by the equations* $B_j = 0$ *by means of the* $\Lambda^k_{ij;1}$ *and the* $\Lambda^k_{ij;2}$.

Let us now return to our problem. In equations (I) and (II), replace the quantities

$\Lambda^{l}_{ijk;h}$ by $\dfrac{\partial\varphi^{l}_{ij}}{\partial\Lambda^{(s)}}$, $\Lambda^{(s)}{}_{;k}$, and the quantities $\Lambda^{l}_{ijk;h}$ by $\dfrac{\partial\varphi^{l}_{ijk}}{\partial\Lambda^{(s)}}$, $\Lambda^{(s)}{}_{;h}$. It should be noted that the quantities $\Lambda^{(s)}{}_{;h}$ are themselves 16 of the functions φ^{l}_{ijk}. By making that substitution, the equations (I) and (II) become a system of equations with *ordinary* partial derivatives and the unknown functions φ^{l}_{ij} and φ^{l}_{ijk} of $\Lambda^{(1)}$, $\Lambda^{(2)}$, $\Lambda^{(3)}$, $\Lambda^{(4)}$. For the rest, one has

$$\varphi^{l}_{ijk} = \frac{\partial\varphi^{l}_{ij}}{\partial\Lambda^{(s)}}\Lambda^{(s)}{}_{;k}\,.$$

Quite certainly, there are $24 - 4 = 20$ φ^{l}_{ij} functions and 96 φ^{l}_{ijk} functions.

One could easily show, but that is not really necessary, that this system of equations with ordinary partial derivatives is in involution. The considerations made concerning the respective numbers of the equations 1, 2, 3, and 4 of system (II') show us tha,t if we arrange the *new* linear equations (II) with respect to $\dfrac{\partial\varphi_{ijk}}{\partial\Lambda^{(s)}}$, with respect to the derivatives $\dfrac{\partial}{\partial\Lambda^{(1)}}, \dfrac{\partial}{\partial\Lambda^{(2)}}, \dfrac{\partial}{\partial\Lambda^{(3)}}, \dfrac{\partial}{\partial\Lambda^{(4)}}$, we will arrive at the same numbers (whose values are always the same, no matter which non-characteristic choice is made for the variables). As a consequence,

1. In the new system (II), there exist 96 linearly independent equations with respect to the 96 derivatives $\dfrac{\partial\varphi^{l}_{ijk}}{\partial\Lambda^{(4)}}$;

2. there are only $96 - 12$ equations which contain none of the derivatives $\dfrac{\partial}{\partial\Lambda^{(4)}}$ and are linearly independent with respect to the $\dfrac{\partial\varphi^{l}_{ijk}}{\partial\Lambda^{(3)}}$.

From the first property, one can deduce that every solution of the differential system which yields the φ^{k}_{ij} and φ^{k}_{ijh} is completely determined by the solution in three dimensions in the section $\Lambda^{(4)} = c^{\text{th}}$.

From the second property, it is found that, in this section, one can arbitrarily define 12 of the unknown functions φ^{k}_{ijh}, *but one cannot give more than 12 of them.*

As a result, *the index of generality being sought is 12.*

In the final analysis, this conclusion is based exclusively on the property of system (1) of containing 24 linearly independent equations ($C_k = 0$) with respect to the $\Lambda^{k}_{ij;4}$ and of containing only $24 - \underline{\underline{12}}$ ($B_j = 0$) linearly independent equations with respect to the $\Lambda^{k}_{ij;3}$ and not containing the $\Lambda^{k}_{ij;4}$. It is this *deficit* of 12 which is maintained when one introduces the second derivatives and which indicates the index of generality sought. This number 12 can thus be definitively obtained by starting from the field equations, <u>*but not taking the functions* $h_{s\alpha}$ *into account*</u>.

It is thus by means of getting to the bottom of things—that is, by trying to recognize just how two Riemannian spaces with absolute parallelism are identical—that one arrives at a surefire method for finding the *essential* degree of generality of those spaces.

In the case of the system $R_{ik} = 0$, which involves Riemannian spaces *without* absolute parallelism, the demonstration is more complicated, because one cannot find the necessary and sufficient conditions to be applied to two Riemannian spaces quite as readily. That will be the topic of a future letter, if you wish.

In any case, please do not be too sparing of my patience. In the end, I am very grateful for having agreed to search, making use of the theorems which I have possessed for quite some time, for a form of presentation which would not be too difficult to assimilate for mathematicians who have no special knowledge of analysis. You are therefore rendering me a great service!

Please accept, dear Sir, my most sincere regards,

E. Cartan

225. To Émile Meyerson[1]

Berlin W, 27 January 1930

Dear Mr. Meyerson.

In the passage sent,[2] something basic is not correct. Namely, it was not inflation which led me to a preoccupation with technical things. As a young man, I was already an examiner at the Swiss Patent Office and, in the years that followed, never completely stopped occupying myself with technical things. This was also advantageous for scientific research. As proof of the paramagnetic atom's centrifugal nature, for example, I was inspired by an evaluation that I had drawn up for a gyrocompass.[3]

Kind regards, your

A. Einstein

P.S. Please give me kind regards to Mr. Metz,[4] whose essay in the *Revue philosophique* of January/February 1929[5] I read with approving interest. I found an inaccuracy that he could probably correct for the better only in one single location. On page 67 in line 10, it should read "Euclidean," not "non-Euclidean." That is to say, if the cells are to be interpreted formally as cubes in the Euclidean sense through formal introduction of a corresponding metric, so this metric is a Euclidean one (which, however, makes no physical sense).[6]

226. From Eduard Einstein

Zurich, 27 January [1930][1]

Dear Papa,

Difficult to communicate over such distances. Because, in the attempt to lure you south,[2] I seem to have badly distorted the picture. For weeks, I have been hardly distinguishable from a healthy person,[3] at most by the amount of psychological insight gained in the meantime. Beyond that: I took care of a young friend in a crisis. Unfortunately, when I spoke with him the second time, he was completely normal again. But let's say thank God instead of unfortunately… Unemployment is the last model that I have in mind, especially as the lectures were the best medicine for me the entire time.[4] I only rarely go to the psychiatrist:[5] so little worth telling is happening. That I am not as laid-back as before; you are probably right, there. What should I do in order to prove happiness? Tell jokes? So, here are some jokes.

When the Bolsheviks came to power, they tried to teach the farmers communist principles. In a lesson, a commissar asked a farmer: "Now, what would you do if you had two houses?"

Farmer: I would give one to a comrade.

Commissar: And if you had two horses?

Farmer: I would give one to a comrade.

Commissar: And if you had two cows?

Farmer: (silent, concerned)

Commissar: And what would you do if you had two cows?

Farmer: But I have two cows.

Your views on the behavior of those without professions are probably questionable, but the evening hour is probably already too advanced for me to challenge them today. But I will write again soon. Will you be able to make very much use of the mathematician?[6] Warm regards, your

Teddy

227. From Walther Mayer

Vienna, 27 January 1930

Dear Professor,

Many thanks for your kind letter![1] I was worried that my derivation of the most general form of the triad ${}_{(\alpha)}h^i$ in the spherically symmetric case[2] might have seemed excessively complicated to you, owing to its purely mathematical character. But I believe that only a systematic procedure can give us the certainty of not having calculated only a special case.

When we later proceed to the treatment of the "dipole," where we possibly can no longer fall back on what is already known, we will have only the analytic method.

I had already suspected that the term $\varepsilon_{\alpha ij}x_j$ would drop out in the spherically symmetric case. By elimination of the $x_\alpha x_i$ term, the further computations are enormously simplified.

However, the generality is by no means compromised. It might indeed appear problematic that the corresponding coordinate transformation, $\bar{x}_i = \varphi(s)x_i$, $i = 1, 3$, $s^2 = \sum x_i^2$, would transform an axis ${}_{(\alpha)}h^i = \lambda\delta_{\alpha i} + \sigma x_\alpha x_i$, which becomes Euclidean at infinity, into an axis ${}_{(\alpha)}\bar{h}^i = \varphi\lambda\delta_{\alpha i}$ which exhibits no Euclidean behavior at infinity. Since, however, as one can readily prove, φ approaches a finite value at infinity, this problem in fact does not arise.

One could thus say that the tetrad as defined in your letter has the most general physically usable form (in the case of spherical symm[etry]).

Setting up the field equations, thereby establishing the system of ordinary second-order differential equations for the three unknown functions, is simply a matter of paying attention. I hope to be able to carry out that calculation while I am still in Vienna.

Concerning my departure, I would be very pleased to be in Berlin on February 1. Because of the high price of hotel rooms, I, however, will not risk leaving here until I know the address of my apartment in Berlin. I heartily thank your esteemed wife for her trouble,[3] and would ask her to please, considering my genuine lack of special requirements, just choose a place without long deliberations and to simply rent a quiet room for me.

I would be happy to arrive there by February 1, since, here in Vienna, I cannot dedicate myself to my work as I would like. I am, myself, uncommonly interested in finding out if everything is indeed as you, esteemed Professor, believe; and as I am beginning to believe also.

In hopes of greeting you soon in Berlin, yours sincerely,

Walther Mayer

228. To Élie Cartan

[Berlin?,] 30 January 1930

Dear Mr. Cartan,

Once again, I must return to the *Index de généralité*.[1] It seems possible to me that the following change of approach might be appropriate. One does not stop with just *one* number for its characterization, but proceeds as follows:

As you have done, one first eliminates the $\frac{\partial f}{\partial x^4}$ from the n equations (if possible). Thus, one obtains r_4 equations which contain $\frac{\partial f}{\partial x^4}$ and $n - r_4$ which do not contain $\frac{\partial f}{\partial x^4}$. Then, from the $n - r_4$ equations, one eliminates $\frac{\partial f}{\partial x^3}$. Thus one obtains r_3 equations which contain $\frac{\partial f}{\partial x^3}$ (but not $\frac{\partial f}{\partial x^4}$), as well as $n - r_4 - r_3$ equations which contain neither $\frac{\partial f}{\partial x^4}$, nor $\frac{\partial f}{\partial x^3}$. One continues decomposing in your way until one obtains the numbers

$$r_4 \qquad r_3 \qquad r_2 \qquad r_1 \quad \text{(for four dimensions)},$$

where $n = r_4 + r_3 + r_2 + r_1$.

If the system is in involution and p is the number of variables f, then the unique solution of the system is determined by the following independent specification:

$(I_1 = \ )p - r_1$ functions of x_1

$(I_2 = \ )p - r_2$ " " x_1, x_2

$(I_3 = \ )p - r_3$ " " x_1, x_2, x_3

$(I_4 = \ )p - r_4$ " " x_1, x_2, x_3, x_4

With it, the free choice of the functions is restricted only by the condition that the continuation of the solution in the continuum of one higher dimension must always attach itself continuously.

For example, for the system of equations.

$\Lambda^{\alpha}_{\mu\nu} = 0$ (Euclidean with ordinary parallelism)

I_4	I_3	I_2	I_1
4	8	12	16

For my system of equations, we should have

I_4	I_3	I_2	I_1
4	20	34	40

Of course, $I_3 - (I_3)$ Eucl. corresponds to your index I_1, taking general covariance into account.

The only question now remaining is *how does one fix the degree of determination in the most natural way when it comes to comparing the degrees of determination of two systems of equations.*[2]

Should one ask:

"How many functions can be freely chosen on a section $x^4 = \text{const.}$," or should one characterize the degree of determination by means of the series I_4, I_3, I_2, I_1?

The whole question has arisen because we would like to establish whether the system of equations

$$R_{ik} = 0 \qquad \Lambda^{\alpha}_{\mu\alpha} = \frac{1}{\psi}\frac{\partial\psi}{\partial x^{\mu}} \qquad S_{\mu} = \frac{1}{\psi}\frac{\partial\chi}{\partial x^{\mu}}$$

and the system of equations

$$\Lambda^{\alpha}_{\underline{\mu}\nu;\nu} - \Lambda^{\sigma}_{\underline{\mu}\tau}\Lambda^{\alpha}_{\sigma\tau} = 0 \qquad \Lambda^{\alpha}_{\mu\nu;\alpha} = 0$$

are equivalent as regards degree of determination.

I incline to the opinion that the second approach is to be preferred because of its completeness. What do you think?

Kind regards. your

A. Einstein

P.S. I have just now read your very thorough letter and—without yet grasping all the details—I am *completely convinced* of the reliability of the results. Nevertheless, I am sending this letter because its contents have not been rendered completely superfluous.

Translators' note: Translation adapted from *Debever 1979*, pp. 156–159.

229. From August Kopff

Berlin-Dahlem, 30 January 1930

Dear Prof. Einstein,

The possibility of an increase in the distances of the bodies in the solar system, due to the tidal friction of the Sun, has been repeatedly mentioned in cosmological considerations.[1] I, however, have not found any attempts to quote quantitative results on this phenomenon anywhere, and I do not quite know how one should approach that goal, since one is totally dependent on hypotheses. Even for the present state of the solar system, the question of the tidal effects of the Sun remains completely open.[2] Through the investigations of, for example, de Sitter, however, it has become probable that the small remaining empirical discrepancies in the secular progression of the distance for Mercury and Venus can rather be explained by the variations in our empirical time measurements.[3] There are too few observational data which would indicate a notable contribution from tidal friction in the case of the Sun.

Respectfully yours,

A. Kopff

230. To Élie Cartan

[Berlin?,] 31 January 1930

Dear Mr. Cartan,

Several things were wrong in my letter from yesterday,[1] but the basic idea still seems reasonable to me,[2] namely, to characterize the degree of generality by the series of numbers I_4 I_3 I_2 I_1. But the numbers given and the comparison with the Euclidean case were wrong.

The following method of fixing I, in the case of our system of equations, seems simple to me. We choose the quantities $h_{s\nu}$ and $h_{s\nu\alpha}\left(=\frac{\partial h_{s\nu}}{\partial x^{\alpha}}\right)$ as variables. I now look at my system, $G^{\mu\alpha} = 0, F^{\mu\alpha} = 0$. These are now equations of the first order in our variables. I use the two identities

$$\Lambda^{\alpha}_{\mu\nu;\alpha} \equiv \frac{\partial\varphi_{\mu}}{\partial x^{\nu}} - \frac{\partial\varphi_{\nu}}{\partial x^{\mu}} \tag{1}$$

$$2\overline{G}^{\mu\alpha} - F^{\mu\alpha} = S^{\alpha}_{\underline{\mu\nu},\nu} + S^{\alpha}_{\underline{\mu\sigma}}\Delta^{\nu}_{s\nu} \text{ [3]} \tag{2}$$

Then 12 of the 22 equations can be written in the form

$$0 = \frac{\partial\varphi_{\mu}}{\partial x^{\nu}} - \frac{\partial\varphi_{\nu}}{\partial x^{\mu}} \tag{3}$$

$$0 = S^{\alpha}_{\underline{\mu\nu},\nu} + S^{\alpha}_{\underline{\mu\sigma}}\Delta^{\nu}_{\sigma\nu}\,. \tag{4}$$

Of these equations, 6 contain no terms differentiated with respect to x^4, two of them having no derivatives with respect to x^3 either. Thus, one has immediately:

10 + 6 equations which contain derivatives with respect to all variables,
4 " " " " " " " $x_1\ x_2\ x_3$
2 " " " " " " " $x_1\ x_2$

Because of general covariance, derivatives with respect to x^4 can be eliminated from 4 of the 16 equations, something which can also be seen from the fact that, so far, only 2 of the three 4-identities have been used. The 4 equations arising from this elimination contain the indices 1, 2, and 3. The original equations may, therefore, be written:

12 with differentiation with respect to $x_3\ x_2\ x_1$
8 " " " " " $x_3\ x_2\ x_1$
2 " " " " " $x_2\ x_1$

With our choice of variables, the complete system reads

$G^{\mu\alpha} = 0$ $F^{\mu\alpha} = 0$	12	8	2	0
$h_{s\mu,\sigma} - h_{s\mu\sigma} = 0$	16	16	16	16
$h_{s\mu,\sigma,\tau} - h_{s\mu\tau,\sigma} = 0$	48	32	16	0

The numbers written alongside the equations state how many of the equations are differentiated with respect to how many of the variables. We now obtain the numbers r_4, r_3, r_2, r_1 for our system of equations:

r_4	r_3	r_2	r_1
76	56	34	16

The number of variables is $p = 80$. Hence, we obtain for the I

I_4	I_3	I_2	I_1
4	24	46	64

Now we compare this *properly* (and not as in my previous letter!) with the Euclidean case, in which we take $h_{s\nu}$ and $h_{s\nu\alpha}$ as variables. We then obtain, in an analogous notation:

$h_{s\mu,\,\nu} - h_{s\nu,\,\mu} = 0$	12	8	4	0
$h_{s\mu,\,\nu} - h_{s\mu,\,\nu} = 0$	16	16	16	16
$h_{s\mu\sigma,\,\tau} - h_{s\mu\tau,\,\sigma} = 0$	48	32	16	0

r_4	r_3	r_2	r_1
76	56	36	16

I_4	I_3	I_2	I_1
4	24	46	64

When comparing, a difference between the two cases appears only in the third column (I_2)! Thus, the determination of the manifold *also* appears to be almost uncanny; a true paradox. Do you think I have "shot a buck" (i.e., committed a blunder)?

Kind regards. your

A. Einstein

Postscript

(1) I find exactly the same situation with your system of equations

$$R_{ik} = 0 \qquad \frac{\partial\varphi_i}{\partial x_k} - \frac{\partial\varphi_k}{\partial x_i} = 0 \qquad \frac{\partial S_i}{\partial x_k} - \frac{\partial S_k}{\partial x_i} = 0$$

It is interesting, too, that the second and third system yield additional conditions for the g_{ik}, so my objection to this system was not totally justified. *Even Maxwell's equations come out in the first approximation*. Thus, it will be necessary to take account of this system.

(2) The *pure* Riemannian theory is also easily compared with Euclid by this method. The results are as follows:

$R_{ik} = 0$

I_4	I_3	I_2	I_0
4	16	30	40

$R_{ik,lm} = 0$

I_4	I_3	I_2	I_1
4	12	24	40

Thus, according to these results, the pure Riemannian theory of relativity is more weakly restricted than the new theory.

I am very eager to find out whether you approve of the method and if you can find any errors in these results. In all honesty, it is scarcely believable that the determination should be so strong in the new theory, the less so since it is not so in the first approximation.

Translators' note: Translation based on *Debever 1979*, p. 160–165.

231. To August Kopff

Berlin, 31 January 1930

Dear Colleague,

I thank you most kindly for your letter.[1] So far as I remember, in the case of Venus, there are unexplained discrepancies with respect to the *geometrical* orbital elements. But I no longer know exactly which ones. The question is whether these could be explained by a force whose moment of torque relative to the midpoint of the Sun has the same direction as the solar axis of rotation.[2] If that were indeed the case, one could calculate the corresponding constant of the tidal friction and, from it, the effects on all the planets. It would be very kind of you if you would check on this.

With kind regards, your

A. Einstein

Not on Mon. or Thurs, mornings.[3]

232. To Josef Strasser[1]

[Berlin,] 31 January 1930

Dear Sir,

To question 1: Decisions have played a very minor role in my life. Everything happened directly out of necessity and without plan.

This answers questions 2 and 3.[2]

Respectfully yours,

233. To Cornelio L. Sagui[1]

[Berlin, after 31 January 1930][2]

[Not selected for translation.]

234. To Élie Cartan

[Berlin,] 2 February 1930

Esteemed Mr. Cartan,

I have now myself found the error which is to blame for my paradoxical result.[1] I thought that

$$\Lambda^{\alpha}_{\mu\nu} = 0 = h_{s\mu,\,\nu} - h_{s\nu,\,\mu}$$

were the equations to be added to

$$h_{s\mu,\,\nu} - h_{s\mu\nu} = 0$$

in order to form a third system consisting of all of them, which would then be formally analogous to a second order system of equations. Strangely enough, I had overlooked the fact that one can eliminate all the derivatives in the first equation by using the second! The comparison of my system with the Euclidean one is, therefore, wrong. But I can find the needed form very simply. I will give the last part of the analysis in an improved manner.

By means of considerations which I indicated to you in my letter, a simple ordering of the equations leads to the following three results

	I_4	I_3	I_2	I_1
$R^i_{klm} = 0$	4	12	24	40
$R_{ik} = 0$	4	16	30	40
My system	4	24	46	64

(The number of identities is always $r_3 + 2r_2 + 3r_1$)

One can see from the first row that, in a Euclidean space, 12 of the $g_{\mu\nu}$ and $g_{\mu\nu\alpha}$ can be freely chosen in a space of three dimensions by a suitable choice of coordinate system. Indeed, this will hold for any given Riemannian space. Furthermore, if one has a space with a Riemannian metric and distant parallelism, then, instead of the variables $h_{s\nu}$, one can choose the $g_{\mu\nu}$ and 6 further variables (e.g., 6 of the $h_{s\nu}$), together with their first derivatives, as field variables (and their first derivatives). Since 12 of the $g_{\mu\nu\alpha}$ can be freely selected by a choice of coordinates on a "three dimensional surface," then, in a $h_{s\nu}$ space, 12 variables are also freely specifiable (by means of a choice of coordinates) on a "surface."

Thus, it follows that, apart from a choice of coordinates,

in the case of $R_{ik} = 0$ $\quad 16 - 12 = 4$

" " " " my equations, $24 - 12 = 12$

field variables (apart from the choice of coordinates) may be prescribed.

This is in exact agreement with your results.[2]

I would like to add a remark. The equations $G^{\mu\alpha} = 0$ alone form a system in involution because of the identity $G^{\mu\alpha}{}_{;\mu} + \Lambda^{\alpha}_{\sigma\tau}G^{\sigma\tau} \equiv 0$.

A system in involution again arises from adding on the equations $F^{\mu\alpha} = 0$. Is it certain that there exists no (second order) system in involution which might arise from the addition of still further equations to my system of equations? In particular, it seems to me that the index of generality of my system of equations might still be too high.[3] Naturally, this would be of the greatest importance.

Kind regards, your

A. Einstein

235. From Élie Cartan

Le Chesnay, 2 February, 1930

Dear Sir,

I have indeed received your two letters from January 30 and 31.[1] The integers I_1, I_2, I_3, I_4 can, in fact, all be considered; unfortunately, only the last of these numbers which is nonzero has a precise mathematical meaning. Here is what, in fact, happens:

I keep your notation: p is [the number of] unknown functions, $r_1 + r_2 + r_3 + r_4$ the linear equations of first order. Let us take

$$s_1 = p - r_1$$
$$s_2 = p - r_2$$
$$s_3 = p - r_3$$
$$s_4 = p - r_4$$

We now substitute, for the given system, that system obtained by adjoining to it the partial derivatives of first order as new unknown functions (*extended* system). The new values of the integers s_i are

$$s'_1 = s_1 + s_2 + s_3 + s_4$$
$$s'_2 = s_2 + s_3 + s_4$$
$$s'_3 = s_3 + s_4$$
$$s'_4 = s_4$$

If $s_4 > 0$, the new value of s_4 remains the same, the number of functions that one can give arbitrarily in x^1, x^2, x^3, x^4 always stays the same. But *the value of s_3 has increased.* In this case ($s_4 > 0$), the integer s_3 thus has no precise meaning at all, and all the more so the integers s_1, s_2.

If, on the contrary, $s_4 = 0$, the integer s_3 keeps its value for all the extended systems and it truly has an intrinsic meaning. But it remains the same for s_2 and for s_1 (unless $s_3 = 0$, in which case, it is the integer s_2 which defines the degree of generality).

All of that can be stated more precisely. The ideal would be to demonstrate that, if $s_4 = 0$ and $s_3 = I > 0$, the integer s_3 is the same for two *equivalent* differential systems, that is systems that can establish a biunique correspondence which is continuous between all the solutions of the first and all the solutions of the second. In this general form, the problem seems to me at present to be intractable. But one can—as I have done in the theory of infinite groups—give a more restrictive definition, which is however still very broad, of two equivalent differential systems. Let a system Σ with p unknown $f_1, f_2, \ldots, f_p$ be adjoined to the system of new equations containing the functions $f_1 \ldots, f_p$ and of new unknown functions $f_{p+1}, \ldots, f_{p+q}$. One thus obtains a new system Σ'. I would say that Σ' is a *holohedral extension* of Σ if every solution of Σ yields values of $f_1 \ldots, f_p$ which can be associated with the values of $f_{p+1}, \ldots, f_{p+q}$ in such a manner as to obtain a solution of Σ'. I would thus maintain that two systems Σ_1 and Σ_2 are equivalent if there exist two holohedric extensions Σ'_1 and Σ'_2 of Σ_1 and Σ_2 which can be derived back from each other by a change of variables and of the unknown functions (the two systems Σ'_1 and Σ'_2 subsume as a result the same number of unknown functions). Under these conditions, the following theorem holds:

If, for the first system Σ_1, the integers $s_1, s_2, \ldots, s_n$ are such that

$$s_n = s_{n-1} = \ldots = s_{\nu+1} = 0 \qquad s_\nu > 0,$$

then one would have for the second system

$$s'_n = s'_{n-1} = \ldots = s'_{\nu+1} = 0 \qquad \underline{s'_\nu = s_\nu}.$$

But let us leave aside these generalities. If you take $I_4 = 4$ to compare two relativistic systems, that removes any precise meaning from I_3. In order that the index of generality I_3 have any significance, it would be necessary to consider the differential system which gives the *invariant* properties of the spaces, that is to say, the one that I indicated to you in my last letter—at least, one must particularize the $h_{s\alpha}$ in such a manner as to render I_4 zero (and to give I_3 its true value).

I was greatly interested in the postscript of your second letter, in which you did not absolutely reject the system $R_{ik} = 0, \dfrac{\partial \varphi_i}{\partial x_k} - \dfrac{\partial \varphi_k}{\partial x_i} = 0 \ldots.$ etc. ... You had played at searching for the expression of $R_{\alpha\beta}$ by means of the $\Lambda^\gamma_{\alpha\beta}$? One finds

$$2R_{\alpha\beta} = \Lambda^{\underline{\beta}}_{\alpha\underline{\mu};\mu} + \Lambda^{\underline{\alpha}}_{\beta\underline{\mu};\mu} - \varphi_{\alpha;\beta} - \varphi_{\beta;\alpha} + \Lambda^{\rho}_{\alpha\underline{\mu}}\Lambda^{\underline{\beta}}_{\mu\rho} + \Lambda^{\rho}_{\beta\underline{\mu}}\Lambda^{\underline{\alpha}}_{\mu\rho}$$
$$+ S_\alpha S_\beta - (\Lambda^{\underline{\beta}}_{\alpha\underline{\rho}} + \Lambda^{\underline{\alpha}}_{\beta\underline{\rho}})\varphi_\rho + 2\Lambda^{\sigma}_{\alpha\rho}\Lambda^{\underline{\sigma}}_{\beta\underline{\rho}} - g_{\alpha\beta}S_\mu S_{\underline{\mu}}$$

Please accept, dear and honored Sir, my most sincere regards,

E. Cartan

This morning, I found a photograph in Le Populaire, a very good likeness, of you holding your violin on the way out of a concert![2]

236. To [?] Melcher

[Berlin,] 3 February 1930

[Not selected for translation.]

237. To Élie Cartan

[Berlin,] 4 February 1930

Dear Mr. Cartan,

I ought to explain my notation again; for I believe you have misunderstood me regarding the I^4 I^3 I^2 I^1 .[1] But first, I am afraid I have not understood what you mean by an "extended system";[2] do you mean

$$a_{ikl}\frac{\partial f_k}{\partial x^l} + b_{ik}f_k = 0 \qquad \text{(original system)}$$

$$\left.\begin{aligned} a_{ikl}\overline{\varphi}_{kl} + b_{ik}f_k &= 0 \\ \frac{\partial f_k}{\partial x^l} - \varphi_{kl} &= 0 \end{aligned}\right\} \qquad \text{(extended system)}$$

The second system would not fit into the scheme, since it contains equations which contain no derivatives of the unknown functions. Every single unknown would then be eliminated from such equations.

My notation was as follows: n (algebraically independent) differential equations are given. One eliminates all the $\frac{\partial f}{\partial x^4}$ from as many of them as possible, and one ends up with

$$r^4 \text{ equations which contain } \frac{\partial f}{\partial x^4}$$

$$n - r^4 \text{ equations which do not contain } \frac{\partial f}{\partial x^4}$$

Now one eliminates the $\frac{\partial f}{\partial x^4}$ from the $n - r^4$ equations. Then there remain r^3 equations which contain the $\frac{\partial f}{\partial x^4}$ and $n - r^4 - r^3$ which contain neither the $\frac{\partial f}{\partial x^4}$ nor the $\frac{\partial f}{\partial x^3}$. One continues in this way and, finally, obtains the quadruple

$$r^4 \; r^3 \; r^2 \; r^1$$

where

$$n = r^4 + r^3 + r^2 + r^1$$

In order that the system be in involution, $r^3 + 2r^2 + 3r^1 \; \left(\sum_\nu \nu r^{l-\nu}\right)$ identities of the kind given by you must hold.

By putting $\overset{4}{I} = p - \overset{4}{r}$ (p = number of variables f), etc., $I^4\; I^3\; I^2\; I^1$, completely characterize the nature of the determination of the system, in which, in a (part of a) continuum of α dimensions, I^α functions are freely specifiable.

An example:

$$\frac{\partial f_\alpha}{\partial x_\beta} - \frac{\partial f_\beta}{\partial x_\alpha} = 0$$

We have, as a result of the elimination procedure,

$$\frac{\partial f_1}{\partial x_4} - \frac{\partial f_4}{\partial x_1} = 0 \quad \frac{\partial f_1}{\partial x_3} - \frac{\partial f_3}{\partial x_1} = 0 \quad \frac{\partial f_1}{\partial x_2} - \frac{\partial f_2}{\partial x_1} = 0$$

$$\frac{\partial f_4}{\partial x_2} - \frac{\partial f_2}{\partial x_4} = 0 \quad \frac{\partial f_2}{\partial x_3} - \frac{\partial f_3}{\partial x_2} = 0$$

$$\frac{\partial f_3}{\partial x_4} - \frac{\partial f_4}{\partial x_3} = 0$$

Number of identities

$$\not{2} \cdot r^3 + 2r^2 = 4$$

this agrees with:

$$\frac{\partial}{\partial x_\gamma}\left(\frac{\partial f_\alpha}{\partial x_\beta} - \frac{\partial f_\beta}{\partial x_\gamma}\right) + \cdot + \cdot \equiv 0.$$

$$r^4 = 3 \qquad r^3 = 2 \qquad r^2 = 1 \qquad r^1 = 0$$

$$I^4 = 1 \qquad I^3 = 2 \qquad I^2 = 3 \qquad I^4 = 4.$$

Of course, what I have said is not at all new after your beautiful presentation, but merely a synoptic presentation of results.—

As to the connection of your system of equations with mine, I see it as follows. Let us set

$$\Lambda^{\sigma}_{\mu\sigma} = \varphi_{\mu} = \frac{\partial \lg \psi}{\partial x^{\mu}}.$$

The manifold also possesses a metric,

$$g_{\mu\nu} = \psi h_{s\mu} h_{s\nu}.$$

If we construct $R^{\mu\nu}$ from *these* $g_{\mu\nu}$, then

$$2R^{\mu\alpha} - (G^{\mu\alpha} + G^{\alpha\mu}),$$

in any case, contains no more second derivatives of h. But it appears from your expression that quadratic terms still remain. For the special case $\psi = \text{const.}$, using the identity which you sent to me, we obtain

$$2R_{\alpha\beta} - (G_{\alpha\beta} + G_{\beta\alpha}) \equiv 2\Lambda^{\sigma}_{\alpha\rho}\Lambda^{\sigma}_{\beta\rho}.$$

But the right-hand side does *not* vanish for constant ψ. So it appears that the two systems differ from each other in content as well as form. This also comes out in the spherically symmetric case.

Kind regards, yours,

A. Einstein

Translators' note: Translation from *Debever 1979*, pp. 175–181.

238. To Eduard Einstein

[Berlin,] 5 February 1930

Dear Tete,

I'm happy that you can pursue the usual activities again. Because with people, it is like with bikes. Only when it's moving can it comfortably retain balance. As long as you also handle the work well, everything is good. I am also following this prescription, only that I have had to reduce my dose. The Viennese mathematician[1] is already here. We only have an hour together every couple of days, but this is what matters. If you come in March,[2] let's go alone with a girl to the villa in Caputh. It is a proper village and very nicely situated. I am very happy there. By the way, one can get from there to Berlin in an hour. The theory is progressing slowly. But I have the boldest hopes. One of the finest French mathematicians corresponds with me about it a lot;[3] that takes much effort, but brings still more joy.

In one respect, you must actually be glad about the symptoms of your illness: one cannot know anything as deeply as when one experiences it oneself. If you overcome it, you will have the chance of becoming a particularly good shrink. If you come in March, we will probably also be able to take a lovely excursion with Toni.[4] I am currently reading fragments of Democritus[5] with her. Beyond that, I am reading a small book by Oppenheimer about the national economy.[6]

Best wishes from your

Papa

239. To Anton Kunitzer[1]

[Berlin W.,] 6 February 1930

Dear Sir,

If you convert unemployment insurance into a pension for single women,[2] there will just as quickly no longer be officially married women in the poorer class. In general, it is advantageous in itself that, through the simplification of housekeeping, women's work has become available for the actual productive economy. The fault of the current state of affairs lies in the flawed organization of economic life which brings to pass that one must work too much and the others are excluded from productive work.

I also cannot agree with your opinion in regards to taxes. It is not essential, in the long run, from which vein of economic life the means for public functions are siphoned off. This is more a technical question. (Cheap collection.)

Respectfully yours,

240. From Élie Cartan

Le Chesnay (Seine et Oise), 27 avenue de Montespan, 7 February 1930

Dear Sir,

I readily understood your introducing the new integers I_1, I_2, etc., which, in your last letter, you furthermore explained quite clearly. I myself, in my papers, made use of similar quantities, which I denoted as s_1, s_2, etc. What I wanted to tell you is that only one of those integers has a truly intrinsic significance, while the others depend on the analytic form which one gives to the differential system under consideration. This is what I want to say.

Let us take, as an example, the system that you have given. If one constructs the general solution of this system by the recurrence method employed in my theory of systems in involution, that solution depends indeed on I_1 arbitrary functions of one variable, on I_2 arbitrary functions of 2 variables, on I_3 arbitrary functions of three variables, etc. But first of all, in your example,

$$\frac{\partial f_\alpha}{\partial x_\beta} - \frac{\partial f_\beta}{\partial x_\alpha} = 0,$$

it is quite clear, even for someone who is not a mathematical expert, that the general solution depends upon an arbitrary function φ of 4 variables:

$$f_\alpha = \frac{\partial \varphi}{\partial x_\alpha},$$

and that is all: when that function is given, the unknown functions are perfectly determined, and one no longer finds any traces of the $I_2 = 3$ arbitrary functions of 2 variables, etc. which were necessary in the general method of integration.

But one can also demonstrate that the general solution nevertheless depends on more arbitrary functions of 1, 2, and 3 variables than are indicated by the numbers I_1, I_2, I_3. In fact, we *extend* the system, that is to say, we write it in the following form:[1]

$$\frac{\partial f_\alpha}{\partial x_\beta} - f_{\alpha\beta} = 0 \quad\Big|\quad \frac{\partial}{\partial x_\beta} - f_{\alpha\beta} = 0 \; (f_{\alpha\beta} = f_{\beta\alpha})$$

$$\frac{\partial f_{\alpha\beta}}{\partial x_\gamma} - \frac{\partial f_{\alpha\gamma}}{\partial x_\beta} = 0$$

which leads naturally, for the f_α, to the same solution $f_\alpha = \dfrac{\partial\varphi}{\partial x_\alpha}$, the $f_{\alpha\beta} \equiv f_{\beta\alpha}$ being $\dfrac{\partial^2\varphi}{\partial x_\alpha \partial x_\beta}$. Let us write out these equations in order. We obtain

$\frac{\partial f_\alpha}{\partial x_4} - f_{\alpha 4} = 0$ $(\alpha = 1, 2, 3, 4)$	$\frac{\partial f_\alpha}{\partial x_3} - f_{\alpha 3} = 0$ $(\alpha = 1, 2, 3, 4)$	$\left(\frac{\partial f_\alpha}{\partial x_2} - f_{\alpha 2} = 0\right) b$ $(\alpha = 1, 2, 3, 4)$	$\frac{\partial f_\alpha}{\partial x_1} - f_{\alpha 1} = 0$ $(\alpha = 1, 2, 3, 4)$
$\frac{\partial f_{1\alpha}}{\partial x_4} - \frac{\partial f_{4\alpha}}{\partial x_1} = 0$ $(\alpha = 1, 2, 3, 4)$	$\frac{\partial f_{1\alpha}}{\partial x_3} - \frac{\partial f_{3\alpha}}{\partial x_1} = 0$ $(\alpha = 1, 2, 3, 4)$	$\frac{\partial f_{1\alpha}}{\partial x_2} - \frac{\partial f_{2\alpha}}{\partial x_4} = 0$ $(\alpha = 1, 2, 3, 4)$	
$\frac{\partial f_{2\alpha}}{\partial x_4} - \frac{\partial f_{4\alpha}}{\partial x_2} = 0$ $(\alpha = 2, 3, 4)$	$\frac{\partial f_{2\alpha}}{\partial x_3} - \frac{\partial f_{3\alpha}}{\partial x_2} = 0$ $(\alpha = 2, 3, 4)$	[2]	
$\frac{\partial f_{3\alpha}}{\partial x_4} - \frac{\partial f_{4\alpha}}{\partial x_3} = 0$ $(\alpha = 3, 4)$			

$r_4 = 4 + 4 + 3 + 2 = 13$ $\quad r_3 = 4 + 4 + 3 = 11$ $\quad r_2 = 4 + 4 = 8$ $\quad r_1 = 4$

Furthermore, the number of unknown functions is

$$4 \text{ (for the } f_\alpha) + 10 = 14$$

$$I_4 = 14 - r = \underline{1},\ I_3 = 14 - 11 = \underline{3},\ I_2 = 14 - 8 = \underline{6},\ I_1 = 14 - 4 = \underline{10}$$

instead of

$$I_4 = 1, I_3 = 2, I_2 = 3, I_1 = 4.$$

The value of I_4 did not change: it is the only one of the integers which has an essential significance. The values of the other I do not correspond to any intrinsic property of the system, but only to the particular analytic form that one has given it.

If we had taken $I_4 = 0, I_3 > 0$, we would always find $I_4 = 0$, and the same value for I_3.

You will note that, in agreement with the general formula that I mentioned to you, when I extend my system from the first to the second order, the new values of I_1, I_2, I_3, I_4 are

$$\begin{array}{cccc} I_1+I_2+I_3+I_4 & I_2+I_3+I_4 & I_3+I_4 & I_4 \\ 4+3+2+1 = 10 & 3+2+1 = 6 & 2+1 = 3 & 1 \end{array}$$

I have given three talks to the physicists,[3] which seemed to interest them. On the subject of the conditions for compatibility, Mr. Hadamard[4] made two extremely important remarks which perhaps will interest you.

In my theory of systems in involution, it is essential to presume that one is dealing with *analytic* equations, that is, the solutions in one dimension which determine a solution in two dimensions are *analytic*, etc. Otherwise, the theorem of Cauchy-Kowalewsky, according to which, the existence of a solution to the system

$$\frac{\partial f_\alpha}{\partial z} = a_{\alpha\beta}\frac{\partial f_\beta}{\partial x} + b_{\alpha\beta}\frac{\partial f_\beta}{\partial y} + c_\alpha$$

corresponding to given initial values $f_\alpha = \Phi_\alpha(x, y)$ for $z = 0$ essentially supposes that the given $\Phi_\alpha(x, y)$ were *analytic*: in this case, one is assured of having *analytic* functions of x, y, z for the f_α which can be expanded in convergent series. But this theorem can prove to be incorrect if the given functions are *not* analytic; the system may not have *any* solution corresponding to the givens. A simple example is provided by the equation

$$\frac{\partial^2 f}{\partial z^2} + \frac{\partial^2 f}{\partial x^2} + \frac{\partial^2 f}{\partial y^2} = 0;$$

if one sets, for $z = 0$,

$$f = \Phi(x, y)$$

$$\frac{\partial f}{\partial z} = \Psi(x, y),$$

and if $\Phi(x, y)$ is not an analytic function of x, y, it is evident that the equation cannot allow any solution corresponding to those of this kind given, since every harmonic function is analytic. But the theorem would be true for the equation

$$\frac{\partial^2 f}{\partial z^2} - \frac{\partial^2 f}{\partial x^2} - \frac{\partial^2 f}{\partial y^2} = 0$$

In reality, physics has always only considered equations for which the difficulty indicated by Mr. Hadamard does not occur. But this is nevertheless an important question which he raised.

The second observation of Mr. Hadamard is also very interesting. Let us consider a differential system which would satisfy determinism, i.e., such that knowledge of a solution would be determined by the value of the solution within a section $x^4 = a$. It can, however, happen that, if one takes the solution in that section, the solution in a neighboring section could be practically *unobservable* by the physicist: If one takes an infinitely small variation, *as small as one likes*, of the given functions in the section $x^4 = a$, it can happen that, in the section $x^4 = a + \varepsilon$, in the near neighborhood of the first, it may be impossible to limit the amplitude of the variations suffered by the solution. Mr. Hadamard cited, as an example, the case of the equation

$$\frac{\partial^2 f}{\partial x^2} + \frac{\partial^2 f}{\partial t^2} = 0$$

let us take, for $t = 0$,

$$f = \varepsilon \sin mx,$$

$$\frac{\partial f}{\partial t} = 0,$$

,

ε being a very small constant. One would have here

$$f = \sin mx \sinh mt;$$

one can see that, however small t may be, the observed function f at that instant t can take on arbitrarily large values (due to the $\sinh mt$), although, within the section $t = 0$, f remains quite small everywhere.

There would thus be, in certain cases, a *mathematical* determinism which would, properly speaking, not be a *physical* determinism.[5]

The preceding situation does not arise when one considers the hyperbolic equation

$$\frac{\partial^2 f}{\partial x^2} - \frac{\partial^2 f}{\partial t^2} = 0 .$$

More generally, the equation

$$\frac{\partial^2 f}{\partial x^2} + \frac{\partial^2 f}{\partial y^2} + \frac{\partial^2 f}{\partial z^2} - \frac{1}{c^2}\frac{\partial^2 f}{\partial t^2} = 0$$

plays on the following property: If one takes a section of the universe which does not intersect the cone $dx^2 + dy^2 + dz^2 - c^2 dt^2 = 0$, (e.g., $t = a$), there is, for that section of the universe, a true physical determinism; but there is none if the section of the universe intersects the cone (e.g., $z = c^{\text{th}}$).

If there were in physics an equation of the form

$$\frac{\partial^2 f}{\partial x^2} + \frac{\partial^2 f}{\partial y^2} - \frac{\partial^2 f}{\partial z^2} - \frac{\partial^2 f}{\partial t^2} = 0$$

there would be no physical determinism for any section of the universe. Fortunately, equations of this type play no role in physics.

All of this proves that, in the current state of the analysis, one is obliged, in the discussion of systems that are as complicated as yours, to limit oneself to analytic solutions. For the rest, the characteristics being the same as for the equation of the propagation of light, there should not be any overly great fears on the subject of physical determinism.

I did not quite understand what you were intending to say on the subject of the system $R_{ik} = 0$, etc. I have always considered the R_{ik} as referring to the values

$$g_{\mu\nu} = h_{s\mu} h_{s\nu}$$

and not to

$$g_{\mu\nu} = \psi\, h_{s\mu} h_{s\nu} .$$

Furthermore, the function ψ can be introduced only if one takes

$$\frac{\partial \varphi_\alpha}{\partial x^\beta} - \frac{\partial \varphi_\beta}{\partial x^\alpha} = 0;$$

one could, more generally, take

$$\frac{\partial \varphi_\alpha}{\partial x^\beta} - \frac{\partial \varphi_\beta}{\partial x^\alpha} = a(\varphi_\alpha S_\beta - \varphi_\beta S_\alpha),$$

$$\frac{\partial S_\alpha}{\partial x^\beta} - \frac{\partial S_\beta}{\partial x^\alpha} = b(\varphi_\alpha S_\beta - \varphi_\beta S_\alpha)$$

with two arbitrary constants a and b.

Have you thought about the question raised by Francis Perrin,[6] to wit, the possibility of completing the second members of your equations in the following manner:

$$\Lambda^{\mu}_{\alpha\beta;\mu} = 0$$

$$G_{\alpha\beta} = k g_{\alpha\beta},$$

k being a constant which one can further suppose to be equal to 1 or -1. This modification is equivalent to assuming the existence of a privileged unit of intervals.

One finds nothing of a homogeneous solution providing a finite space with this modification, but F. Perrin has found a regular, nonhomogeneous solution everywhere within a spherical space with infinite time.

Please accept, dear and honored Sir, my most sincere regards,

E. Cartan

241. From Willy Scherrer

Bern, Manuelstr. 76, 8 February 1920

[Not selected for translation.]

242. On Tomáš Garrigue Masaryk[1]

[*Einstein 1930x*]

DATED after 10 February 1930
PUBLISHED February/March 1930

IN: *Urania*, February/March 1930, p. 18.

If men of Masaryk's human and intellectual stature had been in charge of the destinies of the European states since the beginning of this century, Europe would have been better off.

A. Einstein

243. Calculations

[around 10 February 1930][1]

[Not selected for translation.]

244. Statement on the League of Nations

"An Völkermagazin"
[*Einstein 1930k*]

DATED 13 February1930
PUBLISHED 1930

IN: Richard Bölcsey, *Ein Jahrzehnt Völkerbund.* Berlin: Verlag Völkermagazin Marquardt & Co, 1930, p. 8.

I am seldom enthusiastic about the acts and omissions of the League of Nations, but am always grateful for its existence.[1][2]

A. Einstein

245. Obituary for Paul Levi

"Albert Einstein über Paul Levi"
[*Einstein 1930l*]

DATED 13 February 1930
PUBLISHED 17 February 1930

IN: *Hohenzollerische Blätter, Hechinger Tagblatt*, 17 February 1930, p. 1.

Berlin, 13 February 1930[1)]

Today, I was among the many who, deeply shaken, lay to rest Paul Levi, the noble son of Hechingen.[1] He was one of the most just, most brilliant and courageous humans that I have encountered in my journey through life. He devoted all his unflagging strength to the defense of the weak and oppressed. He often did this at the risk of his life, always with that self-evidence such as is only inherent in such natures that act from the inner compulsion of an insatiable need for justice.[2] May his memory never die and his shining example keep alive the sense of responsibility of those who come after for the public good.

1) From Prof. Albert Einstein, the great founder of the theory of relativity, we received on Saturday afternoon through special delivery the obituary for Paul Levi reprinted above. The newspaper *Tagebuch* carries two poignant epitaphs for Paul Levi, the first from his colleague Max Alsberg,[3] [and] the second from the Editor of the *Tagebuch*, Leopold Schwarzschild.

246. To Élie Cartan

[Berlin?,] ca. 13 February 1930[1]

Dear Mr. Cartan,

Your letter[2] was, once again, extremely interesting to me. I would have very much liked to have heard your lectures! Now, I must once again try your patience, because I did not quite understand your explanation about the index of generality.

With the example

$$\frac{\partial f_i}{\partial x_k} - \frac{\partial f_k}{\partial x_i} = 0, \qquad \text{... (1)}$$

you want to illustrate that (besides I_n) only the index I_{n-1} (where n is the number of dimensions), which you call the *index de généralité*, has the significance of an invariant; in contrast, the indices I_{n-2} etc., which refer to manifolds of lesser dimensionality, would have no significance as invariants.

You show this by replacing (1) by a system of lower order (φ = arbitrary) and a system of higher order. One then indeed obtains other I_1, I_2 values. But one also obtains another I_3, which is the quantity that corresponds to your *index de généralité*. I am of the opinion that one can demand invariance of the *index de généralité* only insofar as one requires independence from the coordinate system.

(1) yields directly

r_4	r_3	r_2	r_1
3	2	1	0

Also, for the *I*'s

I_4	I_3	I_2	I_1
1	2	3	4

You will now say: One can replace (1) by an arbitrary φ. That is, however, then no longer a system of equations.

Furthermore, you second replace the system by

$$\frac{\partial f_\alpha}{\partial x^\beta} - f_{\alpha\beta} = 0 \quad (\text{where } f_{\alpha\beta} \equiv f_{\beta\alpha})$$

$$\frac{\partial f_{\alpha\beta}}{\partial x^\gamma} - \frac{\partial f_{\alpha\gamma}}{\partial x^\beta} = 0$$

But this is not a genuine replacement. If one, however, inserts the equation in parentheses, $f_{\alpha\beta} - f_{\beta\alpha} = 0$, then one does not obtain a system of the type that we are considering, but rather equations which do not contain differential quotients. That is, however, not permitted. There has thus been no proof that the quantities I_4, I_3, I_2, I_1 have no significance as invariants.[3]

$$g_{\mu\nu,\alpha\alpha} - g_{\mu\alpha,\nu\alpha} - g_{\nu\alpha,\mu\alpha} + g_{\alpha\alpha,\mu\nu} = -\frac{\partial\varphi_\mu}{\partial x_\gamma} - \frac{\partial\varphi_\alpha}{\partial x_\nu}$$

$$h_{\alpha\mu,\nu\nu} - h_{\alpha\nu,\nu\mu} = 0$$

$$\underbrace{h_{\alpha\mu,\alpha,\nu} - h_{\alpha\nu\alpha\nu}} = 0$$

$$h_{\alpha\mu,\alpha} = \psi_{,\mu} \qquad h_{11,1} = \psi_1$$

$$h_{12,1} + h_{22,2} = \psi_{,2}$$

$$h_{13,1} + h_{23,2} + h_{33,3} = \psi_3$$

$$h_{14,1} + h_{24,2} + h_{34,3} + h_{44,4} = \psi_{,4}\ 0$$

$$\begin{matrix} h_{11} & h_{12} & h_{13} & h_{14} \\ & h_{22} & h_{23} & h_{24} \\ & & h_{33} & h_{34} \\ & & & h_{44} \end{matrix} \left| \begin{array}{l} h_{a4} = \lambda \\ h_{44} = \mu \\ h_{11} = \sigma \\ h_{22} = \sigma + \tau,8 \quad h_{23} = \tau,2 \\ h_{33} = \sigma \end{array} \right.$$

$$\underline{h_{\alpha4,\nu\nu}} = 0$$

h_4 h_{44}

$$h_{am,nn} - h_{an,nm} \qquad \varphi^3$$

$$h_{11,nn} - \overbrace{(h_{11,1} + h_{12,2} + h_{13,3})_1} = 0$$

$$h_{12,nn} - (h_{11,1} + h_{12,2} + h_{13,3})_2 = 0$$

$$h_{13,nn} - (\qquad)_3 = 0$$

$$h_0 - \underbrace{(h_{22,2} + h_{33,3})_1}_{0} = 0$$

$$h_{22,nn} - (\qquad) = 0$$

1

0

$$h_{23,nn} - (\qquad) = 0$$

4

0

0

$$0 \qquad - (h_{33,3})_1 = 0$$

$$2 = 0$$

$$h_{33,nn} - (h_{33,3})_3 = 0$$

$$h_{11,nn} - \varphi_{,1} = 0$$

$$h_{12,nn} - \varphi_{,2} = 0$$

$$\varphi_{,nn} = 0$$

$$h_{22} = \frac{x}{r}\psi \left| x_3\psi \frac{\partial\psi}{\partial r_3} \right.$$

$$- \quad h_{23} = -\frac{x_2}{r^3} \left| -x_2\psi - \frac{\partial\psi}{\partial r_2} \right.$$

$$h_{33} = 0$$

247. To Élie Cartan

[Berlin,] 13 February 1930

Dear Mr. Cartan,

How delighted I would have been to have heard your three lectures![1] But I am lucky to have received such interesting mail from you. Every letter is truly a joy for me. But I am still not content with the *I*. Your argument can, in fact, be turned against you.

For you consider I_3 in a four dimensional space to be a measure of generality (and not I_4). But in the system

$$\frac{\partial\varphi_\alpha}{\partial x^\beta} - \frac{\partial\varphi_\beta}{\partial x^\alpha} = 0,$$

I_3 is equal to 2; i.e., two of the φ_α can be freely chosen on a three dimensional section, in spite of the fact that everything is fixed by one arbitrary function. (In your example, your representation $\begin{matrix}\dfrac{\partial\varphi_\alpha}{\partial x^\beta} = \varphi_{\alpha\beta} \\ \varphi_{\beta,\gamma} = \varphi_{\gamma,\beta}\end{matrix}$ $(\varphi_{\alpha\beta} \equiv \varphi_{\beta\alpha})$ is actually $I_3 = 3$).

But this does not matter. For such formulation will be right in which the $I_4\ I_3\ I_2\ I_1$ are as small as possible. Thus, I am still of the opinion that generality can only be adequately described by all 4 numbers. (Do not be angry with me about my stubbornness; I can do no different.) I am very glad that I now fully understand your theory of systems in involution. The most beautiful part of it, I think, is your proof of the existence of identities. I look forward to your detailed article, which will finally clear up an important question.

The first of Hadamard's remarks interested me very much, but has not fully convinced me.[2] In particular, in appears to me that, by means of a discontinuous distribution of masses *in the surface* x_3 = const., one can produce functions $f\left(\frac{\partial^2 f}{\partial x^2} + \frac{\partial^2 f}{\partial y^2} + \frac{\partial^2 f}{\partial z^2} = 0\right)$ which possess arbitrary discontinuities in the surface x_3 = const. (e.g., by means of double layers

$$\begin{array}{c|c} & 1\ - \\ + & 1\ - \\ + & 1\ - \\ + & 1\ - \\ & 1\ - \end{array} \qquad x_3 = a$$

The second remark did not really surprise me.[3]

I knew, of course, that, in your system of equations, the R_{ik} are to be constructed out of the $h_{s\mu}h_{s\nu}$, and not out of $\psi h_{s\mu}h_{s\nu}$. I only wanted to say that the R_{ik} built of the latter quantities as $g_{\mu\nu}$ s appears to be closely related to my $G^{ik} + G^{ki}$.

Has it become clear to you why your system is not physically plausible to me? It is because the g_{ik} are already deterministically fixed by R_{ik} without the remaining 6 field variables entering in. They appear to be added a posteriori, with no impact on the g_{ik}. Is this clear?

F. Perrin's system with the cosmological term could perhaps earn some support.[4] But, for the present, I have no opinion on it. At this moment, the burning question is that of the existence of singularity-free solutions which could represent electrons and protons.[5] For, without the solution of this difficult problem, I feel no judgment can be passed on the usefulness of the theory. For the moment, this theory seems to me to be like a starved ape who, after a long search, has found an amazing coconut but cannot open it; so he does not even know whether there is anything inside.

Kind regards, your

A. Einstein.

Traslators' note: Translation based on *Debever 1979*, pp. 194–197.

248. To Marie Curie-Skłodowska

[Berlin?, between 14 February and 5 March 1930][1]

Hopefully, you submitted a *formal application* for Mr. Rosenblum with the research corporation.[2] I obviously cannot do this; you alone are the competent personage in this case. I can only *support* the application.

I am not submitting any proposals concerning the reform of the Institute;[3] rather, unfortunately, the German national commission, which ⟨sadly⟩ demonstrated a regrettable lack of political tact with this.[4] Mr. Krüss[5] drafted it and surprised the commission with his suggestions. In my opinion, his goal in itself is actually good, but the matter will cause bad blood. I'll send you the proposal *in confidence* so that you can consider with your friends what is to be done.[6] Mr. Krüss would not have dared to do this *against the express will of the Foreign Office* if he did not have Mr. Dufour[7] behind him and if he would not have known of the gossip against ⟨Mr.⟩ Luchaire in Paris and Geneva.

The fundamental error was that the commission accepted the Paris Institute, which only raised suspicion among the English, Americans, and Germans. The other states would have long since donated the necessary funds for a permanent organ of the commission in Geneva. The vote of the German commission will unfortunately lead to the deplorable prestige point of view being brought to the fore (in France). ⟨Otherwise⟩ A proposal for the relocation of the Institute to Geneva and for an equal participation of the different states in its financing should have come from France.[8]

Kind regards, your

249. From Myron Mathisson

Warsaw, 14 February 1930

[Not selected for translation.]

250. From Paul Ehrenfest

Leiden, 15 February 1930

[Not selected for translation.]

251. Letter to the German Pro-Palestine Committee[1]

"Einstein's Brief"
[*Einstein 1930n*]

DATED 16 February 1930
PUBLISHED 21 February 1930

IN: *Jüdische Rundschau*, 21 February 1930, p. 98.

Berlin W., 16 February 1930

Dear Count,[2]

To my great regret, I am unable to accept to your kind invitation to take part in the rally of the Pro-Palestine Committee in Hamburg.[3] However, I ask you to convey my friendly regards to the assembly.

The Pro-Palestine Committee, which has set for itself the task of organizing the sympathies that the work of the construction of Jewish Palestine enjoys within the German public, is an encouragement and reinforcement to all of us who are concerned about the future of the work in Palestine. The construction of the Jewish national home in Palestine is a colonial, political, and cultural experiment that rightly garners public interest. For us Jews, however, this construction is in no way a purely economic matter, nor an object of charity, but rather an event of the greatest importance within Jewish history. In the two millennia since its dispersal, but especially in the last hundred years, the Jewish people have experienced a radical change in shape. The weakening of the Jewish community, which has led to the isolation and atomization of the individual,[4] has robbed the Jew of the inner stability that the community offers. However, the desire to assimilate and the inner dependence arising from it have not protected Jews from embittered attacks. As the Jews once again become aware of their nationality and build a home for themselves, they will acquire the very self-respect that is necessary to undertake collective cultural tasks.[5]

The Jews today face a cultural task of the first order: With the consent of the nations organized in the League of Nations,[6] they have begun to establish a Jewish national home in Palestine. I do not need to enumerate in detail for you what was achieved in Palestine in recent years. It is a work that no nation of culture should be ashamed of.

Jewry has not closed themselves from the demands that this work places upon them. At the time of Theodor Herzl, at the end of the previous century, only the small flock of Zionists, often derided at the time, headed undeterred for the goal of the national home; in August 1929, representatives from almost all of the world's Jewish groups have congregated in Zurich in order to form the organ of the enlarged Jewish Agency that, in the future, will be responsible for the creation of the Jewish national home.[7] Because the work of colonization in the future, in order to be able to settle as many immigrants as possible, must proceed at a faster pace, it also requires stronger support for the work on the part of the world's Jewry. The desperate situation of Jews in most countries in eastern Europe[8] and the total ban on immigration to the majority of countries overseas make increasing Palestine's capacity for absorbing Jewish immigration a vital necessity.

Recently, bloody riots[9] have sparked political debate. The Jewish people are well aware that one of the most difficult, but perhaps also the most important, tasks consists of reconciling their national aspirations with the rights and needs of the land's population. In comparison to other countries, Jewish immigration has brought Palestine great prosperity and a high level of civilization,[10] but the Palestinian Arabs, like the other peoples of the Orient today, are also making rapid progress and setting national goals for themselves. Of course, Jewish integration can only flourish in an atmosphere of calm and good will, and we Jews therefore demand that the mandate authorities fulfill their duties,[11] above all the maintenance of order, better in the future than up to now. However, we are, at the same time determined, to find a path of understanding with Arab neighbors, a path that replaces enmity and rivalry with a system of creative cooperation that is in the interest of both peoples and therefore must come about sooner or later.

As I wish your rally the best success and hope that it will win new friends for the work in Palestine, I remain, dear Count,

Your completely devoted,

A. Einstein

252. To Issai Schur

[Berlin,] 17 February 1930

[Not selected for translation.]

253. From Élie Cartan

Le Chesnay (Seine et Oise), 27 avenue de Montespan, 17 February 1930

Dear Sir,

I am very proud that my letters can be of interest to you. You may be sure that for my part, I regard it as a privilege that you are willing to devote to me some of your moments, so precious to science.

I think that, by a series of successive approximations, we shall arrive at the point of understanding each other, even on the subject of the *I*. The argument that you have sent me against myself was quite amusing, but I refuse to accept it![1] In fact, I_3 measures, for me, the degree of generality *in the case that* I_4 *is zero*, and that is the case in your system and in analogous systems. Otherwise, it is the entire I_4 that indicates the number of arbitrary functions of four variables which measure the degree of generality of the solution: if $I_4 > 0$, the entire I_3 no longer has any essential significance. If $I_4 = 0$, it is I_3 that I take as the measure (I_3 arbitrary functions of 3 variables). If $I_4 = I_3 = 0$, it is I_2: the general solution then depends upon I_2 arbitrary functions of two variables and so on. If $I_4 = I_3 = I_2 = I_1 = 0$, the solution of the system depends on a certain number I_0 of arbitrary *constants*, and that number, in this case, is also the measure of the degree of indetermination of the solution.

For the rest, I am rather tempted to admit that you are right if it is a question of comparing two systems of a similar analytic form, including, for example, the same number of equations with partial derivatives of the same order and the same number of unknown functions, as is the case for the two possible systems of 22 equations.

I have clearly understood the reasons for which you reject the system with the $R_{ik} = 0$; but I believe that, in one of your recent letters, you had revised your opinion to some extent, due to the presence of the equations $\frac{\partial S_\alpha}{\partial x^\beta} - \frac{\partial S_\beta}{\partial x^\alpha} = \ldots$ which cause the metric and the parallelism to enter. But evidently, this latter circumstance does not in any way weaken the arguments that you had given and have taken the trouble of summarizing again for me.

Hadamard's first comment, concerning the equation $\frac{\partial^2 f}{\partial x^2} + \frac{\partial^2 f}{\partial y^2} + \frac{\partial^2 f}{\partial z^2} = 0$, does not refer to *discontinuous* variables in the section $z = 0$. In a precise sense, there exists *no* function f that satisfies the equation in question and is regular on the interval $-\varepsilon < z < \varepsilon$, if one requires that function to assume the value $f = \varphi(x, y)$ at $z = 0$, where the function φ is *continuous*; and this even allowing that the derivatives of the first two orders are continuous, but *not analytic*. This derives quite simply from the fact that every regular harmonic function f in a given domain is necessarily analytic within that domain (analytic meaning here that it can be expanded in a power series); as a result, the value taken by f on the *analytic* surface $z = 0$ is necessarily an analytic function of x, y and thus cannot be reduced to $\varphi(x, y)$.

I understand quite well that the solution found by F. Perrin,[2] as well as the isolated solutions which I have found and which lack singularities, can be of no help to you. The entire problem consists in finding a solution without singularity that is sufficiently general so that it can be interpreted physically. But is it quite certain that such solutions exist,[3] and that the coconut contains anything in its interior? One is standing in front of a wall, and the mathematicians are having a difficult time in piercing an opening. One can hardly base one's hopes on a miracle of divination; but you have already experienced several of them!

Please accept, dear and illustrious Master, the expression of my most sincere regards,

E. Cartan

254. To the Chairman of the District Court of Sofia

[Berlin,] 20 February 1930

Since I have been proposed as a witness by the defendants in the Anti-War Committee trial, I would like to allow myself to hereby make the follow statement in writing:

(1) I am convinced that a constant danger of war can actually be spoken of as long as no international guarantees against war are won. In particular, as is well known, there is currently a danger of war that has its origin in the conflicting interests of certain governments and circles toward Russia.[1]

(2) The legal fight against the threat of war is permitted in all truly civilized countries, but it is a meritorious beginning everywhere. Persecution of opponents of war is a disgrace to a modern state and a kind of confession that the public authority favors military goals.

P.S. I'm enclosing here a letter from Finland's minister of war[2] in order to show how a culturally superior country deals with the problem of conscientious objection in an exemplary way.

255. Two Rigorous Static Solutions to the Field Equations of the Unified Field Theory

"Zwei strenge statische Lösungen der Feldgleichungen der einheitlichen Feldtheorie"
[*Einstein and Mayer 1930*]

PRESENTED 20 February 1930
PUBLISHED 11 March 1930

IN: *Preußische Akademie der Wissenschaften* (Berlin). *Physikalisch-mathematische Klasse. Sitzungsberichte* (1930): 110–120.

In the following, we deal with two special cases:[1]

(a) The spatially centrally symmetric (rotationally symmetric) case, which exhibits at the same time mirror symmetry.

Considered physically, this case represents the external field of an electrically charged sphere of nonvanishing mass.

(b) The static solution, corresponding to an arbitrary number of uncharged mass points at rest.

Remark: The developments in §1 up to equation (27) contain only the strict mathematical demonstration that, in the case of central symmetry and spatial mirror symmetry, with a suitable choice of the coordinates, the $h_s{}^{\alpha}$ assume the form given in (27).[2]

§1. The Spatially Centrally Symmetric Case.

We are seeking the most general three-dimensional continuum,

$$x_1, x_2, x_3, h_s{}^{\alpha}(x_1, x_2, x_3)\,,\ s, \alpha = 1, 2, 3\,,$$

which shows the property of rotational symmetry, i.e., it exhibits invariance with respect to the group

$$\bar{x}_{\alpha} = a_{\alpha\beta}x_{\beta}, \quad \alpha, \beta = 1, 2, 3\,,\text{[3]} \qquad (1)$$

where $\|a_{\alpha\beta}\|$ is an orthogonal matrix.

Employing (1), the point $P(x_1, x_2, x_3)$ is transformed into the point $P(\bar{x}_1, \bar{x}_2, \bar{x}_3)$, and the normalized triad $h_s{}^{\alpha}(x)$ at the point P is transformed into the triad

$$\bar{h}_s^{\ \alpha}(\bar{x}) = a_{\alpha\beta}h_s^{\ \beta}(x), \quad s, \alpha, \beta = 1, 2, 3 \qquad (2)$$ [4]

at the point $\bar{P}$.

Necessary and sufficient for rotational symmetry now is the existence of a "local rotation" which is the same at all points of R_3 (a rotation of the local triads), through which the triad $h_s^{\ \alpha}(\bar{x})$ is converted back into the original triad:

$$\bar{h}_s^{\ \alpha}(\bar{x}) = A_{st}h_t^{\ \alpha}(\bar{x}), \quad s, t, \alpha = 1, 2, 3\,. \qquad (3)$$

Let R_3 behave at infinity as if it were Euclidean, i.e., if x_1, x_2, x_3 tends toward infinity (at least one of the three coordinates),[5] then $h_s^{\ \alpha}(x)$ converges toward $\delta_{s\alpha}$. For brevity, we write $h_s^{\ \alpha}(\infty) = \delta_{s\alpha}$.

At infinity, it follows from (2) that $\bar{h}_s^{\ \alpha}(\bar{x}) = a_{\alpha s}$, and thus, from (3), that $a_{\alpha s} = A_{s\alpha}$. Instead of (3), we then have

$$\bar{h}_s^{\ \alpha}(\bar{x}) = a_{ts}h_t^{\ \alpha}(\bar{x}), \quad s, t, \alpha = 1, 2, 3\,, \qquad (3')$$

which, compared to (2), yields

$$a_{\alpha\beta}h_s^{\ \beta}(x_1, x_2, x_3) = a_{ts}h_t^{\ \alpha}(a_{1j}x_j, a_{2j}x_j, a_{3j}x_j)\,, \alpha, \beta, t, s = 1, 2, 3, 4 \qquad (4)$$

as the functional equation for the triad components that we are seeking. The relations (4) are identities for the quantities $x_1, x_2, x_3, a_{\alpha\beta}$, as long as the matrix $\|a_{\alpha\beta}\|$ is orthogonal.

We now consider the point $P(x_1, x_2, x_3)$ and choose for $a_{\alpha\beta}$ the triad:

$$a_{\alpha\beta} = {}_{(\alpha)}\xi_\beta\,, \ \alpha, \beta = 1, 2, 3\,, \qquad (5)$$

which, due to $a_{\alpha\beta}a_{\alpha\gamma} = {}_{(\alpha)}\xi_{\beta(\alpha)}\xi_\gamma = \delta_{\beta\gamma}$, has a Euclidean normalization, whereby we posit

$${}_{(1)}\xi_\alpha = \frac{x_\alpha}{s}, \ s^2 = x_\alpha x_\alpha, \quad \alpha = 1, 2, 3\,. \qquad (5')$$

For this choice of the matrix $\|a_{\alpha\beta}\|$, relation (4) gives

$${}_{(\alpha)}\xi_\beta h_s^\beta(x_1, x_2, x_3) = {}_{(t)}\xi_s h_t^\alpha(s, 0, 0), \quad s, t, \alpha = 1, 2, 3\,. \qquad (6)$$

We shift (6) with ${}_{(a)}\xi_\gamma$ and obtain

$$h_s^{\ \gamma}(x_1, x_2, x_3) = {}_{(t)}\xi_{s(\alpha)}\xi_\gamma h_\tau^{\ \alpha}(s) = {}_{(1)}\xi_{s(1)}\xi_\gamma h_1^{\ 1}(s) \qquad (7)$$
$$+ {}_{(1)}\xi_{s(\alpha)}\xi_\gamma h_1^{\ \alpha}(s) + {}_{(1)}\xi_{\gamma(t)}\xi_s h_t^{\ 1}(s) + {}_{(t)}\xi_{s(\alpha)}\xi_\gamma h_t^{\ \alpha}(s),$$

where the summation in the second line refers only to the indices 2 and 3.

Instead of $h_t^{\alpha}(s, 0, 0)$, we posit $h_t^{\alpha}(s)$. Given (5), we can write (7) also as

$$h_s^{\gamma}(x_1, x_2, x_3) = \frac{x_s x_\gamma}{s^2} h_1^{\ 1}(s) + \frac{x_s}{s} {}_{(\alpha)}\xi_\gamma h_1^{\ \alpha}(s) \quad . \tag{8}$$
$$+ \frac{x_\gamma}{s} {}_{(t)}\xi_s h_t^{\ 1}(s) + {}_{(t)}\xi_{s(\alpha)}\xi_\gamma h_t^{\ \alpha}(s)$$

We now make use of the indeterminacy of the specification of the vectors, ${}_{(2)}\xi_\alpha$, ${}_{(3)}\xi_\alpha$ which must form a Euclidean-normalized triad with ${}_{(1)}\xi_\alpha$.

If in (8), instead of the chosen dyad ${}_{(2)}\xi_\alpha$, ${}_{(3)}\xi_\alpha$, we introduce a dyad which is rotated by an angle ϕ, ${}_{(2)}\eta_\alpha$, ${}_{(3)}\eta_\alpha$

$$\left.\begin{aligned} {}_{(2)}\xi_\alpha &= \cos\phi_{(2)}\eta_\alpha + \sin\phi_3\eta_\alpha \\ {}_{(3)}\xi_\alpha &= -\sin\phi_{(2)}\eta_\alpha + \cos\phi_{(3)}\eta_\alpha \end{aligned}\right\} \tag{9}$$

then we obtain a new representation of the triad $h_s^{\gamma}(x_1, x_2, x_3)$, which contains the arbitrary angle ϕ. This representation has the form:

$$\begin{aligned} h(x_1, x_2, x_3) &= P_{(s\gamma)} + Q_{(s\gamma)}\sin\phi + R_{(s\gamma)}\cos\phi \\ &+ S_{(s\gamma)}\cos^2\phi + T_{(s\gamma)}\sin\phi\cos\phi. \end{aligned} \tag{10}$$

Since (10) holds for any arbitrary value of ϕ, it follows that]

$$h_s^{\gamma}(x_1, x_2, x_3) = P_{(s\gamma)}, Q_{(s\gamma)} = R_{(s\gamma)} = S_{(s\gamma)} = T_{(s\gamma)} = 0. \tag{11}$$

If we carry out this simple calculation, $Q_{(s\gamma)} = 0$, $R_{(s\gamma)} = 0$ yields

$$h_2^{\ 1}(s) = h_3^{\ 1}(s) = h_1^{\ 2}(s) = h_1^{\ 3}(s) = 0. \tag{12}$$

From $S_{(s\gamma)} = 0$, $T_{(s\gamma)} = 0$, it again follows that

$$h_2^{\ 2}(s) = h_3^{\ 3}(s),\ h_2^{\ 3}(s) = -h_3^{\ 2}(s). \tag{13}$$

Due to (12) and (13), (8) becomes

$$\begin{aligned} h_s^{\gamma}(x_1, x_2, x_3) &= \frac{x_s x_\gamma}{s^2} h_1^{\ 1}(s) + h_2^{\ 2}(s)[{}_{(2)}\xi_{s(2)}\xi_\gamma] \\ &+ h_2^{\ 3}(s)[{}_{(2)}\xi_{s(2)}\xi_\gamma]. \end{aligned} \tag{14}$$

Now, ${}_{(2)}\xi_{s(2)}\xi_\gamma + {}_{(3)}\xi_{s(3)}\xi_\gamma = \delta_{(s\gamma)} - {}_{(1)}\xi_{s(1)}\xi_\gamma$, independently of the particular choice of the normalized dyad ${}_{(2)}\xi_s$, ${}_{(3)}\xi_s$. In contrast, on commutation of the vectors ${}_{(2)}\xi_s$, ${}_{(3)}\xi_s$, the quantity ${}_{(2)}\xi_{s(3)}\xi_\gamma - {}_{(3)}\xi_{s(2)}\xi_\gamma$ changes its sign.

If, however, we admit only the transformations (1) for which the determinant of the matrix $\|a_{ik}\|$ has the value +1, then ${}_{(2)}\xi_{s(3)}\xi_{\gamma} - {}_{(3)}\xi_{s(2)}\xi_{\gamma}$ is also independent of the particular choice of the dyad. (Then, $|{}_{(\alpha)}\xi_{\beta}|$ must equal 1, with $\alpha, \beta = 1, 2, 3$, which determines which of the two vectors ${}_{(2)}\xi_{\alpha}$, ${}_{(3)}\xi_{\alpha}$ counts as second or as third, respectively. Such transformations are in fact called rotations.

If we now introduce the alternating tensor $\varepsilon_{\alpha\beta\gamma}$ with $\varepsilon_{123} = 1$, then we have ${}_{(2)}\xi_{s(3)}\xi_{\gamma} - {}_{(3)}\xi_{s(2)}\xi_{\gamma} = \varepsilon_{s\gamma\tau(1)}\xi_{\tau}$, and, instead of (14), we can write

$$h_s^{\ \gamma}(x_1, x_2, x_3) = x_s x_\gamma\, A(s) + \delta_{s\gamma}\, B(s) + \varepsilon_{s\gamma\tau} x_\tau C(s)\,, \tag{15}$$

where

$$A(s) = \frac{1}{s^2}(h_1^{\ 1}(s) - h_2^{\ 2}(s))\,,\ B(s) = h_2^{\ 2}(s)\,,\ C(s) = h_2^{\ 3}(s)\cdot\frac{1}{s} \tag{15'}$$

are arbitrary functions of s, which must only correspond to the condition $h_s^{\ \gamma}(\infty) = \delta_{s\gamma}$.

This necessary form (15) of the triad components is, as a simple calculation shows, also sufficient for the rotational symmetry of R_3.

However, we must also set $C(s) = 0$ whenever also irregular rotations (1) ($|a_{\alpha\beta}| = -1$, "mirror reflections") are allowed.

In the following, this will be the only case that we consider, so that (15) with $C(s) = 0$ will then represent the most general form of the triad components.

We now extend our continuum $x_1, x_2, x_3, h_s^{\ \alpha}(x_1, x_2, x_3)$ to a four-dimensional space, by associating the points x_1, x_2, x_3, x_4 to a tetrad $h_s^{\ \alpha}(x_1, x_2, x_3, x_4)$ with $s, \alpha = 1, \ldots 4$ in such a way that

$$h_s^{\alpha}(x_1, x_2, x_3, x_4) = h_s^{\alpha}(x_1, x_2, x_3), \quad s, \alpha = 1, 2, 3\,, \tag{16}$$

and the other vector components which likewise only depend upon x_1, x_2, x_3 are to be determined in such a manner that R_4 is invariant with respect to the group

$$\bar{x}_\alpha = a_{\alpha\beta} x_\beta, \quad \alpha, \beta = 1, 2, 3 \quad \bar{x}_4 = x_4\,. \tag{17}$$

This R_4 is taken to have a *pseudo-Riemannian* structure, i.e., the metric tensor $g^{\alpha\beta}$ in the normalized tetrad $h_2^{\ \alpha}(x_1, \ldots x_4)$ has the representation[1)]

[1)]Here, we dispense with the introduction of imaginary quantities to procure a definite metric tensor.

$$g^{\alpha\beta} = h_1{}^{\alpha} h_1{}^{\beta} + h_2{}^{\alpha} h_2{}^{\beta} + h_3{}^{\alpha} h_3{}^{\beta} - h_4{}^{\alpha} h_4{}^{3}, \quad \alpha, \beta = 1, \ldots 4 \ . \quad (18)$$

At infinity, $h_s{}^{\alpha}(\infty) = \delta_{s\alpha}$ would hold again.

The transformation (17) takes the tetrad $h_s{}^{\alpha}(x)$ at the point $P(x_1, \ldots x_4)$ into the tetrad

$$\bar{h}_s{}^{\alpha}(\bar{x}) = a_{\alpha\beta} h_s{}^{\beta}(x), \ \alpha, \beta = 1, 2, 3, \ \bar{h}_s{}^{4}(x) = h_s{}^{4}(x), \ s = 1, \ldots 4(19)$$

and now we suppose that there is a local back-rotation, so that

$$\bar{h}_s^{\alpha}(\bar{x}) = B_{st} h_t^{\alpha}(\bar{x}), \qquad s, t, \alpha = 1, 2, 3, 4 \qquad (20)$$

holds, where the B_{st} are constants.

From the behavior at infinity, it follows (19) $\bar{h}_s{}^{\alpha}(\infty) = 0$, $\alpha = 1, 2, 3$, and, furthermore, $\bar{h}_s{}^{4}(\infty) = h_s{}^{4}(\infty) = \delta_{24}$. Setting this into (20) yields

$a_{\alpha s} = B_{s\alpha}$, $s, \alpha = 1, 2, 3$, $B_{4\alpha} = 0$, $\alpha = 1, 2, 3$, and $\delta_{s4} = B_{s4}$.

Because of the choice of the tetrad (16), the relations (19) and (20) are fulfilled apart from

$$\bar{h}_s{}^{4}(\bar{x}) = h_s{}^{4}(x) = a_{ts} h_t{}^{4}(\bar{x}), \quad s, t = 1, 2, 3 \ , \qquad (21)$$

$$\bar{h}_4{}^{4}(\bar{x}) = h_4{}^{4}(x) = h_4{}^{4}(\bar{x}) \ , \qquad (21')$$

$$\bar{h}_4{}^{\alpha}(\bar{x}) = a_{\alpha\beta} h_4{}^{\beta}(x) = h_4{}^{\alpha}(\bar{x}), \quad \alpha = 1, 2, 3 \ , \qquad (21'')$$

which are the functional equations for the remaining tetrad components.

These functional equations are solved using the method that was applied to (6), obtaining

$$h_s{}^{4}(x_1, x_2, x_3) = D(s) x_s, \quad s = 1, 2, 3 \ , \qquad (22)$$

$$h_4{}^{\alpha}(x_1, x_2, x_3) = E(s) x_{\alpha}, \quad \alpha = 1, 2, 3 \ , \qquad (22')$$

$$h_4{}^{4}(x_1, x_2, x_3) = F(s) \ . \qquad (22'')$$

Since, at infinity, $\bar{h}_s{}^{\alpha}(\infty) = \delta_{\alpha s}$ must still hold, for the functions occurring in (15) and (22), at $s = \infty$, the following expansions apply:

$$\left.\begin{aligned} A(s) &= \frac{K}{s^a}(1+(\bullet)),\ a>2,\ B = 1+(\bullet),\ F = 11+(\bullet), \\ C, D, E &= \frac{K}{s^b}(1+(\bullet)), \qquad (b>1), \end{aligned}\right\} \quad (23)$$

where the parentheses $(\bullet)$ denote the factor $\frac{1}{s}$.

In the coordinate system

$$\bar{x} = \phi(s)x_i, \quad i = 1, 2, 3. \quad \bar{x}_4 = x_4 \quad (24)$$

in which the triad $x_1, x_2, x_3, h_s^{\ \alpha}(x_1, x_2, x_3)$, $a = 1, 2, 3$ likewise exhibits rotational symmetry, by a suitable choice of the function ϕ:

$$\phi = e^{-\int \frac{As}{B+As^2}ds}, \quad (25)$$

the term corresponding to the function A in (15) can be made to vanish. Since ϕ approaches a finite value at infinity, the conditions defined in (23) hold for the new values $\bar{B}(s), \bar{F}(s), \bar{C}(s), \bar{D}(s), \bar{E}(s)$, as a simple calculation shows.

By means of the additional coordinate change

$$\bar{x}_4 = x_4 + \psi(s), \quad (26)$$

we can cause the function $D(s)$ which occurs in (22) to again vanish, whereby the relations (23) remain valid in the new coordinate system.

Thus, without a loss of generality, we can assume that $A(s) = D(s) = 0$.

Furthermore, we assume that our triad $x_1, x_2, x_3, h_s^{\alpha}(x_1, x_2, x_3)$ is also invariant with respect to a mirror reflection; this assumption entails the vanishing of the function $C(s)$ in (15). The tetrads (15) and (22) then obtain their most general form

$$\left.\begin{aligned} h_s^{\ \alpha} &= \lambda(s)\delta_{s\alpha},\ \alpha, s = 1, 2, 3, \qquad & h_s^{\ 4} &= 0,\ s = 1, 2, 3, \\ h_4^{\ \alpha} &= \tau(s)x_\alpha,\ \alpha = 1, 2, 3, & h_4^{\ 4} &= \mu(s), \end{aligned}\right\} \quad (27)$$

where we have renamed the functions which still occur here.

We will now search for the solutions of the field equations $G^{\mu\alpha} = 0$ *and* $F^{\mu\alpha} = 0$ [6] *within the unified field theory having the form of (27).*

We denote the covariant tetrad adjunct to $h_s^{\ \alpha}$, defined by the system

$$h_s^{\ \alpha} k_{s\beta} = \delta_\beta^{\ \alpha}, \quad s, \alpha, \beta = 1, \ldots 4, \tag{28}$$

by $k_{s\beta}$, $s, \beta = 1, \ldots 4$, (with $k_{s\alpha} = h_{s\alpha}$ for $s = 1, 2, 3$ and $k_{4\alpha} = -h_{4\alpha}$); then, from (27), it has the components:

$$\left.\begin{array}{ll} k_{s\alpha} = \frac{1}{\lambda}\delta_{s\alpha}, \ \alpha, s = 1, 2, 3, & k_{s4} = -\frac{\tau}{\mu\lambda}x_s, \ s = 1, 2, 3 \\ k_{4\alpha} = 0, \ \pi = 1, 2, 3 & k_{44} = \frac{1}{\mu} \end{array}\right\} \tag{29}$$

Now, we write down the formulas that will be used in the following:

The $\sum_{s=1}^{3} \frac{\partial h_s^{\ l}}{\partial x_k} k_{si} - \frac{\partial h_4^{\ l}}{\partial x_k} k_{4i}$ [7] are to be calculated as follows:

$$\left.\begin{array}{l} \Delta_{i4}^{\ l} = 0, \quad i, l = 1, \ldots 4 \\ \Delta_{ik}^{\ l} = -\frac{\partial ln\lambda}{\partial x_k}\delta_{il}, \quad i, k, l = 1, \ldots 3, \quad \Delta_{ik}^{\ 4} = 0, \quad i, k = 1, \ldots 3, \\ \Delta_{4k}^{\ l} = -\frac{\lambda}{\mu}\frac{\partial}{\partial x_k}\left(\frac{\tau}{\lambda}x_l\right) = -\frac{\lambda}{\mu}\frac{\partial}{\partial x_l}\left(\frac{\tau}{\lambda}x_k\right), \quad k, l = 1, \ldots 3 \\ \Delta_{4k}^{\ 4} = -\frac{\partial ln\lambda}{\partial x_k}\mu, \quad k = 1, \ldots 3. \end{array}\right\} \tag{30}$$

From this, it follows for the quantities $\Lambda_{ik}^{\ l} = \Delta_{ik}^{\ l} - \Delta_{ki}^{\ l}$, $i, k = 1, \ldots 4$:

$$\left.\begin{array}{l} \Lambda_{ik}^{\ l} = \frac{\partial ln\lambda}{\partial x_k}\delta_{kl} - \frac{\partial ln\lambda}{\partial x_k}\delta_{il}, \quad i, k, l = 1, \ldots 3, \\ \Lambda = \frac{\lambda}{\mu}\frac{\partial}{\partial x_i}\left(\frac{\tau}{\lambda}x_l\right), \quad i, l = 1, \ldots 3, \\ \Lambda = 0, \quad i, k = 1, \ldots 3, \\ \Lambda = -\frac{\partial ln\mu}{\partial x_k}, \quad k = 1, \ldots 3, \end{array}\right\} \tag{31}$$

Furthermore, we require the contravariant metric tensor, whose system of components has the following form:

$$\left.\begin{aligned} g^{\alpha\beta} &= \lambda^2\delta_{\alpha\beta} - \tau^2 x_\alpha x_\beta, \quad \alpha, \beta = 1, \ldots 3, \\ g^{4\alpha} &= -\mu\tau x_\alpha, \\ g^{44} &= -\mu^2. \end{aligned}\right\} \tag{32}$$

We first solve the system of the field equations $F^{\mu\nu} \equiv \Lambda^{\alpha}_{\underline{\mu}\underline{\nu};\alpha} = 0$ or $\phi_{\mu,\nu} - \phi_{\nu,\mu} = 0$,[8] where ϕ_μ denotes $\Lambda^{\alpha}_{\mu\alpha}$.

Due to (31), we have

$$\phi_i = \Lambda^{\alpha}_{i\alpha} = \xi_i\left(\frac{\mu'}{\mu} + 2\frac{\lambda'}{\lambda}\right), \quad \xi_i = \frac{x_i}{s}, \quad i = 1, 2, 3, \tag{33}$$

and

$$\phi_4 = \Lambda^{\alpha}_{4\alpha} = -\frac{\lambda}{\mu}\frac{\partial}{\partial x_\alpha}\left(\frac{\tau}{\lambda}x_\alpha\right), \quad \alpha = 1, 2, 3. \tag{33'}$$

The system $\phi_{i,k} - \phi_{k,i} = 0, \quad i, k = 1, 2, 3$ is fulfilled identically; only $\phi_{i,4} - \phi_{4,i} = 0$ (or, since $\phi_{1,4} = 0$, $\phi_{4,i} = 0$) remains, that is

$$\phi_4 = \text{constant}. \tag{34}$$

According to (33′), this yields the equation

$$\frac{\lambda}{\mu}\left[\left(\frac{\tau}{\lambda}\right)' s + 3\frac{\tau}{\lambda}\right] = k, \ k = \text{constant}, \tag{35}$$

which can also be written as

$$\left(\frac{\tau}{\lambda}s^3\right)' = k\frac{\mu}{\lambda}s^2 \tag{36}$$

and integrated, giving:

$$\frac{\tau}{\lambda}s^3 = k\int\frac{\mu}{\lambda}s^2 ds + k_1, \quad k_2 = \text{const.} \tag{36'}$$

At infinity, λ, μ, and τ have the expansions $\lambda = 1 + (\bullet)$, $\mu = 1 + (\bullet)$, and $\tau = \frac{c}{s^b}(1 + (\bullet))$, $b > 1$ with which, according to (36′), it follows that $k = 0$, i.e.,

$$\tau = e\frac{\lambda}{s^3}, \quad e = \text{const.} \tag{37}$$

(We have set e for k_1 here.)

This completes the system $F^{\mu\nu} = 0$.

We now treat the other system of field equations,

$$G^{\mu\alpha} \equiv \Lambda^{\alpha}_{\underline{\mu}\underline{\nu};\nu} - \Lambda^{\sigma}_{\underline{\mu}\underline{\tau}}\Lambda^{\alpha}_{\sigma\tau} = 0 \text{ [9]} \tag{38}$$

which, by a simple rearrangement, can be put in the form:

$$G_{\sigma}^{\ \alpha} \equiv g^{\nu\rho}\left[\frac{\partial\Lambda^{\alpha}_{\sigma\rho}}{\partial x_{\nu}} - \Delta^{j}_{\sigma\nu}\Lambda^{\alpha}_{j\rho} - \Delta^{j}_{\rho\nu}\Lambda^{\alpha}_{\sigma j} + \Delta^{\alpha}_{\nu j}\Lambda^{j}_{\sigma\rho}\right],\ \alpha, \mu = 1, \ldots 4. \text{[10]} \tag{39}$$

We first deal with the partial system $\alpha = 4, \sigma \neq 4$. For it, we obtain from (39)

$$0 = g^{4\rho}\left[\frac{\partial\Lambda^{4}_{\sigma 4}}{\partial x_{\rho}} - \Delta^{j}_{\sigma\rho}\Lambda^{4}_{j4} - \Delta^{4}_{\sigma 4}\Lambda^{4}_{\rho 4} + \Delta^{4}_{4j}\Lambda^{j}_{\sigma\rho}\right] + g^{44}\Delta^{4}_{4j}\Lambda^{j}_{\sigma 4}. \tag{40}$$

Carrying out the calculation, considering (31) and (32), gives ξ_{σ} times a factor:

$$\tau\left(\frac{\mu'}{\mu}\right)'s + s\frac{\lambda'\mu'}{\lambda\mu}\tau - \frac{\mu'\lambda'}{\mu}\left[\left(\frac{\tau}{\lambda}\right)'s + \frac{\tau}{\lambda}\right] + \left(\frac{\mu'}{\mu}\right)^{2}s\tau = 0\,. \tag{41}$$

Due to (35), ($k = 0$!), we have $\left(\frac{\tau}{\lambda}\right)'s + \frac{\tau}{\lambda} = -\frac{2\tau}{\lambda}$.

For $\tau = 0$, (41) is fulfilled if we presuppose that $\tau \neq 0$; then we can shorten (41) through τ and thus obtain

$$\left(\frac{\mu'}{\mu}\right)'s + s\frac{\lambda'\mu'}{\lambda\mu} + 2\frac{\mu'}{\mu} + s\left(\frac{\mu'}{\mu}\right)^{2} = 0\,, \tag{41'}$$

which leads immediately to

$$\mu'\lambda s^{2} = \text{constant} \tag{42}$$

and, furthermore, to

$$\mu = k\int\frac{ds}{\lambda s^{2}} + k_{1}\,. \tag{43}$$

Since λ and μ tend to 1 at infinity, we have $k_{1} = 1$, and thus

$$\mu = 1 + m\int\frac{ds}{\lambda s^{2}},\quad m = \text{constant.}. \tag{44}$$

Now, we consider the partial system $\alpha \neq 4, \sigma = 4$ of the system (39). For it, we have

$$g^{4\rho}\left[\frac{\partial\Lambda^{\alpha}_{4\rho}}{\partial x_{\nu}} - \Delta^{j}_{4\nu}\Lambda^{\alpha}_{j\rho} - \Delta^{j}_{\rho\nu}\Lambda^{\alpha}_{4j} + \Delta^{\alpha}_{\nu j}\Lambda^{j}_{4\rho}\right] + g^{4\rho}[\Delta^{4}_{4j}\Lambda^{j}_{4\rho}] = 0\,. \tag{45}$$

Carrying out the calculations yields ξ_σ times

$$\left(1-\frac{e^2}{s^4}\right)\left[2\lambda^2 e\left(\frac{\lambda}{\mu s^3}\right)' + 2\frac{\mu'\lambda^3 e}{\mu^2 s^3}\right] + \frac{6\lambda^3 e}{\mu s^4} - \frac{4e^3\lambda^3}{\mu s^8} = 0. \tag{46}$$

Due to

$$\frac{6\lambda^3 e}{\mu s^4} - \frac{4e^3\lambda^3}{\mu s^8} = \frac{2\lambda^3 e}{\mu s^4} + \frac{4\lambda^3 e}{\mu s^4}\left(1 - \frac{e}{s^4}\right),$$

it follows from (46) that

$$\left(1-\frac{e^2}{s^4}\right)\left[\lambda^2\left(\frac{\lambda}{\mu s^3}\right) + \frac{\mu'\lambda^3}{\mu^2 s^3} + \frac{2\lambda^3}{\mu s^4}\right] + \frac{\lambda^3}{\mu s^4} = 0. \tag{47}$$

Here, by means of reducing $2e$, equation (46) already fulfills $e = 0$. By means of an elementary rearrangement, we obtain from (47)

$$\left(1-\frac{e^2}{s^4}\right)\left(ln\frac{\lambda}{s}\right)' + \frac{1}{s} = 0, \text{ and thus} \tag{48}$$

$$ln\frac{\lambda}{s} = -\int\frac{ds}{s\left(1-\frac{e^2}{s^4}\right)} + k = -ln\sqrt[4]{s^4 - e^2} + k,$$

or, finally,

$$\lambda = c\frac{s}{\sqrt[4]{s^4 - e^2}}. \tag{49}$$

Since λ is 1 at infinity, we set $c = 1$, and obtain the final result:

$$\lambda = \frac{1}{\sqrt[4]{1 - \frac{e^2}{s^4}}}. \tag{50}$$

With (37), (44), and (50), we thus already know the functions λ, μ, and τ which characterize the rotationally symmetrical case.

Those relations in (39) which we have not yet used are those that apply to the cases with $\alpha = \sigma = 4$ and $\alpha, \sigma \neq 4$; they must be identically satisfied by the functions (37), (44), and (50).

For $\alpha = \sigma = 4$, (39) becomes

$$g^{\nu\rho}\left[\frac{\partial\Lambda^4_{4\rho}}{\partial x_\nu} - \Delta^4_{4\nu}\Lambda^4_{4\rho} - \Delta^j_{\rho\nu}\Lambda^4_{4j}\right] + g^{4\rho}\Delta^4_{4j}\Lambda^j_{4\rho} = 0 \tag{51}$$

or

$$(\lambda^2\delta_{\nu\rho} - \tau^2 x_\nu x_\rho)\left[-\frac{\partial}{\partial x_\nu}\left(\frac{\mu'}{\mu}\xi_\rho\right) - \left(\frac{\mu'}{\mu}\right)^2 \xi_\nu \xi_\rho - \frac{\lambda'\mu'}{\lambda\mu}\xi_\nu\xi_\rho\right] \tag{51'}$$
$$-\tau\mu x_\rho \frac{\mu'\lambda}{\mu^2}\xi_j \frac{\partial}{\partial x_\rho}\left(\frac{\tau}{\lambda}x_j\right) = 0\,.$$

This equation is indeed satisfied by (37), (44), and (50). For $\alpha, \sigma \neq 4$, (39) becomes

$$\left.\begin{aligned}
&(\lambda^2\delta_{\nu\rho} - \tau^2 x_\nu x_\rho)\Big[\frac{\partial}{\partial x_\nu}\Big(\frac{\lambda'}{\lambda}(\xi_\sigma\delta_{\rho\alpha} - \xi_\rho\delta_{\sigma\alpha})\Big) + \left(\frac{\lambda'}{\lambda}\right)^2 \xi_\nu \delta_{j\tau}(\xi_j\delta_{\rho\alpha} - \xi_\rho\delta_{j\alpha}) \\
&\qquad + \left(\frac{\lambda'}{\lambda}\right)^2 \xi\delta(\xi\delta - \xi\delta) - \left(\frac{\lambda'}{\lambda}\right)^2 \xi\delta(\xi\delta - \xi\delta)\Big] \\
&+ \lambda'\tau x_\rho \frac{\partial}{\partial x_j}\left(\frac{\tau}{\lambda}x_\alpha\right)(\xi_\sigma\delta_{j\rho} - \xi_\rho\delta_{j\sigma}) \\
&-\mu\tau x_\nu\Big[\frac{\partial}{\partial x_\nu}\Big(\frac{\lambda}{\mu}\cdot\frac{\partial}{\partial x_\sigma}\Big(\frac{\tau}{\lambda}x_\alpha\Big)\Big) + \frac{\lambda'}{\mu}\xi_\nu\delta_{j\sigma}\frac{\partial}{\partial x_j}\left(\frac{\tau}{\lambda}x_\alpha\right) + \frac{\lambda'}{\mu}\frac{\partial}{\partial x_\sigma}\left(\frac{\tau}{\lambda}x_\alpha\right)(\xi_\sigma\delta_{j\alpha} - \xi_j\delta_{\sigma\alpha}) \\
&\qquad + \frac{\mu'}{\mu}\xi_\nu\frac{\lambda}{\mu}\frac{\partial}{\partial x_\sigma}\left(\frac{\tau}{\lambda}x_\alpha\right) - \frac{\lambda'}{\mu}\xi_j\delta_{\nu\alpha}\frac{\partial}{\partial x_\sigma}\left(\frac{\tau}{\lambda}x_j\right)\Big] \\
&+ \lambda\frac{\partial}{\partial x_j}\left(\frac{\tau}{\lambda}x_\alpha\right)\frac{\partial}{\partial x_j}\left(\frac{\tau}{\lambda}x_\sigma\right) = 0.
\end{aligned}\right\}$$

(52)

This system, also, is satisfied by the functions (37), (44), and (50). Carrying out the corresponding calculation, which requires only a certain attention to detail, is left for the reader.

We note the result: The tetrad

$$\left.\begin{aligned}
&h_s{}^\alpha = \frac{\delta_{s\alpha}}{\sqrt[4]{1 - \frac{e^2}{s^4}}}, \quad \alpha, s = 1, 2, 3, \qquad h_s{}^4 = 0, \\
&h_4{}^\alpha = \frac{e}{\sqrt[4]{1 - \frac{e^2}{s^4}}}, \quad \alpha = 1, 2, 3 \qquad h_4{}^4 = 1 + m\int\sqrt[4]{1 - \frac{e^2}{s^4}}\frac{ds}{s^2}
\end{aligned}\right\} \tag{53}$$

is the most general solution of the centrally symmetric (mirror-symmetric) case. As far as its physical interpretation goes, in our opinion, e can be interpreted as the electric charge and m as the ponderomotive mass. This interpretation is, in itself, arbitrary, apart from the fact that it conforms to the given interpretation of the field in considering the field equations to the first approximation. A notable feature is the occurrence of two and only two constants, which, from a consideration of experience, must be expected.

§2. A Pure, Static Gravitational Field.

From equations (53), we can see that for a vanishing charge e, all the $h_s^{\ \alpha}$ up to $h_4^{\ 4}$ become constant, while the latter is given by $h_4^{\ 4} = 1 - \frac{m}{s}$. This result suggests that there are more general static solutions for which $h_4^{\ 4}$ is variable.

Pursuing this idea, we posit

$$h_s^{\ \alpha} = \delta_{s\alpha}, \quad s = 1, 2, 3 \qquad h_4^{\ \alpha} = \delta_{4\alpha} \cdot \sigma(x_1, x_2, x_3), \tag{1}$$

so that all the $\Delta_{\alpha\beta}^{\gamma}$ are zero, up to

$$\Delta_{4\beta}^{4} = \Lambda_{4\beta}^{4} = -k_{44} h_4^{\ 4}{}_{,\beta} = -\frac{\partial ln\sigma}{\partial x_\beta}. \tag{2}$$

Then all the field equations are satisfied identically, except for

$$G_4^{\ 4} = g^{\nu\rho}\left[\frac{\partial \Lambda_{4\rho}^{4}}{\partial x_\nu} - \Delta_{4\nu}^{4}\Lambda_{4\rho}^{4}\right] = 0. \tag{3}$$

or

$$0 = \sum_\rho \frac{\partial \Lambda}{\partial x_\rho} - \Delta_{4\rho}^{4}\Delta_{4\rho}^{4} = \sum_\rho \frac{\partial ln\sigma}{\partial x_\rho^{\ 2}} + \frac{\partial ln\sigma}{\partial x_\rho} + \frac{\partial ln\sigma}{\partial x_\rho}. \tag{3$'$}$$

This yields for σ:

$$\sum_\rho \frac{\partial^2 \sigma}{\partial x_\rho^{\ 2}} = 0. \tag{4}$$

that is, σ is a potential.

Since σ converges to 1 at infinity, the solution (for a finite number of mass points) is

$$\sigma = 1 + \sum_j \frac{m_j}{r_j}, \quad m_j = \text{const.} \tag{5}$$

This strict result is important for the physical interpretation of the theory for the following reason: The formula (59) shows that there is a strict solution which corresponds to the case that two or more unbound electrically neutral masses are at rest at arbitrary distances from one another. Such a case does not occur in nature. One could be inclined to think that this would indicate a failure of the theory with respect to experience. That would indeed be the case if the equation of motion for these kinds of singularities were allowed by the field equations, which was, in fact, the case in the original version of the theory. It, however, appears not to be the case in the present theory.[1)]

Thus, no argument against the applicability of the theory can be derived from the existence of the static solution considered here. Indeed, however, one can see that, in the new theory, a freedom from singularities must be required of each solution which is supposed to represent the elementary particles of matter.

Without finding such solutions, it would seem not to be possible to draw conclusions about the law of motion for particles from the field equations.

[1)]The derivability of the law of motion in the earlier version of the theory was based upon the fact that there was a field equation in the form of a symmetric tensor equation, whose divergence vanished identically.[11]

That condition is not met in the present theory.

256. To Myron Mathisson

[Berlin,] 20 February 1930

Dear Mr. Mathisson,

I have looked over your manuscript with great interest,[1] though, thus far, only superficially. It seems to me to indicate that you have an unusual talent for formalism and that you are destined for scientific work. The work that you have sent me would, in my opinion, be very well suited to serve as a doctoral dissertation, since it represents an excellent demonstration of your qualifications. If you agree to submit the work, in that sense, I would ask you to inform me about your personal history, your studies, profession, etc. I will then, first of all, write on your behalf to Prof. Loria[2] in Lemberg, or to some other colleague in Poland whom you might suggest.

When all that has been accomplished, I would be willing to make some efforts to obtain a fellowship for you, so that you could work here for a certain period. What do you think about this?

The publication of your work in a scientific journal seems to me to be not too appropriate, since it, in terms of the field forces, does not go beyond special relativity in that it completely neglects the effects of the gravitational field on the mass point.[3] The forces that you have calculated with great care refer to the electrical reaction of the electron on itself. Once the work has been accepted as a dissertation, a shorter extract could nevertheless be published in some scientific journal.

I am sending you a new paper on the unified field theory,[4] since I am presently of the opinion that it is the natural continuing development of the general theory of relativity and because, in this theory, the problem of motion is modified in essential ways compared to the existing one.

Awaiting your prompt answer with interest, I am, with friendly greetings, yours,

257. To Wilhelm Miklas[1]

[Berlin,] 20 February 1930

Esteemed Mr. President,

Today, I received a letter from three leading French scholars, Messrs. Richet, Hadamard, and Langevin, regarding the Halsmann affair.[2] In that letter, the situation is compared with the Dreyfuss affair,[3] and, quite rightly, the thesis is put forth that intellectual Europe has a duty to do everything possible to promote the triumph of justice. The letter was sent to me so that I could make use of it with those who could serve this end most effectively. Since pardoning Halsmann lies in your power, it seems to me to be essential to inform you of that letter and the duty which it imposes on me.[4]

A pardon originating with you would by no means completely satisfy the conscience of those who think and feel justly, but it would indeed provide a proof that, in your nation, there is a reserve of justice which is independent of the passions of the masses.

Through such an act, you would doubtless pay a great service to your country and, moreover, to justice in general.

With the greatest respect,

258. "New Investigation of the Field Equations of the Unified Field Theory"

[after 20 February 1930][1]

New Investigation of the Field Equations of the Unified Field Theory[2]

A. Einstein and W. Mayer[3]

Investigations into the law of motion[4] have led us to the view that the field equations are still in need of modification. The principle for choosing the field equations consists in searching for the strongest possible determination of the field via an "overdetermination". In the following, we show that the previous field equations are capable of a generalization without reducing the degree of determination.[5]

The modification of the field equations developed here results in a still closer connection to the gravitation theory based on the Riemann metric alone than was possible with the previous field equations.[6]

§1. Field Equations and Identities.

Let $L^{\alpha}_{\mu\nu}$ be a tensor which is antisymmetric in its lower indices, which is to be expressed in terms of the $h_s{}^{\nu}$ and their first derivatives. The commutation rule for differentiation then yields the following identity:

$$L^{\alpha}_{\underline{\mu}\underline{\nu};\nu;\alpha} - L^{\alpha}_{\underline{\mu}\underline{\nu};\alpha;\nu} + L^{\alpha}_{\underline{\mu}\underline{\nu};\sigma}\Lambda^{\sigma}_{\nu\alpha} \equiv 0 \,. \tag{1}$$

Positing[7]

$$G^{\mu\alpha} = L^{\alpha}_{\underline{\mu}\underline{\nu};\nu} - L^{\nu}_{\underline{\mu}\underline{\alpha};\nu} - L^{\sigma}_{\underline{\mu}\underline{\tau}}\Lambda^{\alpha}_{\sigma\tau} \tag{2}$$

and

$$F_{\mu\nu} = \Lambda^{\alpha}_{\mu\nu;\alpha} = \varphi_{\mu,\nu} - \varphi_{\nu,\mu} \tag{3}$$

then allows us to write (1) in the form

$$G^{\mu\alpha}{}_{;\alpha} + L^{\sigma}_{\underline{\mu}\underline{\tau}}F_{\sigma\tau} \equiv 0 \,. \tag{1a}$$

As field equations, we take

$$G^{\mu\alpha} = 0 \tag{4}$$

and

$$F_{\mu\nu} = 0 \,. \tag{5}$$

Between these field equations, the identity (1a) holds.

The field equations (5) can, owing to (3), also be replaced by

$$F_{\mu} = \varphi_{\mu} - \frac{\psi_{,\mu}}{\psi} = \Lambda^{\sigma}_{\mu\sigma} - \frac{\psi_{,\mu}}{\psi}, \text{[8]} \tag{5a}$$

where ψ is a scalar, and the comma (as always) denotes ordinary differentiation.

The field equation (4) obtains a particular meaning only when the L are expressed in terms of the h and their (first) derivatives. We are led to an answer for $L^{\alpha}_{\mu\nu}$ which entails that the necessary identities exist which ensure the compatibility of the field equations by defining $2\underline{G^{\mu\alpha}} = G^{\mu\alpha} - G^{\alpha\mu}$.

For brevity, we introduce a tensor which is antisymmetric in all three of its indices

$$\Sigma^{\alpha}_{\underline{\mu\nu}} = L^{\alpha}_{\underline{\mu\nu}} + L^{\mu}_{\underline{\nu\alpha}} + L^{\nu}_{\underline{\alpha\mu}}, \tag{6}$$

and take the following rearrangements into account:

$$L^{\alpha}_{\underline{\mu\nu}} - L^{\mu}_{\underline{\alpha\nu}} = \Sigma^{\nu}_{\underline{\alpha\mu}} - L^{\nu}_{\underline{\alpha\mu}}, \tag{6a}$$

$$L^{\sigma}_{\underline{\mu\tau}} - \Lambda^{\alpha}_{\sigma\tau} = \frac{1}{2}(\Sigma^{\mu}_{\underline{\tau\sigma}} - L^{\mu}_{\underline{\tau\sigma}})\Lambda^{\alpha}_{\sigma\tau}; \tag{6b}$$

by calculation from (2), we initially obtain

$$2\underline{G^{\mu\alpha}} \equiv \left(\Sigma^{\nu}_{\underline{\alpha\mu};\nu} - \frac{1}{2}\Sigma^{\mu}_{\underline{\tau\sigma}}\Lambda^{\alpha}_{\sigma\tau} + \frac{1}{2}\Sigma^{\alpha}_{\underline{\tau\sigma}}\Lambda^{\mu}_{\sigma\tau}\right) - L^{\nu}_{\mu\alpha;\nu} + \frac{1}{2}L^{\mu}_{\underline{\tau\sigma}}\Lambda^{\alpha}_{\sigma\tau} - \frac{1}{2}L^{\alpha}_{\underline{\tau\sigma}}\Lambda^{\mu}_{\sigma\tau} \tag{7}$$

Evaluation of the expression in parentheses yields

$$\Sigma^{\nu}_{\underline{\alpha\mu},\nu} + \Sigma^{\sigma}_{\underline{\alpha\mu}}\Delta^{\nu}_{\sigma\nu}$$

or

$$\Sigma^{\nu}_{\underline{\alpha\mu},\nu} + \left(\varphi_{\sigma} + \frac{h_{,\sigma}}{h}\right)\Sigma^{\sigma}_{\underline{\alpha\mu}},$$

where h denotes the determinant $|h_{s\nu}|$. Due to (5a), we can also write this as

$$\Sigma^{\nu}_{\underline{\alpha\mu},\nu} + \left(\frac{\psi_{,\sigma}}{\psi} + \frac{h_{,\sigma}}{h}\right)\Sigma^{\sigma}_{\underline{\alpha\mu}} + F_{\sigma}\Sigma^{\sigma}_{\underline{\alpha\mu}}. \tag{8}$$

The structure of the last three terms on the right-hand side of (7), together with (6b), have led us to the answer

$$L^{\alpha}_{\mu\nu} = \lambda\Lambda^{\alpha}_{\mu\nu} + \beta(\lambda_{,\mu}\delta_{\nu}{}^{\alpha} - \lambda_{,\nu}\delta_{\mu}{}^{\alpha}), \tag{9}$$

where λ is a scalar, β a constant, and $\delta_{\nu}{}^{\alpha}$ denotes the mixed tensor whose components are equal to 1 or 0, depending on whether their indices are the same or different. We initially obtain[9]

$$\Sigma^{\alpha}_{\underline{\mu}\underline{\nu}} = \lambda S^{\alpha}_{\underline{\mu}\underline{\nu}} = \lambda(\Lambda^{\alpha}_{\underline{\mu}\underline{\nu}} + \Lambda^{\mu}_{\underline{\nu}\underline{\alpha}} + \Lambda^{\nu}_{\underline{\alpha}\underline{\mu}})\ldots \tag{10}$$

$$\begin{aligned} L^{\nu}_{\underline{\mu}\underline{\alpha};\nu} &= \lambda F^{\mu\alpha} + \lambda_{,\nu}\Lambda^{\nu}_{\underline{\mu}\underline{\alpha}} + \beta(\lambda_{;\underline{\mu}}\delta_{\underline{\alpha}}{}^{\nu} - \lambda_{;\underline{\alpha}}\delta_{\underline{\mu}}{}^{\nu}) \\ &= \lambda F^{\mu\alpha} + \lambda_{,\nu}\Lambda^{\nu}_{\underline{\mu}\underline{\alpha}} + \beta(\lambda_{,\underline{\mu};\alpha} - \lambda_{,\alpha;\mu}) = \\ &= \lambda F^{\mu\alpha} + \lambda_{,\nu}\Lambda^{\nu}_{\underline{\mu}\underline{\alpha}} - \beta\lambda_{,\sigma}\Lambda^{\nu}_{\underline{\mu}\underline{\alpha}}\cdots \end{aligned} \tag{11}$$

$$\begin{aligned} \frac{1}{2}(L^{\mu}_{\underline{\tau}\underline{\sigma}}\Lambda^{\alpha}_{\sigma\tau} - L^{\alpha}_{\underline{\tau}\underline{\sigma}}\Lambda^{\mu}_{\sigma\tau}) &= \frac{\beta}{2}[(\lambda_{,\tau}\delta_{\sigma}{}^{\mu} - \lambda_{,\sigma}\delta_{\tau}{}^{\mu})\Lambda^{\alpha}_{\underline{\sigma}\underline{\tau}} - (\lambda_{,\tau}\delta_{\sigma}{}^{\alpha} - \lambda_{,\sigma}\delta_{\tau}{}^{\alpha})\Lambda^{\mu}_{\underline{\sigma}\underline{\tau}}] \\ &= \beta\lambda_{,\tau}(\Lambda^{\alpha}_{\mu\tau} - \Lambda^{\sigma}_{\alpha\tau}) \\ \cancel{= \beta\lambda_{,\tau}(S_{\alpha}} &= \beta\lambda_{,\tau}(S^{\tau}_{\underline{\alpha}\underline{\mu}} - \Lambda^{\tau}_{\underline{\alpha}\underline{\mu}})\ldots \end{aligned} \tag{12}$$

Taking equations (8) through (12) into account, (7) becomes:

$$\left.\begin{aligned} \frac{2}{\lambda}\underline{G}^{\mu\alpha} \equiv -F^{\mu\alpha} - F_{\sigma}S^{\sigma}_{\underline{\mu}\underline{\alpha}} + \frac{1}{\lambda}\left[\Sigma^{\nu}_{\underline{\alpha}\underline{\mu},\nu} + \left(\frac{\psi_{,\sigma}}{\psi} + \beta\frac{\lambda_{,\sigma}}{\lambda} + \frac{h_{,\sigma}}{h}\right)\Sigma^{\sigma}_{\underline{\alpha}\underline{\mu}}\right] \\ + (-1+\beta+\beta)\frac{\lambda_{,\sigma}}{\lambda}\Lambda^{\sigma}_{\underline{\mu}\underline{\alpha}}\,. \end{aligned}\right\} \tag{13}$$

In order to make the last term on the right-hand side of (13) vanish, we set

$$\beta = \frac{1}{2} \tag{13a}$$

and obtain

$$\frac{2}{\lambda}\underline{G}^{\mu\alpha} + F^{\mu\alpha} + F_{\sigma}S^{\sigma}_{\underline{\mu}\underline{\alpha}} \equiv \frac{1}{h\psi\lambda^{\frac{3}{2}}}\left(h\psi\lambda^{\frac{3}{2}}S^{\nu}_{\underline{\alpha}\underline{\mu}}\right)_{,\nu}. \tag{14}$$

From (14), the following identity follows, which is linear and homogeneous in the left-hand sides of the field equations:

$$\left[h\psi\lambda^{\frac{3}{2}}\left(\frac{2}{\lambda}\underline{G}^{\mu\alpha} + F^{\mu\alpha} + F_{\sigma}S^{\sigma}_{\underline{\mu}\underline{\alpha}}\right)\right]_{,\alpha} \equiv 0\,. \tag{15}$$

The existence of the identities (1a) and (15) allows us to conclude the compatibility of the field equations (4) and (5a), independently of the choice of the scalar λ (which we can take tentatively as given). Proof: Certainly, those field equations are compatible which correspond to setting equal to zero

$$\left.\begin{array}{llll} G^{11} & G^{12} & G^{13} & \\ G^{21} & G^{22} & G^{23} & \\ G^{31} & G^{32} & G^{33} & \\ F_1 & F_2 & F_3 & F_4 \end{array}\right\} \quad \ldots(\alpha)$$

since these are 13 equations for the 17 quantities $h_{s\nu}$ and ψ. The remaining field equations, which correspond to setting the expressions G^{14}, G^{24}, G^{34}, G^{41}, G^{42}, G^{43}, and G^{44} equal to zero, are furthermore fulfilled on a section defined by $x_4 = \text{const.}$ The identities (1a) and (15) are, due to the vanishing of the quantities (α), differential equations for the terms $G^{\alpha 4} - G^{4\alpha}$, $G^{\alpha 4} + G^{4\alpha}$, $\alpha = 1, \ldots 4$, with the property that each system of solutions is completely determined by the values of $G_0^{\alpha 4} - G_0^{4\alpha}$, $G_0^{\alpha 4} + G_0^{4\alpha}$ on the section $x_4 = \text{const.}$

Now, $G^{\alpha 4} - G^{4\alpha} = 0$, $G^{\alpha 4} + G^{4\alpha} = 0$ is that system of solutions which vanishes on the section (by assumption). Thus, it follows from the vanishing of the quantities (α) and the quantities G^{14}, G^{24}, G^{34}, G^{41}, G^{42}, G^{43}, G^{44} on the section $x_4 = \text{const.}$ that this last group of quantities vanishes as such.

We now come to the determination of the scalar λ. This itself should, once again, be determined by the $h_s{}^{\nu}$. The only scalar with this property which we have thus far encountered is the scalar ψ that occurs in the field equations (5a). It is self-evident to assume that λ should be a function of ψ. A glance at (4), (2), and (9) tells us that, in the field equations, only the logarithmic derivative $\frac{\lambda_{,\sigma}}{\lambda}$ enters, while considering (5a) shows us that $\frac{\psi_{,\sigma}}{\psi}$ expresses itself only through $h_{s\nu}$ and its first derivatives (and in the latter, linearly). The quantity λ can thus be eliminated from the field equations by setting:

$$\frac{\lambda_{,\sigma}}{\lambda} = a\frac{\psi_{,\sigma}}{\psi} = a\varphi_{\sigma},\ a = \text{const.}, \qquad (16)$$

which is ⟨equivalent⟩ to $\lambda = \psi^{\alpha}$. For $a = 0$, we obtain the existing field equations.[10]

Remark: The subsequent fixing of the quantity λ by the field-quantity ψ ($\lambda = \psi^a$) does not disturb the compatibility proof of the system

$$G^{\mu\alpha} = 0, F^{\mu\alpha} = 0 \text{ (resp. } F_\mu = 0), \lambda = \psi^a \tag{17}$$

Now $\lambda = \psi^a$ is also added to the quantities of the (α) system. The rest of the proof is the same as above. After successful ⟨elimination⟩ substitution of (16) in $G^{\mu\alpha} = 0$, the $\lambda = \psi^a$ can be eliminated in system (17), since this equation only defines the scalar λ.[11]

It indeed develops that, especially in the case of the four-dimensional space, the determination (16) is not the only one that allows the expressions of $\frac{\lambda_{,\sigma}}{\lambda}$ by the $h_s{}^\nu$ and their first derivations (namely in the latter linearly). It follows, namely, from the field equations that there exists a second scalar determined by the $h_s{}^\nu$. Because of (14), the field equations (4), (5) have, as a consequence, the disappearance of

$$\left(h\psi\lambda^{\frac{3}{2}}S^\nu_{\underline{\alpha\mu}}\right)_{,\nu} \quad \ldots\ldots \tag{18}$$

If we set ($\varepsilon_{\alpha\mu\nu\tau}$ the known antisymmetric expression with $\varepsilon_{1234} = 1$)

$$h\varepsilon_{\alpha\mu\nu\tau}\psi\lambda^{\frac{3}{2}}S^\nu_{\underline{\alpha\mu}} = \chi_\tau, \tag{19}$$

then

$$h\psi\lambda^{\frac{3}{2}}S^\nu_{\underline{\alpha\mu}} = \frac{1}{6}\chi_\tau\varepsilon_{\alpha\mu\nu\tau}. \tag{19′}$$

(This follows because $\varepsilon_{\alpha\mu\nu\rho}\varepsilon_{\alpha\nu\mu\tau} = 6\delta_{\rho\tau}$).

The disappearance of is equivalent with $\chi_{\tau,\nu}\varepsilon_{\alpha\mu\nu\tau} = \frac{1}{2}(\chi_{\tau,\nu} - \chi_{\nu,\tau})\varepsilon_{\alpha\mu\nu\tau} = 0$, i.e., with $\chi_{\tau,\nu} - \chi_{\nu,\tau} = 0$. Therefore,

$$\psi\lambda^{\frac{3}{2}}S_\tau = \chi_{,\tau} \tag{20}$$

is valid, whereby it means that

$$S_\tau = h\varepsilon_{\alpha\mu\nu\tau}S^\nu_{\underline{\alpha\mu}}. \tag{21}$$

χ is the scalar whose existence we claimed above.

It now follows very simply that the *answer*

$$\frac{\lambda_{,\sigma}}{\lambda} = -\frac{2}{3}\frac{\Psi_{,\sigma}}{\psi} + bS_{\sigma}, \qquad (22)$$

in which b is an arbitrary constant when substituted in (4), leads to a compatible system (4), (5) of the same degree of determination as the one found so far. [12]

§2. First Approximation: Choice of the Constants.

Of the two possibilities given by (16) and (23a) for the choice of the scalar λ, the first is the formally more natural; it alone will be considered in the following. The choice of the constant a which occurs in (16) can be made only on the basis of physical aspects.

The field equations (4) are, in the first approximation, given by

$$L^{\alpha}_{\underline{\mu\nu},\nu} - L^{\nu}_{\underline{\mu\alpha},\nu} = 0.$$

whose main term is, again,

$$\left(\Lambda^{\alpha}_{\mu\nu} + \frac{a}{2}\varphi_{\mu}\delta_{\nu}{}^{\alpha} - \frac{a}{2}\varphi_{\nu}\delta_{\mu}{}^{\alpha}\right)_{,\nu} - \left(\Lambda^{\nu}_{\mu\alpha} + \frac{a}{2}\varphi_{\mu}\delta_{\alpha}{}^{\nu} - \frac{a}{2}\varphi_{\alpha}\delta_{\mu}{}^{\nu}\right)_{,\nu} = 0$$

or
$$\Lambda^{\alpha}_{\mu\nu,\,\nu} + \frac{a}{2}\varphi_{\mu,\,\alpha} - \frac{a}{2}\varphi_{\nu,\,\nu}\delta^{\alpha}_{\mu} - \Lambda^{\nu}_{\mu\alpha,\,\nu} - \frac{a}{2}(\varphi_{\mu,\,\alpha} - \varphi_{\alpha,\,\mu}) = 0$$ [13]

If we form a scalar equation through contraction with respect to α and μ, we obtain

$$\varphi_{\nu,\nu}\left(-1 + \frac{a}{2} - 4\frac{a}{2}\right) = 0$$

Thus, if $a \neq -\frac{2}{3}$, as we assume, then $\varphi_{\nu,\,\nu}$, vanishes, so that (4) in the first approximation takes the form

$$\bar{h}_{\alpha\mu,\nu,\,\nu} - \bar{h}_{\alpha\nu,\nu,\,\mu} + \frac{a}{2}(\bar{h}_{\nu\mu,\nu,\,\alpha} - \bar{h}_{\nu\nu,\alpha,\,\mu}) = \cancel{\Lambda^{\nu}_{\mu\alpha,\nu}}\ 0, \qquad (4a)$$

where we leave off the terms vanished due to $F_{\mu\alpha} = 0$.[14]

Then, in the usual manner, we have posited

$$h_{\alpha\mu} = \delta_{\alpha\mu} + \bar{h}_{\alpha\mu}$$

with an "infinitesimally small" $\bar{h}_{\alpha\mu}$.[15]

Writing the equations which are symmetric to (4a) by positing

$$\bar{h}_{\alpha\mu} + \bar{h}_{\mu\alpha} = s_{\alpha\mu},$$

one then obtains for

$$a = -2$$

the relation

$$s_{\alpha\mu,\nu,\nu} - s_{\alpha\nu,\nu,\mu} - s_{\mu\nu,\nu,\alpha} + s_{\nu\nu,\alpha,\mu} = 0, \quad (23)$$

[16]

that is, in first approximation, the equations $R_{ik} = 0$ from the theory that is based upon the metric alone. It thus suggests itself that we tentatively give this case ($a = -2$) priority and compare it with what is known empirically.

Because of $F^{\mu\alpha} = 0$, the additional part that is multiplied by $\frac{a}{2}$ has no influence upon the antisymmetric equations, $G^{\mu\alpha} - G^{\alpha\mu} = 0$.[17] Here, too, they are

$$(a_{\mu\nu,\sigma} + a_{\nu\sigma,\mu} + a_{\sigma\mu,\nu})_{,\sigma} = 0 \quad (24)$$

when we set $\bar{h}_{\mu\nu} - \bar{h}_{\nu\mu} = a_{\mu\nu}$.

It can be proven that, as a result of the existence of the two field equations in the first-order problem, the coordinate system can be so chosen that

$$\left(s_{\mu\nu} - \frac{1}{2}\delta_{\mu\nu}s_{\alpha\alpha}\right)_{,\nu} = 0 \quad (28)$$

and

$$a_{\mu\nu,\nu} = 0, \quad (29)$$

which satisfies equations (5) in the first approximation. Equations (23), (24) can then also be written in the form

$$s_{\alpha\mu,\nu,\nu} = 0, \quad (23a)$$

$$a_{\mu\nu,\sigma,\sigma} = 0. \quad (24a)$$

Equations (23a), (25) express the law of the gravitational field, while equations (24a), (26) express the law of the electromagnetic field in the simplest form to the first approximation.

259. "An Addition to the Field Equations of the Unified Field Theory"

[after 20 February 1930][1]

The completion of the system of field equations of the unified field theory, which is given in the following, is based on an identity which is only developed here for the first time ⟨in the following⟩. For brevity, I refer to the detailed description of the theory that appeared in the *Mathematische Annalen* (102.5, pp. 685–697).[2]

The identity (24) l.c. expresses the property of the Λ field which corresponds to the integrability of infinitesimal parallel displacements.[3] We imagine that the index ι is lowered in that equation and that it is then multiplied by the antisymmetric tensor density $\delta^{\iota\kappa\lambda\mu}$, whose components are all equal to ± 1; this presupposes that the dimensionality is four, which we will continue to assume in what follows.

First of all, we obtain

$$(\Lambda^{\iota}_{\kappa\lambda;\mu} + \langle 3\rangle \Lambda^{\iota}_{\kappa\alpha}\Lambda^{\alpha}_{\lambda\mu})\delta^{\iota\kappa\lambda\mu} \equiv 0$$

or

$$\left(\Lambda^{\iota}_{\kappa\lambda,\,\mu} - \frac{1}{2}\Lambda_{\iota\kappa}\Lambda^{\alpha}_{\lambda\mu}\right)\delta^{\iota\kappa\lambda\mu} \equiv 0\,.$$

We transform the second term, making use of a "geodesic" local coordinate system ($g_{\mu\nu} = \delta_{\mu\nu}$). This term then takes on the form:

$$-4(\Lambda^{\alpha}_{12}\Lambda^{\alpha}_{34} + \Lambda^{\alpha}_{13}\Lambda^{\alpha}_{42} + \Lambda^{\alpha}_{14}\Lambda^{\alpha}_{23})\delta^{1234}\,.$$

These are, in all, 12 individual terms which, by rearrangement, appear in the following form:

$$-4(\varphi_1\Lambda^{2}_{34} + \varphi_2\Lambda^{3}_{41} + \varphi_3\Lambda^{4}_{12} + \varphi_4\Lambda^{1}_{23})\delta^{1234},$$

or, returning to a generalized coordinate system,

$$-6\,.$$

260. Calculations

[after 20 February 1930]

[Not selected for translation.]

261. Addendum to the Paper "Compatibility of Field Equations in the Unified Field Theory"

[Berlin?, after 20 February 1930][1]

The results of an unpublished investigation which I have carried out with Mr. W. Mayer[2] have led to the supposition that the field equations of the ⟨general⟩ unified field theory, whose compatibility was demonstrated in a previous article (in *Berichte* 1930, I),[3] can be further generalized. This will be shown in the following.

Let $L_{\mu\nu}^{\alpha}$ denote a tensor which is antisymmetric with respect to the indices μ and ν, and which depends upon the $h_s{}^{\nu}$ in a manner that will be explained below. Applying to it the commutation rule[4] for differentiation yields

$$L_{\underline{\mu}\underline{\nu};\nu;\alpha}^{\alpha} - L_{\underline{\mu}\underline{\nu};\alpha;\nu}^{\alpha} + L_{\underline{\mu}\underline{\nu};\sigma}^{\alpha}\Lambda_{\nu\alpha}^{\sigma} \equiv 0 , \tag{1}$$

or

$$(L_{\underline{\mu}\underline{\nu};\nu}^{\alpha} - L_{\underline{\mu}\underline{\alpha};\nu}^{\nu} - L_{\underline{\mu}\underline{\tau}}^{\sigma}\Lambda_{\sigma\tau}^{\alpha})_{;\alpha} - \cancel{L_{\underline{\mu}\underline{\nu};\alpha;\nu}^{\alpha}} - L_{\underline{\mu}\underline{\nu}}^{\alpha}\Lambda_{\nu\alpha;\sigma}^{\sigma} \equiv 0. \tag{1a}$$

As the field equations, we take[5]

$$G^{\mu\alpha} = L_{\mu\nu;\nu}^{\alpha} - L_{\mu\tau}^{\sigma}\Lambda_{\sigma\tau}^{\alpha} = 0 \text{ and} \tag{2}$$

$$F_{\mu\nu} = \varphi_{\mu,\,\nu} - \varphi_{\nu,\,\mu} = \Lambda_{\mu\nu;\sigma}^{\sigma} = 0 . \tag{3}$$

This system goes over into the earlier one if L is replaced by Λ. The tensor $L_{\mu\nu}^{\alpha}$ can now be chosen in such a way that equation (3) results in the vanishing of $L_{\underline{\mu}\underline{\nu};\alpha}^{\alpha}$; ⟨we⟩ ⟨correspondingly⟩ ⟨we set⟩ we accomplish this by making use of the ansatz

$$L_{\mu\nu}^{\alpha} = \lambda\Lambda_{\mu\nu}^{\alpha} - \lambda_{,\mu}\delta_{\nu}{}^{\alpha} + \lambda_{,\nu}\delta_{\mu}{}^{\alpha} \tag{4}$$

where λ is a scalar that will be determined later, and the comma denotes ordinary differentiation. Then we obtain

$$L_{\mu\nu;\alpha}^{\alpha} \equiv (\lambda_{,\alpha}\Lambda_{\mu\nu}^{\alpha} + \lambda\Lambda_{\mu\nu;\alpha}^{\alpha}) - (\lambda_{,\mu,\,\nu} - \lambda_{,\alpha}\Delta_{\mu\nu}^{\alpha}) + (\lambda_{,\nu,\,\mu} - \lambda_{\alpha}\Delta_{\nu\mu}^{\alpha})$$

or

$$L_{\mu\nu;\alpha}^{\alpha} \equiv \lambda\Lambda_{\mu\nu;\alpha}^{\alpha} \text{ [6]} \tag{5}$$

Considering equation (2) shows us that $\frac{1}{\lambda}G^{\mu\alpha}$ depends only upon the first and second derivatives of log λ but does not contain λ in any other form.[7] On the other hand, owing to (3), we can posit

$$\varphi_\alpha = \Lambda^\sigma_{\alpha\sigma} = \frac{\partial \log\psi}{\partial x^\alpha} ,$$

where ψ is a scalar. We thus obtain the result that (2) depends only upon the $h_s^{\ \nu}$ and both their first derivatives if we take

$$\lambda = \psi^n$$

or

$$\log\lambda = n\log\psi , \tag{6}$$

where n is a constant. The earlier field equations can be obtained by setting $n = 0$ or $\lambda = 1$.

We now have the task of proving the compatibility of the system of equations (2) and (3). We find from (2)

$$2\overline{G}^{\mu\alpha} = G^{\mu\alpha} - G^{\alpha\mu} \equiv (L^\alpha_{\underline{\mu\nu};\nu} - L^\mu_{\underline{\alpha\nu};\nu}) - (L^\sigma_{\underline{\mu\tau}}\Lambda^\alpha_{\sigma\tau} - L^\sigma_{\underline{\alpha\tau}}\Lambda^\mu_{\sigma\tau}) .$$

Making use of the abbreviation

$$\Sigma^\alpha_{\underline{\mu\nu}} = L^\alpha_{\underline{\mu\nu}} + L^\mu_{\underline{\nu\alpha}} + L^\nu_{\underline{\alpha\mu}} = \lambda(\Lambda^\alpha_{\underline{\mu\nu}} + \Lambda^\mu_{\underline{\nu\alpha}} + \Lambda^\nu_{\underline{\alpha\mu}}) = \lambda S^\alpha_{\underline{\mu\nu}} ,$$

the first expression in parentheses takes on the form

$$\Sigma^\nu_{\underline{\alpha\mu};\nu} - L^\nu_{\underline{\alpha\mu};\nu} ,$$

If we expand $\Sigma^\nu_{\alpha\mu;\nu}$ and combine the result with the second parenthesis, we obtain which one can rewrite in the form

$$\cancel{L^\sigma_{\mu\tau}\Delta^\alpha_{\sigma\tau} + L^\tau_{\mu\sigma}\Lambda}$$

$$(L^\sigma_{\underline{\mu\tau}} - L^\tau_{\underline{\mu\sigma}})\Delta^\alpha_{\sigma\tau} - (L^\sigma_{\underline{\alpha\tau}} - L^\tau_{\underline{\alpha\sigma}})\Delta^\mu_{\sigma\tau} ;$$

then, by making use of the relation $\Delta^\alpha_{\nu\alpha} = \frac{\psi_{,\nu}}{\psi} + \frac{h_{,\nu}}{h}$, we get:

$$2\overline{G}^{\mu\alpha} \equiv -L^\nu_{\underline{\alpha\mu};\nu} + \frac{1}{\psi h}(\psi h \Sigma^\nu_{\underline{\alpha\mu}})_{,\nu} \text{ [8]} \tag{7}$$

and also

$$F_{\mu\nu} \equiv F_{\mu,\nu} - F_{\nu,\mu} \tag{8b}$$

holds. Taking account of (8), we can write (6) in the form

$$2\underline{G^{\mu\alpha}} + F^{\mu\alpha} \equiv \frac{1}{\psi\lambda h}(\cancel{\Sigma^\nu_{\alpha\mu}}\psi\lambda h S^\nu_{\underline{\alpha\mu}})_{,\nu} . \tag{6a}$$

From (6a), the identity

$$[\psi\lambda h(2\underline{G^{\mu\alpha}}+F^{\mu\alpha})]_{,\alpha}\equiv 0 \tag{9}$$

follows. From the identities (1a) and (9), the compatibility of the field equations (7), (8) can be deduced, whereby the scalar λ may initially taken to be arbitrary.

Proof: Certainly, those field equations are compatible which correspond to setting

$$\begin{matrix} G^{11} & G^{12} & G^{13} & \\ G^{21} & G^{22} & G^{23} & \\ G^{31} & G^{32} & G^{33} & \\ F_1 & F_2 & F_3 & F_4 \end{matrix}$$

equal to zero; for these are 13 equations for the ⟨16 equ⟩ 17 quantities $h_s^{\ \nu}$ and ψ. Suppose now that on a section $x_4 = \text{const.}$, the remaining field equations are also fulfilled, corresponding to setting the expressions G^{14}, G^{24}, G^{34}, G^{41}, G^{42}, G^{43}, G^{44}, ~~F_4~~ equal to zero. From the identities (1a) and (9), it then follows that on this section, the derivatives of these expressions with respect to x^4 must also vanish. The equations must then all vanish on the infinitesimally neighboring section. The equations are thus all also satisfied on neighboring x^4 sections, out to infinity.

Repeating this procedure then ⟨thus⟩ yields the compatibility proof for all the equations.

[9]Now, instead of giving the scalar λ directly, one can define it through an equation. Since in the field equations, only the $h_s^{\ \nu}$ and their derivatives should occur, while in $L^{\alpha}_{\mu\nu}$ however only the logarithmic derivatives of λ enter, we must see how the latter can be expressed ⟨in terms of the *h*⟩ in terms of the $h_s^{\ \nu}$. We require that the $\frac{\lambda_{,\sigma}}{\lambda}$ contain only linear, first derivatives of the $h_s^{\ \nu}$, and no higher derivatives at all.

Before applying this ansatz for λ, we must carry out an auxiliary calculation. According to (6a), as a result of the field equations, we have the equation

$$(\psi\lambda h S^{\nu}_{\underline{\alpha}\underline{\mu}})_{,\nu} = 0\,. \tag{10}$$

which can also be written in the form

$$\psi\lambda h S^{\nu}_{\underline{\alpha\mu}} \equiv \mathfrak{A}^{\alpha\mu\nu\sigma}{}_{,\sigma},$$

where $\mathfrak{A}$ denotes a tensor density that is antisymmetric with respect to all 4 of its indices. There now exists a covariant tensor $h\varepsilon_{\alpha\mu\nu\tau}$ which is antisymmetric in all its indices, whose components all have magnitudes equal to h. If we multiply the above equation by that tensor, we initially obtain

$$\psi\lambda h S_{\tau} = h\chi_{,\tau}, \tag{11}$$

where

$$S_{\tau} = h S^{\nu}_{\underline{\alpha\mu}}\varepsilon_{\alpha\mu\nu\tau} \tag{12}$$

is a covariant vector, and

$$\chi = \frac{1}{4}\mathfrak{A}^{\alpha\mu\nu\sigma}\varepsilon_{\alpha\mu\nu\sigma} \tag{13}$$

is a scalar. Instead of (11), we write

$$\psi\lambda \cancel{h} S_{\tau} = \chi_{,\tau}. \tag{11a}$$

Thus, in addition to ψ and λ, a third scalar χ appears as a result of the field equations in this theory. ±

We now assume that λ is a function of only ψ and χ:

$$\lambda = \Phi(\psi, \chi), \tag{14}$$

$$\frac{\lambda_{,\sigma}}{\lambda} = \psi\frac{\partial\log\Phi}{\partial\psi}\frac{\psi_{,\sigma}}{\psi} + \frac{\partial\log\Phi}{\partial\chi}\chi_{,\sigma}. \text{[10]} \tag{14a}$$

From (8a), $\frac{\psi_s}{\psi}$, i.e., φ_{μ} or $\Lambda^{\sigma}_{\mu\sigma}$ has been expressed in terms of the h. According to (14a) and (11a), we have

$$\frac{\lambda_{,\sigma}}{\lambda} = \psi\frac{\partial\log\Phi}{\partial\psi}\frac{\psi_{,\sigma}}{\psi} + \psi\frac{\partial\Phi}{\partial\chi}S_{\sigma}, \tag{14b}$$

where $\frac{\psi_{\sigma}}{\psi}$ and S_{σ} are linear functions of the first derivatives of the h. In order for $\frac{\lambda_{\sigma}}{\lambda}$ to be expressible in terms of the h alone, we must have

$$\psi\frac{\partial\log\Phi}{\partial\psi} = A = \text{const.}, \tag{15}$$

$$\psi\frac{\partial\Phi}{\partial\chi} = B = \text{const.} \tag{16}$$

It follows from (15) that $\log\Phi = A\log\psi + \log a$,

or
$$\Phi = \chi a\psi^{A}, \tag{17}$$
where a is a function of χ.
From (17) and (16), we find
$$\psi^{A+1}\frac{\mathrm{d}a}{\mathrm{d}\chi} = B. \tag{18}$$
This equation can be satisfied in two ways:

(1) $B = 0.$ $\quad \frac{\mathrm{d}a}{\mathrm{d}\chi} = 0.$ $\quad A$ arbitrary.

(2) $A+1 = 0.$ $\quad a = c\chi + \beta$ (α and β are constants.)

The first case contains the earlier system of equations as a special case for ($A = 0$). The second case is more interesting to us, since it—as will soon be seen—corresponds to a system of equations with a higher degree of determinacy than the earlier system. According to (17), it gives
$$\lambda = (\beta + c\chi)\psi^{-1}$$
or, with no loss of generality,
$$\lambda = (1 + \alpha\chi)\psi^{-1}. \tag{19}$$
Now (14b)[11] and (19) result in
$$\frac{\lambda_{,\sigma}}{\lambda} = -\frac{\psi_{,\sigma}}{\psi} + cS_{\sigma}. \tag{20}$$
This equation, however, can be valid[12] only in connection with
$$S_{\sigma,\tau} - S_{\tau,\sigma} = 0. \tag{21}$$
From (4), (20), and (21), we find the final result of this investigation. We must set
$$L^{\alpha}_{\mu\nu} = \Lambda^{\alpha}_{\mu\nu} + (\varphi_{\mu} - cS_{\mu})\delta_{\nu}{}^{\alpha} - (\varphi_{\nu} - cS_{\nu})\delta_{\mu}{}^{\alpha}, \tag{22}$$
where S_{σ} must satisfy equation (21).

In the case that $c \neq 0$, one thus finds it necessary to add the system (21) to the field equations, which implies an increase in the degree of their determinacy. The new system of field equations, (7), (8), and (21) is thus superior to the earlier equations from a methodological point of view, from which it was derived by means of a generalization. The occurrence of the arbitrary numerical constant c would appear to be significant from the physical point of view.

§2. The Field Equations in First Approximation.[13][14]

[p. 1]

$$\underbrace{(\lambda^{\alpha}_{\underline{\mu}\underline{\nu};\nu} + Q^{\mu\alpha})}_{G^{\mu\alpha}}{}_{/\mu} = 0 \qquad G^{\mu\alpha}{}_{;\mu} + G^{\sigma\tau}(c_1\Lambda^{\alpha}_{\sigma\tau} + c_2\Lambda^{\tau}_{\sigma\alpha} + c_3\Lambda^{\sigma}_{\tau\alpha}) \text{ [15]}$$

$$\Lambda^{\alpha}_{\mu\nu;\mu} + \Lambda^{\tau}_{\sigma\nu;\nu}(c_1\Lambda^{\alpha}_{\sigma\tau} + c_2\Lambda^{\tau}_{\sigma\alpha} + c_3\Lambda^{\sigma}_{\tau\alpha}) + Q^{\mu\alpha}_{;\mu} + \ldots \equiv 0 \quad \text{[16]}$$

$$(1)\ \frac{1}{2}(\Lambda^{\alpha}_{\sigma\tau}\Lambda^{\mu}_{\sigma\tau})_{;\mu}$$

$$(2)\ \Lambda^{\tau}_{\sigma\nu;\nu}\Lambda^{\alpha}_{\sigma\tau}\Big|(\Lambda^{\tau}_{\sigma\nu}\Lambda^{\alpha}_{\sigma\tau})_{;\nu} + \underbrace{\Lambda^{\tau}_{\sigma\nu}(\Lambda^{\alpha}_{\tau\nu,\sigma} + \Lambda^{\alpha}_{\nu\sigma;\tau})}$$

$$(\Lambda^{\tau}_{\underset{\mu\sigma}{\sigma\nu}}\Lambda^{\alpha}_{\tau\nu})_{\underset{\mu}{\sigma}} + \cancel{\Lambda^{\alpha}_{\tau\langle\nu\rangle}\Lambda^{\tau}_{\nu\sigma;\sigma}} + (\Lambda^{\overset{\mu}{\tau}}_{\sigma\nu}\Lambda^{\alpha}_{\nu\sigma})_{;\underset{\mu}{\tau}}$$

$$(3)\quad \Lambda^{\tau}_{\sigma\nu;\nu}\Lambda^{\tau}_{\sigma\alpha}\Big|(\Lambda^{\tau}_{\sigma\nu}\Lambda^{\tau}_{\sigma\alpha})_{;\nu} + \Lambda^{\tau}_{\sigma\nu}\underbrace{(\Lambda^{\tau}_{\alpha\nu,\sigma} + \Lambda^{\tau}_{\nu\sigma;\alpha})}$$

$$(\Lambda^{\tau}_{\underset{\mu}{\sigma}\nu}\Lambda^{\tau}_{\alpha\nu})_{\underset{\mu}{\sigma}} + \Lambda^{\tau}_{\sigma\nu}\Lambda^{\tau}_{\nu\sigma;\sigma} - \frac{1}{2}(\Lambda^{\tau}_{\nu\sigma}\Lambda^{\tau}_{\nu\sigma})_{;\alpha}$$

$$(4)\ \Lambda^{\tau}_{\sigma\nu;\nu}\Lambda^{\sigma}_{\tau\alpha}\Big|(\Lambda^{\tau}_{\sigma\nu}\Lambda^{\sigma}_{\tau\alpha})_{;\nu} + \Lambda^{\tau}_{\sigma\nu}(\Lambda^{\sigma}_{\alpha\nu;\tau} + \Lambda^{\sigma}_{\nu\tau;\alpha})$$

$$(\Lambda^{\tau}_{\sigma\nu}\Lambda^{\sigma}_{\alpha\nu})_{;\tau} + \frac{1}{2}(\Lambda^{\sigma}_{\nu\tau}\Lambda^{\tau}_{\nu\sigma})_{;\alpha}$$

$$+\frac{1}{2}(\Lambda^{\alpha}_{\sigma\tau}\Lambda^{\mu}_{\sigma\tau})_{\mu} + \frac{c_1}{2}\Lambda^{\tau}_{\mu\sigma}\Lambda^{\alpha}_{\tau\sigma} \qquad + \frac{c_2}{2}\Lambda^{\tau}_{\mu\nu}\Lambda^{\tau}_{\alpha\nu}$$

$$Q^{\mu\alpha} + \frac{c_1}{2}(\Lambda^{\tau}_{\mu\sigma}\Lambda^{\alpha}_{\tau\sigma}) + \frac{c_1}{2}\Lambda^{\mu}_{\sigma\nu}\Lambda^{\alpha}_{\nu\sigma} + \frac{c_2}{2}\Lambda^{\tau}_{\sigma\mu}\Lambda^{\tau}_{\sigma\alpha} + \frac{c_2}{4}\Lambda^{\tau}_{\nu\sigma}\Lambda^{\tau}_{\nu\sigma}\delta_{\alpha\mu}$$

$$-\frac{c_3}{\cancel{4}}\Lambda^{\tau}_{\nu\sigma}\Lambda^{\tau}_{\nu\sigma}\delta_{\alpha\mu} + \frac{c_3}{\cancel{2}}\Lambda^{\tau}_{\sigma\mu}\Lambda^{\sigma}_{\tau\alpha} + \frac{c_3}{2}\Lambda^{\mu}_{\sigma\nu}\Lambda^{\sigma}_{\alpha\nu} - \frac{c_3}{2}\Lambda^{\sigma}_{\nu\tau}\Lambda^{\tau}_{\nu\sigma}\delta_{\alpha\mu}$$

$$Q^{\mu\alpha} + \underset{0}{\left(\frac{1}{2} - \frac{c_1}{2}\right)}\Lambda^{\alpha}_{\sigma\tau}\Lambda^{\mu}_{\sigma\tau} + \underset{1}{c_1}\Lambda^{\sigma}_{\mu\tau}\Lambda^{\alpha}_{\sigma\tau} + \frac{c_2}{\cancel{2}}\Lambda^{\sigma}_{\mu\tau}\Lambda^{\sigma}_{\alpha\tau} + \underset{1}{c_3}\Lambda^{\tau}_{\mu\sigma}\Lambda^{\sigma}_{\alpha\tau}$$

$$+\underset{1}{c_3}\Lambda^{\mu}_{\sigma\nu}\Lambda^{\sigma}_{\alpha\nu} - \underset{0}{\frac{c_2}{4}}\Lambda^{\nu}_{\sigma\tau}\Lambda^{\nu}_{\sigma\tau}\delta_{\alpha\mu} - \underset{-1/2}{\frac{c_3}{2}}\Lambda^{\sigma}_{\nu\tau}\Lambda^{\tau}_{\nu\sigma}\delta_{\mu\alpha} = 0$$

$$Q^{\alpha\alpha} + \left(\frac{1}{2} - \frac{c_1}{2} + c_2 \cancel{+ 2c_2}\right)\Lambda^{\nu}_{\sigma\tau}\Lambda^{\nu}_{\sigma\tau} + (c_1 + c_3 \cancel{+ c_3 + 2c_3})\Lambda^{\sigma}_{\mu\tau}\Lambda^{\tau}_{\mu\sigma} = 0$$

$$Q^{\alpha\alpha} \neq 0.$$

$$Q^{\mu\alpha} - Q^{\alpha\mu} = 0 \qquad \frac{1}{h}\varepsilon^{\mu\alpha\sigma\tau}\Big| h\varepsilon_{\mu\alpha\sigma\tau}$$

$$\underline{G_{\sigma\tau}L^{\tau}_{\kappa\lambda}} \qquad G_{\sigma\tau\rho} \qquad G_{\sigma}\delta^{\alpha}_{\mu} \qquad \frac{1}{h}(hS^{\alpha}_{\underline{\mu\nu}})_{,\nu}$$

$$\underline{G_{\sigma\tau}\varphi_{\rho}} \qquad (G_{\sigma\tau\rho} - G_{\sigma\rho\tau}). + \ldots + (hS^{\alpha}_{\mu\nu})_{,\nu}\varepsilon_{\mu\alpha\sigma\tau}$$

$$\frac{G_{\sigma\tau}L_{\kappa\lambda}}{\underline{\Lambda} \quad \underline{S}} \qquad G_{\sigma\tau\rho\,,\lambda}\,\varepsilon \qquad S^{\iota}_{\kappa\lambda;\mu;\nu} - S^{\iota}_{\kappa\lambda;\nu;\mu} = S^{\iota}_{\kappa\lambda;\sigma}\Lambda^{\sigma}_{\mu\nu}$$

$$(hS^{\alpha}_{\underline{\mu\nu}}\varphi_{\mu})_{,\nu,\,\alpha}$$

[p. 2]

$$h_{\alpha\mu,\,\nu\nu} - h_{\alpha\nu,\,\nu\mu} = G^{\mu\alpha} = 0$$ [17]

$$G^{\mu\alpha}{}_{,\mu} = 0$$

$$G^{\mu\alpha}{}_{,\alpha} = h_{\alpha\mu\alpha\nu\nu} - h_{\alpha\nu,\alpha\nu\mu}$$

$$G^{\alpha\alpha} = h_{\alpha\alpha,\,\nu\nu} - h_{\alpha\nu,\alpha\nu/\mu}$$

$$\underbrace{G^{\mu\alpha}{}_{,\alpha} - G^{\alpha\alpha}{}_{,\mu}}_{= 0 \text{ when } \varphi_{\mu} = 0} = \varphi_{\mu,\,\alpha\alpha}$$

262. From Eduard Einstein

Zurich, 23 February [1930][1]

Dear Papa,

I'm coming at the beginning of March.[2] But can't stay that long and have to work a lot. Because I will probably already have exams in July in a subject that one would more reasonably assume is never mastered.[3] Besides, during this time, we want to again feed our annual need for theater visits. Do you also go to the cinema? I have often gone to films and balls recently. But we also want to discuss the problems of life. I have written lovely new psychological aphorisms.[4] Perhaps I will type them (I have a typewriter now) and send them to you. I'll suggest an exciting book to you: Class Reunion by Franz Werfel.[5]

For today, many regards (I will write again in more detail) and thanks for your kind letter.

Your

Teddy

I am doing very well.

Enclosing a spirited letter probably meant for you that arrived today. What unfortunate things you receive as letters!

263. From Myron Mathisson

Warsaw, 23 February 1930

[Not selected for translation]

264. To Mileva Einstein-Marić

[Berlin, after 23 February 1930][1]

Dear Mileva,

I am really looking forward to Tetel and will see that he is properly calm and comfortable here.[2] Prof. Maier[3] wrote to me that Tetel should ride the second class sleeping train;[4] I bow to high science and think it would not do much harm. But I see that, in general, the life of the little people, with exactly prescribed routes and the drudgery of everyday life, better preserves the equilibrium of their nerves than the Md.r.F = members of the idle rich class.[5] Removing even the smaller stresses and strains creates nervous weakness and lack of resilience, like much lying [causes] degeneration of the muscles. I'm angry that those in Dortmund will have a child after all.[6] This fate will run its course, as tragic as it is. They are coming in May to visit us in the timber cottage.

Kind regards from your

Albert

P.S. A gifted graphologist[7] has read from your writing that you had suffered from severe back pain. That's an accomplishment!

265. To the Editor of *Falastin*[1]

"Einstein und die Araberfrage"
[*Einstein 1930w*]

DATED [25 February 1930][2]
PUBLISHED 21 March 1930

IN: *Jüdische Rundschau*, 21 March 1930, p. 155.

[See Doc. 267 in the documentary edition].

266. To Deutsche Liga für Menschenrechte[1]

[Berlin,] 25 February 1930

[Not selected for translation.]

267. To the Editor of *Falastin*[1]

[*Einstein 1930s*]

DATED 25 February 1930
PUBLISHED 15 March 1930

IN: *Falastin,* 15 March 1930, p. 1.

[See documentary edition for English text.]

268. To Federigo Enriques[1]

[Berlin,] 27 February 1930

[Not selected for translation.]

269. To Myron Mathisson

[Berlin,] 27 February 1930

Dear Mr. Mathisson,

I have written to Prof. Bialobrzeski about your dissertation.[1] I would be happy to bear the costs related to your completing the doctorate, so that you are not burdened by them. It would be good if you would also write to Prof. Bialobrzeski on your own part, so that you will know whether or not you will have to submit a Polish translation. I have sent the German text to Prof. B. so that he can gain a first impression. If you need that copy also, you can ask him to return it to you. When the matter of your dissertation has been settled, I will try—presuming that you are in agreement—to obtain a fellowship for you, so that you can spend some time working here. Written correspondence is practically impossible for me, since I am much too overloaded with work.

I now see that you have made decisive mathematical progress through the methods introduced in your work, so that its publication in a journal—for example, the Annalen der Physik—would certainly be justified. Would it not be possible to dispense with the frightful Schouten notation?[2] That would be a great boon for the reading mankind.

I must say to you quite openly, that due to its lack of unification, the general theory of relativity in its present form no longer seems at all satisfactory to me. I am sending you my more recent works, from which you can recognize my new path.[3] The remarkable thing is now that the problem of motion in the new theory cannot be treated in analogy to the old one, in that one cannot operate here with singularities.[4] We can discuss all these things when you come here.

Send again news to your

270. To Frida Perlen[1]

[Berlin,] 27 February 1930

Dear Mrs. Perlen,

If three people sign for Germany, no more than one may be a Jew.[2] I am surprised that you do not realize this yourself.

While I regret having to cause you this disappointment, I am yours, with kind regards,

271. From Upton Sinclair

[Pasadena,] February 28, 1930

[See documentary edition for English text.]

272. On Masaryk's Eighthieth Birthday

"Zum achtzigsten Geburtstag Masaryks"
[*Einstein 1930p*]

PUBLISHED 1 March 1930
IN: *Die Wahrheit*, 1 March 1930, p. 4.[1]

Prof. Masaryk is a wonderful example of how love of one's own people can be reconciled with a cosmopolitan attitude and way of thinking.[2]

273. To Eduard Einstein

[Berlin,] Sunday [2 March 1930][1]

Dear Tetel,

Thanks for your letter.[2] I am happy I will now finally see you again and am convinced that you will enjoy your stay. You should not get too many impressions, only so many that you don't get downright bored. You will be best be taken care of if we stay in the city; on the other hand, it is more picturesque in the wooden house in Caputh, but care would be more difficult if we go there alone. Also, it is more inconvenient to come into the city (every half hour by Omnibus and electric tram). You choose.

I am in a very good mood, because my new theory is progressing so beautifully. In other ways as well, I am content, particularly with my health. I am curious to learn from you how the volcanic aspect of your nervous inner life operates.[3] Indeed, I am also meschugge, but only in *my* way. With me, there is much that is interesting, particularly in the correspondence. We can also talk about your impressions of medical studies. I think anatomy and physiology interesting but depressing because of the aloofness of the truly deeper problems. I have played much music, only classical. But I will make far-reaching concessions to you in this regard, if you want. Bring a few surprises with you again.

Warm regards from your

Papa.

274. To Maja Winteler-Einstein

Berlin, 2 March 1930

Dear Maja,

It is incomprehensible to me that I just now suddenly felt the need to write to you—impulses, even the most unbelievable, come over people without one knowing where they came from. Do you remember that our father also never wrote private letters, not even once, to his sisters?[1] A certain graphologist[2] recently got his hands on your writing. After he had spoken excitedly about your goodness, tranquility, and talent, he added "a lot of headaches." Hopefully that belongs to the past. I am happy with my life. Health has improved and I am having great luck with work. What I began back then in Scharbeuz has developed slowly but wonderfully.[3]

To do the calculations, I have the help of a brilliant mathematician, Walter Mayer, from Vienna, a splendid person whom you would really like.

At Planck's,[4] I heard raves of your domesticity and of the care that a lady from Hamburg with a curious name enjoyed at your home.[5] It must also be enchanting to be in this peace and serene life as you are leading it. I always struggle in my machinery of correspondence, which is received by many helpful contemporaries in frantic action. Tetel is coming here soon.[6] Some time ago, he had nervous symptoms, panic attacks, depression. He is better again, but who knows what will happen? Albert is expecting a child with his Donna,[7] which is also terrible. My descendants will considerably exceed me in insanity, I fear. Nothing to be done. You have chosen the right path.

Our Jews reveal themselves in the Palestinian-Arab matter as chauvinistic nationalists without psychological instinct or sense for fairness.[8] Good thing that they are powerless and don't have cannons. I cannot follow along if it does not get better. The collective is and remains an evil beast, worse than the individuals who compose it. I have contacted a leading Arab;[9] we'll see what may come of it? But, in any case, I'm staying here, so you don't need to have any fear that I will get a dagger in the dear stomach.

In the autograph collection of the League of Nations, I have recently provided the following contribution: "What the League of Nations did and what it refrained from doing has seldom inspired me; but I was always grateful for its existence."[10] I have long had a representative in Geneva[11] anyway and do not concern myself much anymore with the Committee.

In April, we are going to Caputh. Then I will sail again and vegetate with Mother Nature, the pleasant one.

Warm regards, also to Pauli,[12] from your,

Albert.

275. From Myron Mathisson

Warsaw, 2 March 1930

Dear Sir,

I thank you infinitely for your two notes and your intercession in my favor with Prof. Bialobrzeski.[1] You have refloated my boat, and I expect that it will now resume its voyage. I will not fail to report my news promptly to you. As for the assistance which you have offered, the thought that you would pay my expenses would be absolutely unbearable for me. The luck of having been noticed by you will serve as a counterweight for the annoyances of the task of writing.

The publication of my work would also make possible the publication of my subsequent ones.[2] In order to justify the Schouten's notation,[3] I permit myself to write *Gauss*'s formula in the form (which is, by the way, incorrect, at least if one does not restrict the permissible coordinates) used by *Laue* (Relativitätstheorie II, 1. edition, page 90):[4]

$$\int \operatorname{Div} P\, d\Sigma = -\int \sqrt{-g} \begin{bmatrix} P^1 & P^2 & P^3 & P^4 \\ d_{\mathrm{I}}x^1 & d_{\mathrm{I}}x^2 & d_{\mathrm{I}}x^3 & d_{\mathrm{I}}x^4 \\ d_{\mathrm{II}}x^1 & d_{\mathrm{II}}x^2 & d_{\mathrm{II}}x^3 & d_{\mathrm{II}}x^4 \\ d_{\mathrm{III}}x^1 & d_{\mathrm{III}}x^2 & d_{\mathrm{III}}x^3 & d_{\mathrm{III}}x^4 \end{bmatrix}$$

I am aware that printing theoretical works is certainly difficult. Accordingly, I have limited myself to what is most necessary.

It seems to me that the method which I have used is not necessarily tied to the concept of particle-singularities,[5] since the singularities which play a role here could be interpreted as purely mathematical constructions, leading to the elimination of the dimensions of the electron. This is quite similar to the Newtonian potential outside a small sphere of variable density. The expansion of that potential, in terms of spherical functions, appears to have poles. But the analytic character of an *infinite* expansion of singularities is different from that which one finds in the case of true singularities. In my work which I sent to you, this distinction does not enter. It played a central role in my research into the uniqueness of the solutions.

I know, Sir, that you have sought repeatedly to establish equations which allow regular solutions for particles. Wave mechanics would seem to favor the concept of continuous matter. Louis de Broglie, by the way, saw a grave defect in that theory, of which he was one of the founders. He is an adherent of singularities. Prof. Schrödinger himself has written that the constancy of the mass and the charge for an energy packet seem completely incomprehensible. I dare cite these opinions of the two pillars of wave mechanics only to underline the close connection that I have found between the truly singular character of particles and the constancy of *e* and *m*. I understand that the ideas which I have quoted, as well as their authors, are quite well known to you. I also understand that my ideas can be discussed only when I will be in the happy situation of seeing you in person. I hope that it will not take long for this moment to arrive, and that fills me with joy.

Please accept, Sir, the assurance of my gratitude and of my profound devotion.

Myron Mathisson

276. To Robert Weltsch

Berlin W, 3 March 1930

Dear Mr. Weltsch,

I have sent the editor of "Falastin" an answer, as far as I am concerned.[1] Since this answer contained concrete suggestions, I did not want to burden your conscience with it because, otherwise, you could have come into an uncomfortable position vis à vis the Zionist organization.[2] Your article interested me extraordinarily. If our people were reasonable and just, they would, at least retrospectively, appreciate this evidence of higher insight. I am returning the article to you with the same letter.

I am also sending you a letter, the content of which has made a downright devastating impression on me.[3] If what is disclosed in it really corresponds to the facts, and the Zionist organization is actually guilty of breach of trust in such measures, for me, there would only be one [thing to do]: to break off every connection and resign from the Agency.[4] I would be indebted to you if you could assuage me somewhat with insight into this matter.

Kind regards, Your

A. Einstein

277. From Mileva Einstein-Marić

[Zurich, between 3 and 9 March 1930][1]

Dear Albert,

Your views on possible mollycoddling are certainly quite correct,[2] but I don't believe you also would recommend that someone with an injured leg dance the Charleston as much as possible, or ask someone whose arm was noticeably painful to play the piano as much as possible. And when this poor Tete longs for you after overcoming an illness, and you don't want to visit him for a while, and he decides to make the long journey himself just to be able to be with you a bit, one cannot then put him in the carriage that is the hardest and most uncomfortable and shakes the most because his nerves are a bit more sensitive at the moment.[3] I beg you to view the matter like this, and don't reproach Teddi. We live so simply, on the whole, and handle everything ourselves, so that Tete must also provide many things for himself and gladly busies himself with everything. At the moment, he only must be cared for. I would like to request another favor that you will surely grant me. It concerns your statement: I am also meshugge, which is placating in your letter to Teddi.[4] I beg you not to say such a thing to Teddi. I believe that you will hardly find any sort of difference in him, compared with earlier, but there also never was one to be found; he has just suffered very, very much. But it would be cruel to say to the boy, even only in jest: You are crazy; I believe that would make him infinitely unhappy. It would perhaps be best if not much is said about his illness, because that causes him to stew over it. I would have found it much more correct if he did not travel for the time being, but he has such a strong desire to see you that one could not refuse him. But I beg you to care for him properly or, if it does not go well for some reason, send him back to me quickly.

I don't understand at all why you gave my handwriting to a graphologist.[5] To what end? Any such interpretation of my handwriting can be of no interest to you.

It would interest me to hear how you find Teddi and how he will get there. Would you please write me a few words about it?

Kind regards,

Mileva

278. To Maurice Solovine[1]

[Berlin,] 4 March 1930

Dear Solovine,

It has taken a little while to finish reading your Democritus[2] because I was obsessed with my own work and also deterred by many other things. Incidentally, the old copy turned up again in the meantime.[3]

I enjoyed your introduction the most. The depiction of Democritus's relationship to his predecessors[4] seems to me to be very well done. For me, at least, a light bulb went off. (Reconciliation of the rigid absolute with the amorphous changing in atom and movement.[5] Worthy of admiration in the original is the treatment of the qualities of the senses. Moving, how he labored with the sense of sight, tenaciously holding on to the fundamental thought.[6] Among the moral aphorisms are some really good ones, but many [are] weirdly bourgeois (herd-animal-moral theory).[7] On the whole, the translation seems to be really well-done, as far I can judge with my poor French. The strong belief in physical causality, which also does not stop for the will of homo sapiens, is admirable.[8] As far as I know, Spinoza[9] was the first to again become so radical and consistent.

My field theory is making good progress. Cartan[10] has worked beautifully in that respect. I myself am working with a mathematician (S. Mayer from Vienna), a splendid chap who would have had a professorship long ago were he not a Jew.[11]

I still think often of the beautiful days in Paris[12] but am happier with my comparatively quiet local existence. Don't hesitate to contact me if you think that I can do anything that you want, and be warmly greeted from your

A. Einstein

279. To Julian Liban[1]

[Berlin?, after 5 March 1930][2]

[Not selected for Translation.]

280. To Allied Jewish Campaign

"An die alliierte jüdische Kampagne"
[*Einstein 1930r*]

PUBLISHED 6 March 1930
IN: *Jewish Daily Bulletin*, 6 March 1930, p. 2.

[See documentary edition for English text.]

281. To Felix Iversen[1]

[Berlin,] 6 March 1930

Dear Prof. Iversen,

By way of thanking you very much for your valuable clarification,[2] I am enclosing a copy of a letter that I sent along with the same message to the Finnish minister of war.[3] I am particularly pleased to find a like-minded peer on these important human areas. How lovely it would be if one could mobilize an imposing phalanx of recognized productive minds for the struggle against the barbarism of the military and warfare.

Kind regards, your

282. To Walter Loeb[1]

[Berlin,] 6 March 1930

[Not selected for translation.]

283. To Juho Niukkanan[1]

[Berlin,] 6 March 1930

Dear Mr. Minister,

I have been enlightened by like-minded friends from Finland that my letter of commendation addressed to you was based upon false assumptions.[2] Then, from the standpoint of morals and fairness, it must be demanded that conscientious objectors are only employed for such services that are not closely linked with military affairs. This condition was not satisfied with your citizen A. Pekurinen,[3] and it therefore cannot be regarded as morally justified that one has imposed a dishonorable punishment on him due to his refusal to serve.

Respectfully yours,

284. To Karl Spiecker[1]

[Berlin,] 6 March 1930

Dear Mr. Spieker,

If I can be certain that the film will not be used to earn money on the basis of the desire for sensation, I would not mind if it is made.[2] Far more has already been done to my person than allows itself to be combined with good taste.[3]

Kind regards, your

285. From Luther Pfahler Eisenhart[1]

[Princeton,] 8 March 1930

[See documentary edition for English text.]

286. To Marie Curie-Skłodowska

[Berlin,] 10 March 1930

Dear Mrs. Curie,

I absolutely share your opinion about pursuing the standardization of bibliographies and similar technical tasks.[1] I can't report anything about the detrimental activity of Mr. Krüss from the standpoint of international rapprochement because I do not have anyone behind me here.[2] My replacement by Krüss has its good side, in spite of everything, in that such a genuine connection of the commission with German intellectual circles is being established,[3] which would have been impossible through my intercession. In the long run, I am convinced that it will be beneficial if I am replaced by Mr. Krüss.

Kind regards, your

287. From Sergei N. Bernstein[1]

[Kharkov,] 10 March 1930

Esteemed Colleague,

From the 24th to the 29th of June [this year], the Mathematicians' Congress of the USSR will take place in Kharkov.[2] Foreign mathematicians are also being invited to this congress, and many of them (e.g., Hadamard, Borel, Lévi-Cività, Blumenthal, Lichtenstein)[3] have accepted the invitation from the organizing committee to participate in it.

Now the organizing committee is contacting you with the same request: we and all our Russian colleagues would be very happy to see you at our congress and to hear a general lecture from you on a topic of your choice.

During your stay in Kharkov, the organizing committee will provide you with accommodations, free of charge, and will pay an honorarium of 250 rubles.[4]

If you will have the desire to make a trip to Moscow, we will make an effort to find you cost-free living quarters there as well.

We hope to receive a positive answer from you as soon as possible, and would ask you to mention the topic of your talk and, if possible, to give a brief summary of its contents (abstract) at the latest by 25 March 1930.

With greatest respect, the chairman of the organizing committee,

S. Bernstein

288. To Heinrich Zangger

[Berlin, 12 March 1930][1]

Dear Zangger,

We are truly happy together, more than could be explained from the fact or consciousness of common blood.[2] The boy has the instinct to not overwind the bobbin again.[3] He did not have the healthy inertia and overloaded himself, be it by simply going along or from ambition, which, incidentally, is not as different as one would like to think. The course of studies—as it has evolved—is actually a serious threat to the nervous balance and intellectual character. I held myself for a true idiot in the early days of study and had difficulty maintaining balance, given the overstimulation. After the conclusion of my studies, I was put off thinking for a whole year, which later became my refuge, my blue heaven.[4] I believe, in all seriousness, that the correct method for the education of a person has not yet been found.

I am well, particularly at work. I believe in my new path, which has again brought me into total intellectual solitude. It is delightful and full of tension. Tetel wrote to you about "clairvoyance."[5] That is a crazy matter. A woman of fifty-five years was sitting there.[6] She was given pieces of jewelry, pencils, pocket watches. She takes an object and touches it. The man or woman concerned comes forward. "You have had gas poisoning ('inhaled poison'). You work in a large house and are feared by your subordinates." So it goes on, with great accuracy. Here fact, here reason, both in hopeless conflict.

Warm regards, your

A. E.

289. To Czesław Białobrzeski

[Berlin,] 13 March 1930

[Not selected for translation.]

290. To Myron Mathisson

[Berlin,] 13 March 1930

[Not selected for translation.]

291. To Margarete Lebach

[Berlin,] 15 March 1930

[Not selected for translation.]

292. From Myron Mathisson

Warsaw, 15 March 1930

[Not selected for translation.]

293. To Cyril Clemens[1]

Berlin W, 18 March 1930

Dear Sir,

Considering that Mussolini[2] is honorary president of your society, I regret not being able to accept your kind offer.[3]

With the utmost respect,

A. Einstein

294. To Martin Meyer[1]

[Berlin,] 18 March 1930

[Not selected for translation.]

295. To Jay Morein[1]

[Berlin,] 18 March 1930

Dear Sir,

In the theoretical sciences, the external cause for discoveries is always based in the current state of empirical and theoretical knowledge. In my work, I cannot specify what could be described as an *external* cause for my theories. The first question was: how does the representation of a beam of light depend on the state of motion of the system of coordinates to which it is being related? The second question: Upon what is the equality of inertial and gravitational mass of bodies based? The third question: Can gravitational fields and electromagnetic fields be understood theoretically in a unified way?[2]

These questions characterize my real life's work. What I came up with, otherwise, was more the work of chance and related to current problems of physics.[3]

Respectfully yours,

296. To Erwin Schrödinger[1]

[Berlin,] 18 March 1930

Dear Schrödinger,

Thank you sincerely for the incredibly punctual shipment of the transcripts.[2] With Marianow, nothing was correct; with Officer Lungenschuss, just as few expeditions, and far more headache than usual.[3] In this case, no one present knew further particulars, in contrast to the two successful cases. In the case of Rathenau,[4] there is suspicion of previous knowledge. In these things, producing some halfway reasonable objectives would probably require a lifetime, such that we must leave it to others.

Warm regards, your

Einstein.

297. To Max Weinreich[1]

[Berlin,] 18 March 1930

I hereby call upon you to remove me as the honorary chairman of your board of trustees, as I have learned that the Yiddish Scientific Institute has acted in an anti-Zionistic sense.[2]

Respectfully yours,

298. To Heinrich Werneke[1]

[Berlin,] 18 March 1930

Dear Sir,

Undoubtedly, the goal of the detoxification and humanization of the German schoolbook is one of the most important and urgent tasks that a true friend of humanity must face in this country.[2] Without a doubt, much too little has happened in this area. With the dreadful barrage of publications that reach me, I have not seen your *Primer on Ethics*.[3]

The guiding star of a new primer would have to be: the fostering of humane and tolerant sentiments and a healthy human understanding, along with the renunciation of polemics and mockery. Moreover, above all, high quality in conjunction with plainness and simplicity.

Unfortunately, I myself am far too overburdened to be able to judge and select myself. But if a recommending judgment is available from pedagogues who have my confidence by virtue of their other statements, the book will not lack my moral support. I lack the means for material support. Finally, I would like to advise you to get in touch with the Franco-German Society, p.A. Mr. Dr. O. Grautoff,[4] Berlin W. 30, Haberlandstrasse 2, as, in my opinion, they are likely to be in contact with the appropriate persons.

Kind regards, Your

299. To Ralph Lyndal Worrall[1]

[Berlin,] 18 March 1930

Dear Sir,

As far as I could see from a superficial survey, I agree with you.[2] All of physics is, in that sense, *realistic*[,] in that it starts from the hypothesis of a reality which is independent of our perceptions and thinking.[3] It is only, to that extent, *no longer materialistic* in that it does not recognize matter as the irreducible conceptual basis of the theoretical system of physics.[4]

Yours respectfully,

300. From Erwin Schrödinger

[Berlin,] 18 March 1930

Dear Mr. Einstein,

Many thanks for your kind letter.[1] It was, for me, a rather great relief and reassurance that these matters, when one approaches them a bit more closely, become, if not really understandable, then at least considerably *more* understandable. Even if we still do not understand everything, we can indeed extrapolate and think to ourselves: if we could investigate more precisely and systematically, then the supernatural nimbus would vanish, more and more, until, finally, nothing would remain except, at most, an enormous amplification of abilities which one knows from ordinary life, e.g., of the ability to feel the unspoken and unconscious expression of will or opinion of someone present, without having to speak a word.

Your letter brought me a second uncommonly fine joy, which a person only very seldom experiences in this form. You, probably completely unconsciously, made use of a different form of address than previously, namely *Du* instead of *Sie*.[2] That little confusion made me very happy. The sympathy which I feel for you is of such a very special sort, so completely different from what I have ever felt for a man before.

I have here not answered, yet, by using your most kind and friendly *Du*, since it might still be possible that you, for some sort of external reasons, do not wish to make the one-time, hurried mix-up into a norm for the future. It is indeed only a superficiality, which is not of great importance. But in any case, it made me happy as a fine, hardly conscious indicator of feelings, very, very much so.

With the heartiest greetings, yours sincerely,

Schrödinger

301. On Progress of the Unified Field Theory

"Über die Fortschritte der einheitlichen Feldtheorie"
[*Einstein 1930t*]

DATED 20 March 1930

IN: *Preußische Akademie der Wissenschaften* (Berlin). *Physikalisch-mathematische Klasse. Sitzungsberichte* (1930): 143.

X. Meeting of the Physical-Mathematical Class. 20 March

Presiding Secretary: MR. PLANCK

*1. MR. EINSTEIN spoke "On Progress of the Unified Field Theory."[1]

Various points of view are described which have led to the field equations. Along with the requirement of general covariance, which is a matter of course in the general theory of relativity, the need for the equations to be compatible plays the most important role, whereby there are more equations than are required by determinism. For the integration, the requirement of freedom from singularities is added. The results obtained thus far are compared with empirical evidence.

302. On the Death of Lord Balfour

"Zum Tode von Lord Balfour"
[*Einstein 1930v*]

DATED 20 March 1920[1]
PUBLISHED 21 March 1930

IN: *Jüdische Rundschau*, 21 March 1930, p. 154.

Einstein sent the J. T. A. the following declaration:

Hardly a person has done so much for the moral recovery of the Jewish people as Lord Balfour. His name will live on among us for as long as there is Jewry.[2]

A. Einstein

303. To Paul Ehrenfest

Berlin W, 21 March 1930

Dear Ehrenfest,

I am coming to you around April 1 in order to present the theories to you before your holiday.[1] I am very much looking forward to it. Schur must have written to you, as he told me and will gladly do what he can.[2] About graphology and things of greater dubiousness, I have never officially expressed myself and have also not gathered enough proper experiences for that purpose either. However, people have readily abused my imprudence of sticking my nose in on such occasions for advertisement.[3] With me, every peep becomes a ⟨trumpet blast⟩ trumpet solo.

Warm regards, your

A. Einstein.

304. From Leo Szilard

London, 22 March 1930

Dear Professor,

I wanted to report to you that I have stayed in London longer than planned.[1]

I send you here the last report from the eastern front.[2] Up to now, the best pump, with a compression ratio of 1:5, had an efficiency of 14%.

The following as an explanation of the symbols:

Machine type *K1* is a machine which we have begun to construct, alongside the large experimental apparatus, and which we want to develop into a complete cooling machine which works with solid potassium. One indeed cannot work with pure potassium in the large machine, since heating the whole apparatus would be too complicated.[3]

In the meantime, I have considered the matter of applying in America for the cooling system for the whole house, and, for reasons of clarity, I applied for the invention in my name alone. As the patent attorney told me, it would present no difficulty, if we should find the matter to be promising, to transfer the patent to several names in common.

The title of the patent is "Refrigerating Plant," and it was applied for in America in March 1930. It will be considered to be common property, like all the cooling inventions.[4]

As regards Bihaly,[5] I should like to suggest that we set up the following: Previously, in fixing [the conditions], we determined a certain participation for Kornfeld;[6] one-seventh of the profits realized are to go to the co-workers, and we will give the first 50,000 reichsmarks (within the framework of this profit share of one-seventh) to Kornfeld, so long as certain preconditions, which we specified at that time, are met. I would now suggest securing the next 25,000 reichmarks, which are earmarked for the coworkers in the same manner for Bihaly, with the same preconditions as in the case of Kornfeld.

I am enclosing a postcard with the request that you use it to confirm the agreement regarding Bihaly, if you agree with the proposal.

So much for the cooling. Concerning myself, I would like to report that the ideas of which I previously gave you an exposé are slowly taking on a definitive form in my head; but there is no danger of their becoming an obsession.[7] I believe that in the course of the coming year, I will divide my time between the AEG. [Allgemeine Elektrizitätsgesellschaft] and a modern religious pilgrimage, in which I will slowly but surely search out one person after another to talk to them about everything and, if an opportunity arises, talk about these ideas.

If thereby one has formed a conception of the attitude of some twenty people in Germany, France, and England, then one can approach them all at the same time and take a first step in bringing about an initially loose organization. Of course, no one can be expected to overcome their reserve and agree to be the first to take part. This would be asking too much. But I think one can proceed as follows: One draws up a list of people who have shown great sympathy and who are assumed to show a certain willingness. The question would then be asked whether they would be willing to take part, provided that at least half of those on the list are also willing to take part.

The preparation for this project is a great deal of work, because it is not enough to know the individual people, but one also has to know what they think about each other and how they relate to each other, otherwise, all will become a fine mess. And when you've got the whole thing, you've only ever got something like a "Society of Friends of..."; but I have a good conscience about it, and that is the safest criterion.

When I'm back in Berlin, I'd like to talk to you about it in peace. I am traveling to Berlin as soon as the work that I brought with me here ends, or when the war bugle of the AEG calls. It probably will happen between April 1 and 10.

With cordial greetings, your very devoted

Leo Szilard

P.S. I would very much like to know if you are well. If you could write half a word about it (that you can conjure up) on the attached postcard, I would be very happy.

305. To Hans Albert Einstein and Frieda Einstein-Knecht

[Berlin,] 24 March 1930

Dear Albert and dear Friedi,

As long as it made sense, I did what I could to turn fate around.[1] Now, however, the fact is there, and I reconcile myself to it as always. I am happy that you are coming and hope[2] that the rural tranquility will do you as much good as [it does] me. If you still have the notion to transition into the field of patents, (1) Albert, you can speak with Seligsohn[3] and another man here, who I have made aware of you. The matter does not seem to be too difficult as it appeared after the threatening letters. I believe that one can achieve more there with healthier intelligence than in your local factory. It is going pretty well with me, both in regard to health as well as with work. The new theory now also finds much lively interest in addition to the skepticism of colleagues. I myself am working diligently, together with a fine Viennese mathematician[4] to whom the local academy has given a fellowship for this purpose.

Best regards to you both, your

Papa

I have spent lovely days with Tetel.[5] He went home again yesterday. He is good and clever, but one must fear for his nerves. In any case, he must be greatly spared.

306. To Luther Pfahler Eisenhart

Berlin, 24 March 1930

Dear Mr. Eisenhart,

I am very happy to hear that you are interested in the new theory,[1] which, from the formal standpoint, is really very satisfying. In the matter of its physical interpretation, I have not yet made much progress, but I have great hopes. The idea of teleparallelism occurred to me two years ago, while I was resting due to illness, without my knowing at the time that this concept had already been considered by you and other mathematicians.[2] For me, the lawful [gesetzliche] determination of such a manifold has been the principal problem, but I am also concerned that proper credit be given to those who thought of connecting the Riemannian metric with teleparallelism before I did. If you want to publish a remark on this topic with respect to Cartan,[3] that would make me quite happy. I would also be very willing to submit such a note to our Academy, since it would be nice if the whole discussion about this point were to appear in our Meeting Reports.[4]

With hearty greetings to you, to Mr. Veblen, and to Mr. Wigner,[5] yours truly,

A. Einstein.

307. On the Bezalel Academy[1]

[*Einstein 1930gga*]

DATED 28 March 1930
PUBLISHED 1 August 1930

IN: *The Palestine Weekly,* 1 August 1930, p. 1

According to my conviction, it is urgently necessary to keep up the Bezalel School of Art. The artistic and scientific spirit must be cultivated in Palestine; for only thus will Palestine be felt for all times by the Jews to be a Spiritual Centre. The Jewish community has survived through the millennia as a cultural, not a political group, this we must bear in mind in our desire that Eretz Israel should be and should remain the embodiment of the Jewish corporate ideal. [2] [3]

Berlin, March 28, 1930.

308. To Chaim Herman Müntz

[Berlin,] 28 March 1930.

Dear Mr. Müntz,

I thank you very much for your interesting letter. You should not regret that you have not been able to dedicate more of your time to purely scientific work in these months. There are other, no less important things to experience, and that is what you are doing. It is unforgivable that I wrote the equations incorrectly for you and that this caused you to lose some time. But you already had the commutation relations from which the equations readily follow. I have now solved the centrally symmetric problem, together with Mr. Mayer.[1] You have made it more difficult for yourself by not having suitably specialized the coordinates right from the beginning (one can make the $h_1{}^4 h_2{}^4 h_3{}^4$ vanish). We are now indeed quite near to a solution of the matter problem (solutions free from singularities). But before everything is consistent, one cannot know if the path taken will lead to success. The question that you have so diligently investigated, namely, that of the compatibility of partial differential equations, was solved in a wonderful way by Cartan, but his results have not yet been published.[2] You will have fun with them. In the meantime, my proof of compatibility will suffice, particularly since it is given in the Academy reports making use of the simpler form of the third identity as shown by Cartan.[3] The theory is now completely satisfying for me. Only the question of its validity remains almost completely unsolved. We are, however,—as it seems—on the right track. I am happy that I was able to persevere. You know best how difficult that was. In hopes of seeing you this summer in Caputh and with hearty greetings to you and your wife, yours truly,

A. E.

309. To Albert Hellwig[1]

[Berlin,] 31 March 1930

Dear Dr. Hellwig,

It was undoubtedly careless of me that I allowed myself to get carried away by curiosity to have a closer look at two so-called clairvoyants who then immediately made dishonest advertisements using my name.[2] With the first one, a so-called "metagraphologist,"[3] I noticed that he often gathered from the writing amazingly sure judgments about character and bodily features, but that, when he does not see or hear the concerned persons, [he] cannot say anything accurate without a specimen of handwriting. With Akkaringa, where, supposedly, information was given over the history of objects, the presence of the concerned person was likewise necessary for moderate success. I could not bring out further details about the way she operated, because I have not been able to carry out enough pure experiments. The most extreme thing that would have come into consideration there, for example, that I have not been able to rule out with certainty, would be the so-called "telepathy." For the moment, however, I do not believe in it either.

Since I, in no way, would like to further encourage superstition, and I lack the time for sufficient careful observation, I have decided to stay away from such events in the future.

Respectfully yours,

310. Foreword to *Reichinstein 1930*

"Geleitwort"
[*Einstein 1930y*]

COMPLETED ca. April 1930[1]
PUBLISHED 1930

IN: D. Reichinstein. *Grenzflächenvorgänge in der unbelebten und belebten Natur*. Barth: Leipzig 1930.[2]

I take the liberty of saying a few words at the start of this book despite the fact that I am not competent to offer a judgment on the system of assumptions posed within it. The excuse for this lies in the conviction that the book appears to me to reveal a crucial gap in the current detection of the mechanism of heterogeneous chemical reactions and develops a possible way to eliminate this gap, whose critical examination I would like to commend to the attention of experts.

If the oxidation rate of the solid phosphorus in oxygen in its gas form rises and, with still further increased concentrations of oxygen, falls again, this fact cannot be reconciled with the law of mass action in the usual interpretation of surface reactions (assuming of course that contamination plays no role.)

The author attempts to bring many related facts from the most diverse subjects under a single viewpoint, the most essential of which are the following:

(1) In the reaction layer,[3] which is treated as a phase, the law of mass action applies, whereby the concentration of the substance forming the solid phase (as well as those of the remaining reactants) is treated as variable (a prior unknown and respectively dependent upon the concentration of the remainder.)

(2) The sum of the concentrations of all the substances in the reaction layer is constant (displacement principle).

The second hypothesis should only claim the character of a rough approximation.

The hypothesis of a variable concentration of the solid reactant is undoubtedly very paradoxical from the standpoint of molecular theory. Whichever way the decision may finally turn out, the present book will stimulate excitement, in any case, through the problems it presents.

A. Einstein

311. To Elsa Einstein

Leiden, 1 April 1930

Dear Else,

I was *very touched* by the flowers. Otherwise, I was really well cared for; only the travel party was lousy, and in Amsterdam, I shuffled around tremendously. But it was unnecessary! The children picked me up here while Ehrenfest was in a colloquium.[1] I am very well, and, this time, I swear on my honor to return in decent condition.

Best regards from your

Albert

312. To Elsa Einstein

[Leiden, 3 April 1930]

Dear Else,

With the flowers and the tobacco, you have touched a soft spot in my heart. I must repeatedly think about how very happy I was upon opening the packet. Wednesday, I lectured, and tomorrow once more, each an hour and a half, roughly like in Paris.[1] However, it is very comfortable. This evening, I'm going to the de Hasens[2] to chat. This afternoon, I was at de Sitter's,[3] yesterday as well because of a scientific question. Ehrenfest's children[4] are delightful. We have done E. a total injustice with the money. I had pleaded not to give me any additional payment. However, the payments were maintained; however, in exchange, the stipulation was made, on the other hand, that smaller expenditures in the service of scholarship in the broader sense may be taken from the fund, the rest [of it], meaning the majority, belongs to me.[5] The distributions for smaller expenses must always be presented to me and Mr. Fokker for approval, which has also happened every time. Everything is also in the best order, and E. owes us absolutely nothing and has not taken anything from us. I am happy to be able to report this to you. Now I can arrange for a separation into different accounts of that which is for me and that which should be available for other purposes. I'm not doing that, however, because it isn't worth it. I'm taking the travel expenses out now, and something over that, but leaving the rest here.

I'm thinking to come home Monday or Tuesday, but I don't entirely know for sure yet. In the meantime, warm greetings to you and Margot[6] from your

Albert

313. To Siegfried Jacoby[1]

[Berlin,] 9 April 1930

Dear Mr. Jacoby,

Unfortunately, the letter I alluded to in my letter is at my country house in Caputh.[2] As soon as I am out there, I will look for it and make the content available to you. It would please me very much if the accusations raised against you in the same [letter] proved unfounded.[3] The assertion of the not entirely appropriate use of your earlier connection to me gained probability for me because, a few years ago, you tried, without informing me, to use this relationship vis à vis the Foreign Office in order to gain access to certain files for yourself.[4]

Friendly greetings, your

314. To Paul Painlevé

[Berlin,] 9 April 1930

Dear Mr. Painlevé,

I am very pleased that, through your letter, you are prompting me to express my opinion over the proposals of the German national delegation which, to my great regret and with the Foreign Office's emphatic disapproval, were raised for a decision in the lap of the national delegation, and precisely in a session in which only about half a dozen members of the committee were present.[1] The editorial work of the proposal comes from the pen of Mr. Krüss, and nothing about the content was announced to us *before* the session; that there was no time to lose was the justification given for this unusually hasty proceeding on such an important matter.[2]

Above all objective questions, there must be the strong will to promote the international cooperation and mutual trust upon which all beneficial cooperation rests. The initiative of the German national delegation is in no way suited to promote the spirit of mutual trust. If one wanted to influence the organization of future work, then one should have started with confidential conversations with the French entities as self-evident acknowledgment of the fact that the burdens of the Institute have thus far laid exclusively on the shoulders of France.[3] This would have been all the more imperative, as, from the German side, not much positive work in the service of the CICI has been achieved yet.

As far as the objective arguments are concerned, I agree with you concerning certain critical remarks which are listed in the memorandum. But on the other hand, I am convinced that [we] *must absolutely endeavor that the Institute remains an undivided unity.* If sub-Institutes were to be established in different countries, for which there is unfortunately already an existing precedent, the value of the Institute for the development of international spirit would be lost.[4] We must not make concessions to national narcissism and jealousy, this evil hereditary disease of European history, if the ICIC wants to remain true to its higher purpose.

However,[5] I must still say one more thing on this occasion: considering the current prevailing mentality, I have always regretted that the Institute was founded in Paris and exclusively with French funds. To be sure, this happened from noble motives, but it could not but provoke powerful suspicion in such a politically turbulent time. This suspicion could not be completely dispelled even by Mr. Luchaire's[6] truly exemplary administration in terms of international objectivity.

I am convinced that, in terms of international understanding, it would be an act of great merit if the proposal came from the French: We want to relocate the Institute as a whole to Geneva, and all states should contribute to its financing according to a fixed rate. After the great sacrifices France has made, such a proposal may not come from the other side;[7] but if it came from France, one would so thankfully welcome this act of self-restraint as a meritorious deed in the service of strengthening the international ideas everywhere.

This is only but a wish in the long term. First and foremost, the requirement of the hour appears to me to be maintaining the unity of the Institute.

Kind regards, your

A. Einstein[8]

315. To Joel Agustus Rogers[1]

[Berlin,] 9 April 1930

Dear Sir,

The quoted interview actually took place, and I said something like what was in the article.[2] Of course, I do not deny [the existence of] racial differences among people altogether, but I am convinced that they are grossly overestimated in our time of nationalistic sentiment.

Respectfully yours,

316. To Otto David Tolischus[1]

[Berlin,] 9 April 1930

Dear Sir,

Prof. Freundlich's claim regarding the deflection of rays is that, through the previous solar eclipse recordings, probably the existence, but not the exact numerical value, of the deflection is ascertained.[2]

It is also true that, for two years, I have been working on a new foundation for the general theory of relativity that treats gravity and electricity as a unified field. The calculation of the implications of this theory has, however, not yet progressed enough to allow for a comparison of the theory with experience.[3]

Respectfully yours,

317. To Caesar Koch

Berlin, 11 April 1930

Dear Uncle, My Dears,

Now the feared and expected has occurred.[1] You, first uncle, have all along not been coddled by destiny for as long as I can remember and must now, in old age, experience the worst thing that can happen to a father. At the same time, you are by nature joyful and thus able to be cheerful and content like virtually no one else in the family. But you will get over this difficult pain; all of us who love you fervently hope for that. It is comforting that the good one did not have to suffer as bitterly, physically and emotionally, as is often the case with this illness. In the end, it comes down not to the length, but rather to the content of our life, and that is, with all of you,—as far as it depends on natural predisposition—a harmonious and pleasant one. This predisposition is the best [thing] that nature can give to us to take with us on life's path.

Things are going well here, even if Elsa and I are not quite alright in the heart. But we have a little wooden house nearby in which we have more peace than in the city. While I sincerely wish that, together, you all will soon overcome the deep pain to some extent; I am yours, with warm regards,

Albert

318. Review of *Weinberg 1930*

[*Einstein 1930aa*]

DATED 12 April 1930[1]
PUBLISHED 30 May 1930

IN: *Die Naturwissenschaften* 18 (1930): 536.

[Not selected for translation.]

319. To Hugo Andres Krüss

[Berlin,] 12 April 1930

Dear Mr. Krüss,

As I see from your report that significant differences of opinion exist between us regarding the future of the International Institute for Intellectual Cooperation; I have decided not to allow myself to be represented by you this summer, but rather to take part in the meetings myself.[1]

Yours with kind regards,

320. To Upton Sinclair[1]

Berlin, 12 April 1930

Dear Mr. Sinclair,

Your letter causes me no small embarrassment.[2] I have dealt with two people, so-called "metagraphologists."[3] The result was that, other than certain uncanny intuitive determinations from the character of handwriting, I have not been able to establish anything reasonably certain, and the people have made an improper advertisement with the help of my name.[4] Because I don't want to encourage the current rampant superstition and lascivious sensationalism of the public, and I do not have at my disposal a competent opinion of the issues, I do not want further use to be made of my name in this context in public.

However, this does not preclude me from recommending the book of such an important author, as you are, to a publisher after examination if I were assured that no use would be made of my name publicly in this context. Under this condition, I am happy to comply with your request.

Kind regards, your

A. Einstein.

321. To Max Weinreich

[Berlin,] 12 April 1930

[Not selected for translation.]

322. From Joseph Sauter

Wabern, 12 April 1930

Dear Mr. Einstein,

I was very pleased that you kindly sent (in mid-December) your article, "A Unified Field Theory, Based on the Riemannian Metric and Teleparallelism."[1] That I am only now expressing my thanks for it is due to the following circumstance: Only now have I succeeded in finding all those "eggs of Columbus" and breaking them open, which were hidden in the mathematical proofs.

Formula (17) caused me some difficulties; I would rather write it as follows:[2]

$$A^{\sigma\sigma'\sigma''\dots}_{\tau\tau'\tau''\dots;\rho} = A^{\sigma\sigma'\sigma''\dots}_{\tau\tau'\tau''\dots,\rho} + \sum \Delta^{\sigma^*}_{\mu\rho} A^{\sigma\sigma'\sigma''\dots\mu\dots}_{\tau\tau'\tau''\dots\tau^*\dots} - \sum \Delta^{\mu}_{\tau^*\rho} A^{\sigma\sigma'\sigma''\dots\sigma^*\dots}_{\tau\tau'\tau''\dots\mu\dots}$$

The chiefly interesting point appears, to me, to be that, instead of the shortest lines (or of the Γ ∴), the loxodromes, or lines of constant direction, become important. I was curious enough to find out directly whether your basic equations (29), (30) are fulfilled in spherical geometry, i.e., in geography, if one defines any two arbitrary worms as "parallel" (for example, one in India, the other in Argentina) which make the same angle with the meridian of their location; in the Mercator projection, those worms would appear to be parallel in the usual sense and would be the loxodromes projected as straight lines. To that end, I set x_1 for the geographical longitude, x_2 for the geographical latitude, and $h_1{}^1 = \frac{\sec x_2}{R}$, $h_2{}^2 = \frac{1}{R}$, with R

as the radius of the Earth. The result was

$$g^{11} = \frac{\sec^2 x_2}{R^2}, \quad g^{22} = \frac{1}{R^2}, \quad h_{11} = R\cos x_2, \quad h_{22} = R, \quad h = R^2 \cos x_2,$$

$$g_{11} = R^2 \cos^2 x_2,\ g_{22} = R^2,\ g = R^4 \cos^2 x_2,\ h_{11,2} = -R\sin x_2,\ \Delta^1_{12} = -\mathrm{tg}x_2,$$

$$\Lambda^1_{12} = -\mathrm{tg}x_2, \qquad \Lambda^1_{21} = \mathrm{tg}x_2, \qquad \Lambda^{112} = -\frac{\sec^2 x_2 \mathrm{tg}x_2}{R^4} = -\Lambda^{211},$$

$$\Lambda^{112}{}_{,2} = \frac{1}{R^4}(2\sec^2 x_2 - 3\sec^4 x_2) = -\Lambda^{211}{}_{,2},\ \Lambda^{112}{}_{;2} = -\frac{\sec^4 x_2}{R^4},\ \Lambda = -\frac{\sec^4 x_2}{R^4},$$

$F^{\mu\nu} = 0$, $G^{\mu 2} = 0 = G^{21}$ and, oh dear! $G = -\frac{\sec^2 x_2}{R^4}(\sec^2 x_2 + \mathrm{tg}^2 x_2)$.

The meaning of the lovely theorem at the end of §4 as an element of the argumentation is not apparent to me.[3] Mr. Besso is of the opinion that the third paragraph on page 10 should not begin with, "First, we have to show...," but instead with, "It only remains to show..."[4]

Your first article, "A New Possibility for a Unified field Theory,"[5] gave 4 equations, $\phi_{\alpha,\beta,\beta} = 0$, and one equation, $\phi_{\alpha,\alpha} = 0$, for the electromagnetic field, and, in addition, the relation $R_{\alpha\beta} = 0$ for the gravitational field. Now, we have for the electromagnetic field 6 equations (46) plus 4 equations (47); furthermore, $2R_{\alpha\beta} = -g_{\mu\mu,\alpha,\beta} = -2\bar{h}_{\mu\mu,\alpha,\beta}$ is found, which in general does not vanish. Should we now throw Maxwell's and Poisson's equations on the dust heap?

Equation (48), as a result of the orthogonality, becomes equivalent to $x^\sigma = \alpha_{\mu\sigma} x^{\mu'}$; but only the latter form serves as an element of the argumentation.

We will not speak of the replacement of Δ by Λ in (34).[6]

I fear that my enjoyment of your whole undertaking using teleparallelism is of a similar nature as the enjoyment that I had from Abraham's theory of the rigid electron,[7] from which you wanted to discourage me! In the fact that the laws of physics, as far as observations have investigated them to date, can be so finely described in a unified way using teleparallelism, there is perhaps an objection to those previous observations; am I not correct?

With hearty thanks and respectful greetings, yours truly,

J. Sauter

PS. I have your article, "Compatibility,"[8] but will read it only after I have finished writing the present letter!

323. From Hugo Andres Krüss

Geneva, 16 April 1930

Dear Mr. Einstein,

I learned from your letter of the twelfth of this month, which I received here, that you intend to personally participate in this year's meeting of the [International] Committee for Intellectual Cooperation.[1] I understand that, given the great importance of the decisions to be made this year, you want to exercise your mandate yourself.[2] However, I am genuinely saddened that you are spurred [to make] this decision because significant differences of opinion exist between us.[3]

When you speak of my report, I assume that you mean the statement that the German delegation for Intellectual Cooperation has submitted on the question of reorganization and for which I prepared a draft, which was discussed in your presence at the meeting of the German delegation on February 11of this year and was adopted with some amendments.[4]

However, as you will remember vis à vis this statement of the German delegation, I reserved for myself complete discretion for my conduct in the study committee currently meeting, where I attend not as a representative of the German delegation, but rather am appointed on an individual basis, and, second, that I fundamentally do not think it right to enter into this kind of negotiations with hands tied, but rather to be free in the course of the negotiation to find what is correct and feasible together with the other participants.

For this reason, I sincerely regret not having been informed of your view early enough that I then would have been in the position to bring it to bear in the negotiations of the inquiry committee.

With best regards,

Krüß

324. From Oswald Veblen

[Princeton,] 17 April 1930

[See documentary edition for English text.]

325. To Emma Kayser-Sussmann[1]

Berlin, 20 April 1930

Dear Mrs. Kayser!

We are spending the evening here with Rudi and Ilse in a most pleasant fashion, thinking of you with festive feelings, which are called for day by day such as tomorrow.[2] Growing old truly is an art that not everybody can master gracefully, but being a good mother-in-law, being one every day for many a year, that is an achievement of which you have every right to be proud. I am wishing for you, with all my heart, to keep satisfyingly healthy, combined with a well balanced measure of wisdom and sacrifice, without which those of advanced years unfortunately cannot manage. As to our physicians who just love to lecture us continuously, we always have satisfaction that they, too, eventually must taste the bitter fruit, just as we all are destined to do.

My heartfelt wishes for many more happy days and years.

Your,

A. Einstein.

326. To Gewerkschaft Deutscher Geistesarbeiter[1]

[Berlin,] 24 April 1930

I hereby declare that I have complete trust in the current leadership of the Union of German Intellectual Workers.[2] The difficult situation of a large portion of intellectual workers demands strict solidarity and mutual trust. Only then can the interests of intellectual workers be effectively perceived by the outside world.[3] Every formation of sects generates counter currents, and thereby worsens the situation of the whole.

327. To Henry Noel Brailsford[1]

[Berlin,] 24 April 1930

[Not selected for translation.]

328. To Eugene Spiess[1]

[Berlin,] 24 April 1930

Dear Sir,

No one is safe from error except the one who does not think at all.[2] However, the community of the last kind cannot support any academic journal. Therefore, if man is not to confront the mysteries of existence blind and deaf, then all those new ideas whose erroneous character has yet to be identified must be conveyed to people.

Kind regards, your

329. Aphorism for Jüdische Altershilfe[1]

[Caputh, 25 April 1930]

In the fate of the elderly, one discerns the culture of the young.

330. To Rudolf Kayser

Berlin, 25 April 1930

Dear Rudi,

You have made me very happy with this book dedication.[1] Those who create love nothing *more* than the products of their work. To be connected with these is therefore the clearest proof of genuine affection. As this is mutual, it is a happy circumstance between us that will survive the years and tribulations.

Warm regards and thanks from your

Albert.

331. From Hans Reichenbach[1]

Berlin, 25 April 1930

Dear Mr. Einstein,

You will certainly be interested to hear that I have now succeeded in establishing a journal to bring to life our exact natural philosophical orientation which will be jointly edited by Mr. Schlick, Mr. Carnap, and myself, and will be published by

Felix Meiner.[2] I would like to devote one of the first issues to the problems of the theory of knowledge with respect to quantum physics. May I ask whether you would be interested in writing a contribution on your views of that topic? We would be most particularly happy to receive a contribution from you.

Or do you perhaps have some other manuscript of a general content which you could place at our disposal? We would find an emphasis on our common fundamental philosophical stance between your direction and ours to be most especially opportune.

With kindest greetings, yours truly,

332. To Alfred Apfel[1]

[Berlin,] 30 April 1930

Dear Doctor Apfel,

Thank you very much for the delivery of your journal, which appears to me as an extremely meritorious enterprise. The radicalization of social struggle, which represents a serious danger for the whole culture, can only be counteracted through unbiased information from educated people. I see, time and again, that understanding of the serious latent danger, even by politicians and lawyers who are of good will, falls far short of necessity. Your journal can become a bulwark against the danger threatening rights, from fascist or imperialist and Bolshevik or communist sides, if you can retain full objectivity toward both of them.[2]

Kind regards, your

333. To Oswald Veblen

[Berlin,] 30 April 1930

Dear Mr. Veblen,

I remember my statement at the time and the occasion on which I made it.[1] I don't mind if you use it in the intended way and consider this as a special token of friendly sentiments.

However, I would like you to consider that this statement could easily seem frivolous to the reader who does not know the occasion of its origin. One could, for example, express the thought as: Nature conceals her secret through the grandeur of her being, but not through guile.

Kind regards, your

334. To Eduard Einstein

Caputh, [before 4 May 1930][1]

Dear Tetel,

You know my stubbornness in writing letters. But everything is going well here. Albert is there with his wife and is enjoying his vacation here.[2] He is a clever, healthy guy and still looks like a boy as he walks the path to fatherhood. The sayings[3] were forwarded to me in Holland. I studied them on the train between Leiden and Utrecht; they are somewhat too clever and not natural and original enough for my taste. You just have a very harsh father, harsh, at least, toward everything on paper, otherwise tolerant. I should quickly ask you something about physiology, as long as you have not forgotten it yet? Do we know how the transmission of stimuli through nerves is possible? Don't let vitalism fool you. The fact, for example, that one can get an entire animal out of one of two cells which resulted from splitting the primordial cells of an embryo is not proof against the clockwork-causality of biological occurrences. This fact only shows that the environment of a cell substantially affects what develops from it. Of course, I also believe that physical-chemical analysis can just as little illuminate the essence of life as the essence of, say, a book, though there, no one doubts the validity of the physical-chemical laws. The meaning lies not in the *parts*, but in a *synthesis* of effects that we confront wholly dumb. Goethe said it wonderfully: You have the pieces in your hand, lacking, unfortunately, only the spiritual bond.[4]

Work with Dr. Mayer[5] is moving forward nicely. I have never before had such a fascinating problem. I think I am now able to grasp why electric particles all have the same electric charge. It is like a simple knotting of space. It is almost too good to be true.

If you come in the summer, we can always be outside. It will do you good. The sailboat is coming into the water these days. Hopefully, Albert still has something of it.

Be with mother,[6] warmly greeted by your

Papa

335. From Lauder W. Jones[1]

Paris, 6 May 1930

[See documentary edition for English text.]

336. To Antonina Vallentin-Luchaire

Caputh, 16 May 1930

Dear Mrs. Luchaire,

I will most certainly come to Geneva in the summer.[1] In my opinion, it would be best if Painlevé,[2] Mrs. Curie,[3] and I were to take unofficial steps with influential divisions of the League of Nations in order to achieve a renewal of the League of Nation's Commission in the sense that only people enter who are genuinely devoted to the international cause. Nothing reasonable can be achieved with people such as [those] with whom you are now forming the commission.[4]

Your husband's[5] enmity comes largely because of a League-Institute tension that must necessarily exist as long as the Institute, through its chair and its one-sided dependence on the French state, is perceived as a foreign element by the Genevois. There will certainly come a time in which you will agree with me in this [matter]. In the meantime, there appears to be nothing to do on this point, because the French are unfortunately turning this matter into a question of prestige.[6] For the time being, it is necessary to do everything to prevent the fragmentation of the Institute, as, otherwise, it would be wholly impossible for the president of the Institute to serve such splintered machinery.[7] Definitely speak to Mrs. Curie, to whom I send warm regards.

Sending warm regards to you and your husband, your

337. To Robert Weltsch

Caputh, 16 May 1930

Dear Mr. Weltsch,

I sent the telegram immediately upon receipt of Buber's letter, and indeed exactly in Buber's formulation.[1] Please also communicate it to Mr. Buber, as well as perhaps the Jewish Telegraphic Agency.

I heard that Ussischkin delivered a scandalous speech on the Arab question.[2] If things continues like this, no respectably minded Jew can be a Zionist anymore.

Kind regards, your

A. Einstein

338. Preface to Upton Sinclair's *Mental Radio*

"Preface"
[*Einstein 1930z*]

DATED 23 May 1930
PUBLISHED 1930

IN: *Sinclair 1930a*.

PREFACE

[1] I have read the book of Upton Sinclair with great interest and am convinced that the same deserves the most earnest consideration, not only of the laity, but also of the psychologists by profession. The results of the telepathic experiments carefully and plainly set forth in this book stand surely far beyond those which a nature investigator holds to be thinkable. On the other hand, it is out of the question in the case of so conscientious an observer and writer as Upton Sinclair that he is carrying on a conscious deception of the reading world; his good faith and dependability are not to be doubted. So if somehow the facts here set forth rest not upon telepathy, but upon some unconscious hypnotic influence from person to person, this also would be of high psychological interest. In no case should the psychologically interested circles [2] pass over this book heedlessly.

[signed] A. EINSTEIN

May 23, 1930

339. To Friedrich Siegmund-Schultze[1]

Caputh, 23 May 1930

Dear Mr. Siegmund-Schultze,

Unfortunately, it is impossible for me to participate in the congress, as much as the gathering of religious organizations for the purposes of peace efforts is dear to my heart.[2] I am too overwhelmed by work and am neither a speaker nor an organizer.

If I had delivered the intended speech, I would have said that, in the course of history, priests brought much endless strife and war on mankind and therefore have a great debt to pay. They have very rarely opposed the organized powers of hate, they have mostly been their servants. I don't beg them, but rather show them the path of their human duty.[3]

With sincere wishes for the success of your congress, I am respectfully yours,

340. To Eduard Einstein

Caputh, [before 27 May 1930][1]

[Not selected for translation.]

341. From Eduard Einstein

Zurich, 27 May [1930][1]

[Not selected for translation.]

342. To Aron Tänzer[1]

Caputh, 28 May 1930

Honored Rabbi,

I thank you sincerely for sending my paternal family tree, which previously was only known to me up to my grandfather.[2] Unfortunately, it is not possible to connect the names with vivid visualizations. Time devours its children with stump and stem, and only a faint name remains.

Friendly greetings, your

A. Einstein

343. From Hugo Bergmann

Jerusalem, 5 June 1930

Dear Professor,

I thank you sincerely for the willingness with which you fulfilled my request and have telegraphed the high commissioner for Palestine regarding a pardon of those convicted.[1] You probably know the result by now: three of the murderers are to be executed; the remaining twenty-two were pardoned.[2] The Arab press published the news of your telegram,[3] which I sent to them with large type so that the fact of your intervention will have been imprinted on the Arab consciousness. It is tragicomically apropos that, immediately after the announcement of the pardons, the Zionist daily newspaper The Palestine Bulletin pointed out that the Jews were the first to have advocated for the pardon.[4] The Arab press rightly replied to this by asking whether the Jews had forgotten the smear campaign which they had launched against Brith Shalom because of its appeal for clemency.[5] Anyway, the fact that one now refers to the clemency petition may be noted as a good omen for the beginning of a change in mood in the Jewish camp. Of course, it is to be deeply regretted that the Zionist leadership missed this brilliant opportunity to change the relationship between the two peoples through an energetic step and to turn it around for the better. I am convinced that a telegram from Weizmann that would have recommended pardon would have worked like a miracle.[6]

Now, that is in the past, and we must endeavor to operate so that new opportunities for rapprochement are not thrown away. The Zionist executive, unfortunately, still has not recognized that the Arab problem is the key point of our whole situation, that a still so benevolent England also can do nothing for us if the East is against us, and that a still so malevolent England can do nothing of substance against us if the East is with us. I am convinced that the possibility of a radical change in the relationship between both peoples still exists, though we have lost importance chances. We must position ourselves with full consciousness on the side of the Arabs. Of course, that does not mean positioning ourselves against England; our politics must always be flexible in this respect, but through a extensive cooperation with the Arab people, we must expel the causes of Jewish-Arab conflict from the world once and for all. We would have to, of our own accord, establish a far-sighted program of Semitic cooperation which must include roughly the following points:

(1) Permanent renunciation of the Jewish state.[7]

(2) The obligation to regulate colonization so that no Fellah will be dispossessed of his soil. Corresponding guarantee for the adherence to this obligation through a neutral court of arbitration or the like.[8]

(3) Jewish aid to the Fellah, be it through the founding of an agricultural bank with Jewish funds, be it through expanding the activity of existing Jewish institutes to Arab villages, be it through corresponding Zionist political pressure on the Palestinian government in the sense of politics friendly toward the Fellah.

(4) Adjustment of Jewish immigration to the economic capacity of the country.[9]

(5) Jewish support for an Arab federation that would unite the three mandate states of Palestine under English sovereignty, Transjordan,[10] and Iraq, whereby, on the one hand, the special position of Palestine would remain safeguarded, on the other hand,

(6) Jews would receive the right to settle in Transjordan and Iraq under the express condition that these two countries remain outside the area of the Jewish national home.[11]

I am convinced that Arab public opinion would also still be ready today to accept such a program which, on the one hand, far widens the borders of Jewish colonization; on the other hand, would guarantee the realization of the Arabs' national goals with the help of the Jews. For this purpose, it would still be the requirement

(7) to get behind a *democratic* parliament in Palestine.

Admittedly, the question of the parliament is not as acute today as [it was] a few months ago, and an Arab leader said to me this week: "Today, the parliament is no longer important to us; we have achieved what we wanted even without the parliament (Immigration, regulation of the sale of land.)"[12] However, this statement of the Arab was unfoundedly optimistic. The English political line runs in a zigzag, handing out blows and sweetmeat, now to the right and now to the left. But that is precisely why we must formulate the demand for a democratically elected parliament that would call the Fellah into the political arena, who, if we consistently pursue farmer-friendly politics, would be more easily brought over to our side than those currently in power in the Arab camp.

I would be very pleased if this program found your agreement and if you would be ready to support it.

With respectful regards to you and your wife,[13] your,

Hugo Bergmann

Admittedly, I assume that you receive and read the Falastin. As a precaution, I am sending you an excerpt from the last issue, where a Jewish-Arab coalition is proposed which should open Arab nationalists from Jaffa to Basra to Jewish colonization.[14]

344. Message to the World Power Conference[1]

"Zur Weltmachtkonferenz"
[*Einstein 1930cc*]

PUBLISHED 8 June 1930

IN: *Vossische Zeitung*, 8 June 1930.

It is certain that, through the development of technology, the great concept of our planet's international political organization has stepped from the realm of otherworldly speculation into that of tangible goals.[2] The international concept also has a fruitful effect on technology. So, at the same time, we see in the conference of world powers an auspicious manifestation of nations' cultural, scientific, and political cooperation.[3]

345. From Joseph John Elliott

[10 Springwood Avenue, Huddersfield England. 1[6] June 1930][1]

[See documentary edition for English text.]

346. To Mileva Einstein-Marić

Caputh, [between 12 and 15 June 1930][1]

Dear Mileva,

The letter sent to you concerned the investment of around 5,000 dollars.[2] Thalmann suggested the purchase of government bonds similar to those you already have.[3] I think, however, that it would be wise to take out a first mortgage on a good house *here* which would earn 8–10%.[4] Even if the local financial conditions are not good, there is hardly any danger that a new inflation is coming. In contrast, foreign investments in Germany offer protection whose endangerment America, for example, will invariably try to prevent.[5] The worst thing would be that a house went to you, but with the low mortgaging of the building, that would not be too bad.

I advise you, therefore, to let me take care of a first mortgage for you. But if you insist on leaving the money in America at the lower interest rate, I give my consent as well, even if I consider it as less advantageous. In this case, I would immediately write to Ladenburg Thalmann. So let me know right away.

I am coming to Geneva in the last third of July and will visit you at this time when you are at home.[6] Then, I can bring Tetel with me here, where he will be wonderfully accommodated here in the cottage.[7] Now, I've just come straight from England, where I received an honorary doctorate from Cambridge and gave a lecture in Nottingham.[8] I am sorry that you have met so much misfortune at home.[9] Only through death is one secure from it, but not before.

With warm greetings to you and Tetel, your

Albert.

Tetel's letter with the wonderful aphorisms[10] pleased me greatly. I see in him a true inner kinship. I would just still like some simplicity and modesty in external things, and some greater order in external things for him. Discuss business matters with him as well, so that he learns something about it.

347. From Mileva Einstein-Marić

Zurich, 15 June 1930

Dear Albert,

Your suggestion regarding that money[1] came to me somewhat unexpectedly, and I had to think it over a little bit first before I could answer you.

The nice thing about the current situation for me is mainly that things function flawlessly and everything goes so impersonally. With an investment like the one you now propose, one does not know who one could end up with; if payment does not happen, the profit that one should have from the higher interest is gone for a long time. In the event of the owner's inability to pay, the house does not naturally fall to you, but rather one has the right to demand the sale of the house. During this time, one does not receive the interest that one should have regularly because one needs it to live, has troubles and also expenses. We had *one* tenant who could not pay,[2] and you cannot imagine how much trouble it caused until we could just get him out; as a result, I know these things.

Of course, not all people are of this sort. If you think it is good that these 5,000 dollars are invested in this way, then it is also ultimately fine with me. Dr. Zürcher[3] thinks you should *definitely* be professionally advised as to whether the house in question also has a corresponding value. I am also concerned that you will have a lot of work with it, which wouldn't be alright with me. Most notably, I would especially regret it if there were to be any disagreements as a result of it; I am so nervous at the moment that I would actually prefer to forgo a percent than to go through any sort of fuss. I trust in you, and hope that this change of placement runs peacefully.

I have found Tete well, but the poor thing must study hard, because the exam is to take place in the middle of July.[4] He could also postpone it until September, but I would much prefer if he could do it now so then, afterwards, to be able to actually have holidays. With your praise of his aphorisms,[5] he was incredibly pleased with himself. I think the most painful thing for him about the whole current toil is that he is not able to think of some new aphorisms.

With kind regards,

Mileva

348. From Eduard Einstein

[Zurich, after 15 June 1930][1]

Dear Papa,

When my friend Antes became so emaciated to the point of being a skeleton from exams preparation (with his somewhat irrational system of gobbling one subject in about two to three days), Engel began to suffer from nervous intestinal issues; I decided to escape these and similar tribulations and, without hesitation, postponed my exams until 8 September.[2] As a result, my availability during the earlier part of the holiday will, naturally, be significantly limited.[3] After the decision, I hurried to the library, borrowed some novels by Ulitz, Hamsun, Heinrich Mann,[4] and others, noticed that I had almost completely forgotten how to read, got in touch with some old friends in order to ask them what possibilities are available for passing free time. Meanwhile, I dedicate most of the time to my girlfriend, Clärchen.[5] I wrote down a sonnet about her, in which I rhymed her name with fairytales, couples, and tiny hairs. In this context, I would like to mention: If you ever again are offered a car, do not foolishly reject it, but rather send that to me; the attempt in dealings with modern young ladies, to out-do those who possess cars, presents such demands regarding the conversation that, even from one of a poetic nature, they are only fulfilled in their best hours. My third collection of aphorisms from this spring is coming soon.[6] Because you liked the previous ones so much, I'm sending you, here, a few more sentences by great writers, some from poems:

A shudder's the truest sign of humanity.
Goethe Faust II.[7]

For the poet, everything that is ephemeral is only a parable of an unknown primal experience searching for memories in him.
Gottfried Benn.[8]

The perfect poetry is a state of being completely outside oneself, the perfect prose is a state of coming completely into oneself.
Hugo von Hofmannsthal.[9]

The man who acts never has any conscience; no one has any conscience but the man who thinks.
Goethe.[10]

Always torn open by us again,
the god is the place that heals.
We are Sharpness, because we will
to know; he's divided and serene.
...
A god can do it. But how shall a man, say,
get to him through the narrow lyre and follow?
His mind's dichotomy. Where two heartways
cross there stands no temple for Apollo.
Rilke (Sonnets to Orpheus)[11]

The sonnets to Orpheus seem almost like a collection of aphorisms expressed as poems, therefore not completely satisfying. However, if you want to read thoughtful poetry's greatest format, then I recommend the little festival production of Pandora by Goethe, as well as Helena from the same author (Faust, 2nd half, 3rd act).[12]

In my actions, as in my poetic utterances, I have become a complete romantic. I forgot the abstract entirely, except for certain psychological skills that always make maximum impact on relations between people.

Best wishes, your

Teddy

How is it going with the theory?
Will you give the student lecture?[13]

349. "The Problem of Space, Field, and Ether in Physics"

"Das Raum-, Feld- und Äther-Problem in der Physik"
[*Einstein 1930dd*]

PRESENTED 16 June 1930
PUBLISHED between 25 June and 31 December 1930

IN: *Verhandlungen der 2. Weltkraftkonferenz, Berlin 1930,* Bd. 19, s. 1–5.

This translation is reprinted from *The New York Times*, 17 June 1930.

4,000 BEWILDERED AS EINSTEIN SPEAKS

Scientist Before World Power Conference in Berlin Traces Rise of Relativity Theories.

FULL TEXT OF THE SPEECH

Professor Says Space, Raised to Scientific Reality by Newton, Is Swallowing Ether and Time.

PHOTOGRAPHERS ANNOY HIM

Physicist on Rostrum Is Patient With Them at First, but Demurs as Huge Light Gets Near.

Special Cable to THE NEW YORK TIMES.

BERLIN, June 16.—Professor Albert Einstein stepped up to the rostrum on the stage of the Kroll Opera House this afternoon and in the most matter-of-fact conversational manner he told the 4,000 delegates to the World Power Conference all about relativity.

Like a professor addressing a class in advanced calculus, the scientist smiled apologetically and launched forth into forty minutes of bewildering differential coordinates, continuums, four-dimensional space, and metrics of various kinds, which it is doubtful more than one-half of 1 per cent of the audience really understood. It was the first time Dr. Einstein had ever consented to speak on Einstein, and it was the first serious public utterance he ever made without recourse to gigantic equations and mystifying mathematics.

Traces Development of Theories.

The author of the theory of relativity traced the development of ideas about geometry and space. After discussing the theories of the Greeks, Descartes and Newton, Dr. Einstein showed how the special theory of relativity had established that space and time must be united in a single four-dimensional continuum.

Then he went on to show the place of the general theory of relativity as a system which could embrace all Gaussian coordinate systems in this four-dimensional space and answer the question:

"What are the simplest mathematical conditions to which Riemannian metric in four-dimensional space can be reduced?"

This theory, Professor Einstein said, would have been ideal if only gravitational fields and not electromagnetic phenomena had existed in nature, but electro-magnetic phenomena could not be represented by Riemannian metric. This necessitated the development of the unified field theory, which he believes has solved the difficulty.

Summing up, Dr. Einstein said:

"Space, brought to light by the material object and raised to scientific reality by Newton, has in the last few decades swallowed up ether and time and is about to swallow up the field theory and the corpuscular theory as well, so that it will remain as the only theory representing reality."

Professor Einstein was introduced by President Oskar von Miller of the power conference, who told the audience, with a twinkle in his eye, that it was about to hear an interesting little discussion of a few points on which Professor Einstein was especially well-qualified to speak and expressed the hope that all would find enlightenment.

As Dr. Einstein rose a corps of cameramen began clicking and grinding, while the batteries of floodlights edged nearer the rostrum. The professor was obviously uncomfortable, but refrained from objecting until one huge floodlight was within six feet of his head. Then he turned toward Herr von Miller and murmured, "This is unbearable." Then the troop of photographers was shooed briskly away.

Dr. Einstein, who is accustomed to speak before the Prussian Academy of Science and other such bodies, maintained the same manner he would have employed in lecturing to a small group of specialists. He gestured sometimes with his hands, indicating how clear and obvious his reasoning was, and occasionally he looked up from his paper to smile upon his intent hearers who, he seemed to assume, were grasping everything.

TEXT OF THE ADDRESS.

The text of Professor Einstein's address on "Space, Field and Ether Problems in Physics" was as follows:

Conceptions and conceptional systems, logically regarded, never originate from sense experiences. But they are always caused, however indirectly, by sense experiences; they are related to sense experiences and in this relationship lies their meaning and their significance.

If we wish to be clear on the meaning of the pre-scientific conception of space we must seek to visualize those characteristics of our world of experience which have given rise to the formation of a conception of space and of geometric conceptions in general. Regarded from this standpoint, the conception of a real world of externals and material bodies undoubtedly preceded the conception of space. [1]

We need not further analyze what characteristics of our world of experience have led to these fundamental conceptions and in what the close linking up of these concepts with the world of experience consists.

Among the many things which are included in the term "material objects" one category plays a particular rôle. This we call "the relative position of solid bodies." Conceptions of space as well as the conceptions of the system of Euclidean geometry are based upon this idea. The most important conceptual element for the comprehension of the law of "the position" of motionless bodies is that of their contacts. On this are based the most important concepts of congruence and measurement.

Significance of Greek Geometry.

The great significance of the geometry of the Greeks lies in the fact that, so far as we know, it represents the first attempt to comprehend a complex of sense experiences through a logical deductive system. Instead of starting from matter with its manifold forms, it is based on a few formal elements: point, line, plane and distance.

From these were constructed material forms and positional relations between bodies which were purely theoretical and were founded on certain established rules: the axioms. These fundamental elements are themselves idealizations of material objects.

The conception of a space continuum does not appear at all in Greek geometry, although it certainly forms a part of pre-scientific thought. It was first introduced into mathematics by Descartes, the founder of modern geometry. The Greeks were satisfied to study reciprocal relationships between their idealized material objects: points, lines, planes and distances.

Their conception of space was based on the idea that it was easier to study the relationships of all bodies as compared with one than as compared with one another. This one body, however, is the fiction of an infinitely extended body or one with which all the others can be brought into contact. It is clear that the existence of a quasi-rigid earth surface, or the existence of drawing paper in a study of plane figures, must have given rise by means of drawn representations to the formation of this conception.

The service which Descartes rendered to mathematics through the introduction of a space continuum cannot be too highly estimated. [2] In the first place, it made possible the study of geometrical figures by means of analysis. Secondly, it strengthened geometry as a science in a decisive manner. Henceforth a straight line and a plane were no longer favored in principle over other lines and surfaces but all lines and surfaces received equal treatment.

One Axiom Replaces System.

A single axiom took the place of the complicated axiom system of Euclidean geometry. This axiom, in the words of today, reads: There are systems of coordinates compared with which the interval as of neighboring points P and G may be expressed by coordinate differentials dx_1, dx_2, dx_3 in the formula:

$$ds^2=dx_1^2+dx_2^2+dx_3^2.$$

From this, i. e., from Euclidean metric, all conceptions and propositions in Euclidean geometry can be deduced. [3]

However, perhaps the most important thing is that, without the introduction of a continuum of space, in the Cartesian sense, a formulation of Newton's mechanics would have been impossible. [4] The fundamental conception of acceleration used in this theory must be supported by the conception of the Cartesian coordinates of space, for accelaration in no wise may be deduced from concepts which only relate to relative positions of bodies, or material points and their time changes.

It may rightly be said that, according to Newton's theory, space plays a rôle of physical reality, as Newton well knew, although this fact was later overlooked.

The Cartesian coordinates of space had therefore to begin, from the point of view of physics, with two independent functions. It established through the Pythagorean theorem possible positions of practically rigid bodies, as well as the inertial movement of material points. It seemed absolute in the sense that it worked but that nothing could work upon it or modify it. It was the infinite, the eternally unchanging repository of all that is and happens.

The frame of the Newtonian theory is distinguished by concepts of space-time and ponderable matter. To this there came in the nineteenth century a new element —ether. [5] As soon as the undulatory character of light had been established by Young and Fresnel it was considered necessary to accept an inert substance which permeates all bodies and completely fills all space—ether, the vibrations of which were supposed to be light.

Faraday-Maxwell Theory.

Newton's theoretical framework was completely destroyed by the Faraday-Maxwell field theory of electro-magnetic phenomena, for the realization gradually grew that electro-magnetic fields, which are also to be found in empty space, could not be regarded in a satisfactory manner as mechanical conditions of the ether without encountering objections.

One became accustomed to regarding electro-magnetic fields as fundamentals of no mechanical nature. Moreover, they were still regarded, as heretofore, as conditions of the ether, which, however, could no longer be regarded as a form analogous to solid matter—all the less, since at the turn of the century the concept of the molecular structure of matter gained more and more ground.

[6] Even though these electro-magnetic fields had established themselves as not mechanically comprehensible. fundamental substances, there still remained the question of the mechanical characteristics of their medium, the ether. H. A. Lorentz answered this by stating that all electro-magnetic facts force us to the conclusion that ether is everywhere motionless as opposed to Cartesian and Newtonian space.

How close was the thought: The fields are conditions of space; space and ether are one and the same. That it was not realized lay in the fact that space, as the basis of Euclidean metric and Galileo-Newtonian inertia, was considered absolute; that is, incapable of being influenced. It was considered a rigid frame of the world, which, so to speak, existed before all physics and could not be the basis of changing conditions.

The next step in the development of the conception of space was that of the special theory of relativity. The law of the spreading of light in empty space in connection with the relativity principle regarding uniform movement resulted in the necessity that space-time be united in a single four-dimensional continuum. For it was recognized that reality did not conform with the conception of simultaneous events.

A Time Coordinate Used.

A Euclidean metric had to be ascribed, as Minkowski was the [7] first clearly to recognize, to this four-dimensional space, which, by the use of an imaginary time coordinate, would be completely analagous to a metric of three-dimensional space of Euclidean geometry.

On the existence of a space structure expressible through Euclidean metric was founded the later development, which has become known under the terms of "the general theory of relativity" and "the unified field theory."

After it has been realized that no absolute character could be ascribed, not only to speed but to acceleration, it was revealed that reality did not conform to the conception of an inertial system in nature.

It was clear that laws must be so formulated that this formulation could claim validity in a four-dimensional space in terms of every Gaussian system of coordinates—a general covariance of equations which express the laws of nature. [8] This is the formal content of the general principle of relativity. Its force lies in the question: What are the simplest general equation systems of covariance?

This question in this generalization has not yet been productive. A statement has still to come as to the character of the structure of space. This is supplied by the special theory of relativity, the validity of which for small areas must be granted. That means: There is a structure of space which for the infinitesimal surroundings of every point can be expressed mathematically through a Euclidean metric. Or: Space possesses a Riemannian metric.

On the physical ground it was clear that this Riemannian metric also formed simultaneously the mathematical expression of the gravitational field.

The mathematical question corresponding to the gravitation problem was, therefore, this: What are the simplest mathematical conditions to which a Riemannian metric in four-dimensional space can be reduced? In this manner the field equations of gravitation of the general theory of relativity were found, which have received the well-known confirmations. [9]

Space Loses Absoluteness.

The significance of this theory for the recognition of the structure of space can be characterized thus: Space under the general relativity theory loses its absolute character. Until that phase of the development, space was accepted as something the inner substance of which was not capable of being influenced and was in no wise changeable. Therefore, a special ether had to be accepted as a basis of the field conditions localized in empty space.

Now, however, the real quality of space, the metric structure, was recognized as changeable and capable of being influenced. The condition of space gained a field character; space became analogous in structure to the electro-magnetic field. Separation of the concepts of space and ether was thus to a certain extent automatically removed after the special theory of relativity had already removed the last bit of substance from ether. [10]

The general theory of relativity in its former shape would have been, from the logical standpoint, an ideal physical theory on account of its completeness, had there been only gravitational fields and no electro-magnetic fields in nature. The latter, however, could not be represented through Riemannian metric.

One had to seek a structure of greater richness of form which would encompass the Riemannian metric structure and at the same time be able mathematically to describe electro-magnetic fields. This task is to be solved through the unified field theory by the establishment of a space structure the mathematical characteristics of which are as follows: [11]

P and P′ are any two points of a continuum. PG and P′G′ are two line elements going out from these points.

The hypothesis of the metric structure states that the quality of the two line elements may be spoken of intelligently; more generally, that line elements are comparable in respect to their size. The Riemannian character of the metric is expressed by the hypothesis that the square of the size of the line element may be expressed by a homogeneous function of the second degree of differential coordinates.

On the other hand, a statement within the frame of Riemannian geometry about a direction relation, for example about the parallelism of the two line elements PG and P′G′, has no meaning. If the hypothesis is added that parallel relations of line elements can be intelligently spoken of, then one attains the formal basis of the unified field theory.

Space Representing Reality.

To attain completeness it is now only necessary to add the hypothesis that the angle between two line elements going out from the same point is not changed by a parallel movement of elements. The mathematical expression of the field law should be the simplest mathematical conditions to which such a space structure can be reduced. [12]

The discovery of these laws appears to have been made and they correspond, in the first degree, in fact, with the known empirical laws of gravitation and electricity. Whether these laws also supply a useful theory for particles of matter and their movements must be shown by further mathematical investigations. [13]

Taken together we can say symbolically: Space, brought to light by the material object and raised to scientific reality by Newton, in the last few decades has swallowed up ether and time and is about to swallow up the field theory and the corpuscular theory as well, so that it will remain as the only theory representing reality. [14]

350. From Myron Mathisson

Warsaw, 17 June 1930

Sir,

I am taking advantage of the permission that you gave me to make a report on my situation.[1]

A representative of the Rockefeller Foundation was in Warsaw on June 12. He told me that the main difficulty in my case for the fellowship selection committee is that they have no idea of my scientific qualifications, since I have not published anything.[2] The decision will be communicated to Prof. Bialobrzeski[3] around the 15th of July.

I have written a German version of my work on the solutions of the equations of gravitation and electromagnetism on the De Sitter surface.[4] The relevance of the second article, like that of the first,[5] could appear to be dubious before clarifying more precisely the relation between my ideas and wave mechanics. Establishing that relationship is the problem that is occupying my attention now. What I am lacking is time.

Please accept, Sir, my humble tribute,

M. Mathisson

351. To Hugo Bergmann

Caputh, 19 June 1930

Dear Mr. Bergmann,

I am in complete agreement with everything that you have written to me,[1] except for one point. A united parliament can have catastrophic effects for the minority in case of conflict.[2] Some way must be found to secure the existence and cultural self-government of our minority. Otherwise, however, I am, as I said, in complete agreement with you. Only direct cooperation with the Arabs can create a dignified and secure existence. If the Jews do not accept this, the entire Jewish position in the Arab countries will gradually become completely untenable. *It makes me less sad that the Jews are not wise enough to grasp this than that they are not righteous enough to want it.*

Kind regards, your

A. Einstein

352. To Paolo Straneo[1]

Berlin W, 19 June 1930

[Not selected for translation.]

353. To Mileva Einstein-Marić

[Berlin or Caputh,] Friday evening [either 27 June or 4 July 1930][1]

Dear Mileva,

I received your letter concerning the house an hour ago.[2] I have confidence in your judgment regarding houses, because you already invested well once.[3] True, it will not give a splendid rate of return straightaway. But *here*, conditions are so uncertain that I would not dare to forcefully interfere in your business; every day here, new charges can be imposed on houses so that one never knows where he stands.[4] If you are also sure that the purchase has *no particular snags* and that you hold on to *reasonably sufficient liquid reserves* so that you don't run into difficulty, then buy the house, if you think it's good. That you are already familiar with the administration and can do it yourself also speaks in favor of that. Tetel[5] can also learn about these things. Do you have good technical and financial experts at hand?

Warm greetings to you and Tetel,

Albert

354. To the Maccabi World Union[1]

Caputh, 28 June 1930

Without a doubt, the "Maccabi" World Association contributes not insignificantly to raising the feeling of solidarity, healthy self-confidence, and social independence of Jews. It thus deserves the support of all Jewish circles, even those who are not as interested in sport.[2]

355. To Muhammad Roshan Akhtar

Caputh, 28 June 1930

Dear Sir,

I am thoroughly sympathetic to the basic idea of the proposal in your article.[1] It would be the most beautiful moral success for both peoples, and would also be practically advantageous if they could come to a *direct* understanding. May the memory of our cooperation in the early Middle Ages deliver the psychological prerequisites for the healing of our relationship.[2]

Respectfully yours,

356. To Karl F. Bargheer

Caputh, 28 June 1930

Dear Sir,

The suggestion that you have made,[1] that the high pressure in the interior of large stars could be the cause of the conversion of matter into radiant energy, is completely new to me. Since we cannot make any predictions about processes at such high pressures, neither experimentally nor theoretically, and we are rather helpless in the face of the fact of the enormous production of heat in the larger stars, your idea would seem to me to be worthy of serious consideration. It would perhaps be a good idea for you to publish your thoughts in a scientific journal, for example, the *Naturwissenschaften*, and subject them to discussion. To be sure, you would then have to formulate your ideas more precisely and, in particular, try to show empirically that the pressure which is present at the centers [of large stars] and the radiation per unit mass produced there are correlated to one another.

With greatest respect,

357. To Manny Strauss[1]

Caputh, 28 June 1930

Dear Mr. Manny Strauss,

I have heard about you, that you have campaigned with extraordinary success for Jewish colonizing efforts.[2] For that reason, I am writing the following to you:

I have been informed by Mr. F. von Maltzahn[3] about a noble project which has as its goal the settlement of a million eastern Jews in an area of ten million hectares in Peru. Prof. Franz Oppenheimer, who is certainly known to you as a notable expert in the field of national economy, has declared the envisaged terrain as quite suitable.[4] Furthermore, you will learn from the enclosed copy of a letter from the American embassy[5] that the government of the United States also looks favorably upon such a plan.

With the great endangerment of the Jewish masses in the East,[6] the immediate scrutiny of this plan for colonization is of great importance. The same cannot be pursued energetically enough from here, however, given the prevailing difficult circumstances. What's more, in the event of a favorable outcome, this great endeavor can only be taken in hand by the Jews of America. Therefore, I turn to you with the plea that you might want to take the scrutiny of the plan and, eventually, its realization in hand. I have urged Mr. v. Maltzahn to submit all currently available documents to you.

Respectfully yours,

358. To Frida Perlen

Caputh, 3 July 1930

Dear Ms. Perlen,

I find that Ms. Schwimmer's[1] conduct is of great value and that this personality deserves the support of all truly humane thinkers.

The governments are the representatives of the people still caught in lingering traditions of military duty. Without the struggle of the spiritually advanced toward defeating those public authorities dependent on powers, the so bitterly necessary step toward world peace is not attainable. They who have convinced themselves of the necessity of this step are also obliged to advocate for this conviction publicly, whereby they must take on the conflict with public power. Only when sufficiently many and influential personalities muster the moral strength for this behavior can success be achieved.[2]

Such action is revolutionary. A blowing up of enslavement that has become unbearable and entrenched in laws has never been achieved other than through revolutionary acts. This path is also unavoidable in this case. Ms. Schwimmer deserves the credit, to have recognized this and acted courageously after this awareness.

With kind regards, yours,

A. Einstein

359. From Eduard Einstein

[Zurich, after 6 July 1930][1]

Dear Papa,

I saw you just now in about ten copies of the Munich Illustrated,[2] all possible faces, just not what you actually look like. Permit me; I'm sending you a further series of aphorisms as an enclosure.[3] So fortunately, these are probably almost at the end. If you don't like them, I won't be cross with you because of it: Aphorisms (I'm talking about serious aphorisms) are hardly the most certain means of conveying thoughts from person to person. One must already be very happy if, here and there, someone who is occupied with something similar is moved when he reads them.

A true source of wisdom, if also very resigned, is Niels Lyhne von Jakobsen,[4] who I am just reading.

This evening, I am playing with a brilliant violinist, one of my colleagues. Otherwise, when my girlfriend[5] is otherwise busy (which unfortunately is far too often the case), I play tennis with a friend. The heat is almost unbearable.

Please excuse me occasionally to acquaintances, I'm thinking of Mrs. Mendel, the Kaysers, Mrs. Michanowsky, the Juliusburgers,[6] and so on, that I never write.

For today, many greetings to all of you.

Teddy

When are you coming to Switzerland?

Have you already received a car as a gift?

360. To Max Warburg

Berlin, 7 July 1930

Mr. Max Warburg,

Mr. von Maltzahn told me you would be interested in the Peru Colonization Project for Eastern Jews,[1] indeed, that you would be ready to actively participate in the matter in some way. Since I know that my name possesses certain publicity value with the Jews, and I consider the project as worthy of consideration, I would like to say to you that I am ready to do everything you want me to in order to promote this concern (or to clarify it, in the meantime). There may be, here, a real possibility for helping a great part of the Jewish people to obtain a healthy existence. Meanwhile, Prof. Oppenheim[2] assured me that the country and the climate are completely suitable.

Kind regards, your

A. Einstein

361. To Mileva Einstein-Marić

[Caputh, before 11 July 1930][1]

Dear Mileva,

Enclosed is the requested note for L. Thalmann.[2] I'm coming to you all around 29 July and will stay a few days.[3] Hopefully, it will be nice, and I can take Tetel[4] with me to Caputh for a while. But think carefully about the house;[5] do you have someone there who is helping you? Tetel has inherited his father's business talents....

To a merry reunion! Greetings to you both, your

Albert

P.S. I want to see to it that I bring a violin with me to leave there.[6] Tetel could, however, get a *bow*, just in case. I only have *one*.

362. To Eduard Einstein and Mileva Einstein-Marić

[Caputh, 11 July 1930]

Dear Tetel and dear Miza,

I am coming to Zurich around 29 July and will stay a few days with you,[1] if you are there. Before that, I am going to the proceedings in Geneva for a week.[2] It would be nice if you, l. Tetel, came to me in Caputh in August or September. You can study splendidly in fresh air there.[3] What is going on with the new house?[4]

To a merry reunion,

Papa

The aphorisms[5] pleased me greatly, for the most part.

363. On Victor Adler[1]

"Ein Besuch bei Victor Adler"
[*Einstein 1968*]

DATED 12 July 1930[2]
PUBLISHED 1968

IN: *Victor Adler im Spiegel seiner Zeitgenossen. Beiträge zahlreicher Autoren. Vienna: Verlag der Wiener Volksbuchgemeinde, 1968.*

[Not selected for translation.]

364. To Emma Adler[1]

Caputh, 12 July 1930

[Not selected for translation.]

365. To Wilhelm Blaschke[1]

Caputh, 12 July 1930

Dear Colleague,

Mr. Grommer[2] has already written to me in the same sense. I, however, hold it in no way advisable to comply with his request. I have had Mr. Grommer as a collaborator with me for ten years. He is always discontent because of his peculiar illness[3] and always seeks the cause of his discontent in external relationships, such that I have truly suffered much with him. His highest goal was always a professorship in Russia. Now he has achieved this goal and is, of course, discontent again. I have no means of providing for him, as long as he stays here. Furthermore, I harbor the apprehension that he no longer wants to return and that he could fall on very hard times as a result. I also genuinely do not need him, because his reliability is not sufficient for truly productive collaboration.

There is nothing else for me to do but to leave him there.

With best regards, your

366. To Bund der Kriegsdienstgegner[1]

Caputh, 12 July 1930

I cannot decide whether or not to cosign the letter addressed to the Norwegian minister of war[2] that was sent to me. We would have to be quite content for the time being if we could establish that, in all countries, actual wartime service and military training can be refused for ethical reasons.[3] But our struggle must not begin with loci minoris resistentiae (places of least resistance).[4]

Respectfully yours,

367. To Ferdinand Müller

Caputh, 12 July 1930

Dear Mr. Müller,[1]

I have read your manuscript[2] and found a lot of true things in it. However, regarding the main point, namely, the way in which good and bad violins differ from one another, this was explained—correctly, in my opinion—by a physics colleague who gave a lecture last Wednesday in the university's physics colloquium. The following was already established: The vibrations of the string are transformed in the vibrations of the bridge, which, on the one hand, through the sound post on the bottom and, on the other hand, through the foot of the bridge facing away from the sound post, are transferred to the top. From bottom and top, the vibrations then are carried over into the air. The following was new for me: The good instruments distinguish themselves from the bad ones mainly in that in the middle pitches, the main part of, on the one hand, the bottom and, on the other hand, the top, vibrate in opposite directions. (For example, the top up and, at the same time, the bottom down). By this means, an optimal transmission into the air is achieved. This finding, which is based on solid experimental studies of vibration from the most diverse points of the violin, seems to me to be quite accurate, if also in no way sufficient for the construction of violins.

The main objection to your attempt at explanation is that it makes the changing string tension responsible for the generation of sound instead of the force of inertia.[3]

Hence, I am not in agreement with the intention to advertise your violin with my statements. Indeed, it is true that the violins shown to me were of thoroughly decent quality. But I know several violin makers who, in this respect, at least, have shown just as good practical results. First and foremost to be named here is a Mr. Lewin,[4] Berlin W. 30, Heilbronnerstrasse 22, who, in two days, refined a violin known to me in a wondrous way. If I were to give you permission to use my name for advertising, several others would be able demand this benefit from me with the same justification, so that an unbearable circumstance would arise for me.

I lent the violin I left behind to Prof. Born,[5] theoretical physicist at the University of Göttingen, for his musically gifted daughter.[6] If you want it, you can get the instrument back. It is a good instrument, but not as good as the violin refined by Mr. Lewin. If you come to Berlin in winter, I could compare one of your new instruments with that one.

Respectfully yours,

368. To Emanuel von der Pahlen[1]

Caputh, 12 July 1930

Dear Colleague,

Eddington's theory of astrophysics also makes no attempt[2] to answer the question of which factors are responsible for the heat production in stars per unit mass.[3] Now, a rank amateur, a Mr. Bargheer, Oberwallweg 1 in Bückeburg, along with all sorts of fantastic and untenable ideas, has made a suggestion which is capable and worthy of being tested.[4] He suggests, namely, that the production of heat (for example, by neutralization of electric charges) is dependent on the pressure to which the matter concerned is subjected. This could, in fact, be tested. Thus, we know the masses M and the diameters $2R$ of the various types of stars, and so also the pressure at their centers $p = \frac{9}{8\pi}\kappa\frac{M^2}{R^4}$ (κ = gravitational constant), where, for simplicity, the density is presumed to be constant ⟨as a function of R⟩. From the absolute magnitude of the star and its color index, the approximate total energy emission E of the star can then be computed using Planck's formula. Then, [Bargheer's] assertion consists of stating that $\frac{E}{M}$ is a function of p. Admittedly, the mass M will be known only for those stars which are part of a binary system. For comparison, we can also consider the Earth, since, from the temperature gradient and the thermal conductivity of the stone, its heat production can also be estimated.

I have already been aware for some time that $\frac{E}{M}$ for the Earth is much smaller than it is for the Sun, but I did not think of attributing that to the pressure.

With friendly greetings, yours truly,

A. Einstein

369. From Emil Rupp

[Göttingen,] 12 July 1930

Dear Professor,

In the Annalen der Physik, a doctoral student in Munich has strongly attacked the experiments on interference with canal rays.[1] His own experiments are so rudimentary that I cannot understand whether he was inept or whether his apparatus was worthless. He reminds me of someone who photographs a diatom through a magnifying glass and claims that an observer using a microscope could not see properly.

I will, in any case, take up the experiments again soon, and hope to be able to show you, too, the interferences with large path differences.[2]

A draft version of a reply is enclosed.[3]

With my sincere greetings, yours truly,

E. Rupp

370. To Leo Szilard

[Caputh,] 13 July 1930

[Not selected for translation.]

371. From Aurel Stodola

Zurich, 13 July 1930

Esteemed Mr. Einstein,

After some weeks, I have finally made time for a matter which has been on my mind for some time: studying your theory of relativity. My interest, suppressed for a long time, has suddenly shot up like a fountain jet, but the rusty thinking-joints are not able to follow, and so I turn to you, considering your kind readiness to provide information, with a request for your friendly elucidation of the following question which is gnawing at my mind in an indescribable fashion:

(1) The coordinate x_4 in your "Foundations of Relativity," 1916, p. 62[1] is indeed the "*cosmic time*,"[2] is it not, which one can read off a clock whose dial must be divided ("calibrated"?) from the clock measuring the proper time s, according to eq. (72)

$$dx_4 = \left(1 + \frac{\kappa}{8\pi}\int\frac{\rho d\tau}{r}\right) ds \qquad \ldots(1)^{[3]}$$

(2) I send a light beam from S_1 to S_2

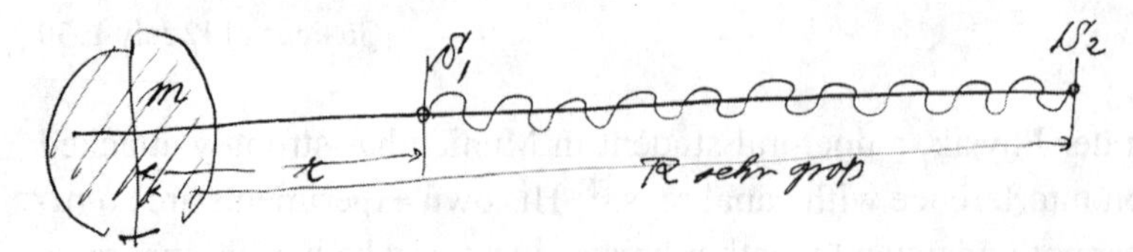

According to the article of Von Laue (Physikalische Zeitschrift 1921),[4] if its frequency is ν_1 when it was at the point S_1, it will still be the same at $S_2 = \nu_1$,

(3) I now apply the principle of your fundamental article of 1911,[5] that the number of light waves between S_1 and $S_{2(1)}$ should be independent of the time, in the following form:

Let t_1 be that *cosmic time interval* in S_1 which *corresponds* to the cosmic time interval t_2 in S_2. From the relative size of these two, one could obtain information about the light signals sent from S_1 at regular intervals ("x_4 time-seconds"), whose arrivals at S_2 are marked by the x_4 clocks there. The number of waves sent from S_1 into the beam is then $t_1\nu_1$, and, at S_2, a number $t_2\nu_1$ are received, and since these must be equal, it follows that

$$t_1 = t_2 \qquad \ldots(2)$$

(4) However, these "*corresponding*" time intervals are related, according to the exposition given by Pauli in the *Encyclopedia of Mathematics*, due to the "*transverse*" Doppler effect (as he calls it) by the equation[6]

$$t_1 = t_2\left[1 - \frac{\phi}{c^2}\right], \qquad \ldots(3)$$

where ϕ is the *increase* of the gravitational potential between 1 and 2. I have arrived here at a contradiction which I am not in a position to resolve.

(5) In your fundamental review article of 1911, you did not make use of the transverse Doppler effect but rather of the ordinary [linear] effect, and required that one measure the time (that is indeed the cosmic time) at S_2, with a clock that ticks by a factor $\left[1+\frac{\phi}{c^2}\right]$ more slowly (compared at the same position) than at S_1. This means that the clock at S_1 must tick more rapidly in the ratio

$$t_1 \cong t_2\left[1+\frac{\phi}{c^2}\right] \qquad \ldots(4)$$

than that at S_2. The various derivations (2), (3), and (4), in my interpretation, give differing results for the ratio of the cosmic times at S_1 and S_2.

I can find no way at all, avoiding (2), purely from the general integrals of the gravitational equations, to make the connection between the cosmic times at S_1 and S_2, since eq. (1) compares only the *cosmic and the proper times* at a single *particular* position.

I have also written to Profs. Weyl and Pauli,[7] but those gentlemen have so many other things to think about and express themselves so abstractly that I, in my desperation, could see no other means of rescue than to turn to the Master of the Masters. My good friend Besso,[8] who visits me regularly in Zurich, declared himself unable to resolve the discrepancy.

With the greatest suspense and my warmest esteem, yours sincerely,

A. Stodola

If the solution of Laue were to allow that ν_2 at S_2 need not be equal to ν_1 at S_1, then, due to the equation $t_1\nu_1 = t_2\nu_2$, everything would clear up satisfactorily.

372. “On the Nature of Reality”

“Über das Wesen der Realität”
[*Einstein and Tagore 1931*]

DATED 14 July 1930

IN: *The Modern Review* January 1931, 42–43.

42 THE MODERN REVIEW FOR JANUARY, 1931

its own Executive Council;—also its own Secretariat on the League of Nations plan; and it will have the power to appoint Committees and *rapporteurs*.

As on the one hand, the British Indian Federation becomes more and more decentralized, and on the other, the Indian States become more and more homogeneous and conformable in political structure to the Provincial Governments, they will become closely integrated into the one Central all-India Federation.

This the dynamic aspect, the principle of growth and evolution, latent in the above scheme,—otherwise called the logic of events, the dynamic of history.

BANGALORE,
21st March, 1928

The Nature of Reality

Authorized Version]

[1] (A conversation between Rabindranath Tagore and Professor Albert Einstein in the afternoon of July 14th, 1930, at the Professor's residence in [2] Kaputh)

E. Do you believe in the Divine as isolated from the world?

T. Not isolated. The infinite personality [3] of Man comprehends the Universe. There cannot be anything that cannot be subsumed by the human personality, and this proves that the truth of the Universe is human truth.

I have taken a scientific fact to explain this. Matter is composed of protons and electrons, with gaps between them, but matter may seem to be solid without the links in spaces which unify the individual electrons and protons. Similarly humanity is composed of individuals, yet they have their inter-connection of human relationship, which gives living unity to man's world. The entire universe is linked up with us, as individuals, in a similar manner;—it is a human universe.

I have pursued this thought through art, [4] literature, and the religious consciousness of man.

E. There are two different conceptions about the nature of the Universe:

(1) The world as a unity dependent on humanity.

(2) the world as a reality independent of the human factor.

[5] *T.* When our universe is in harmony with man, the eternal, we know it as truth, we feel it as beauty.

E. This is the purely human conception of the universe.

T. There can be no other conception. This world is a human world—the scientific view of it is also that of the scientific man. Therefore, the world apart from us does not [6] exist; it is a relative world, depending for its reality upon our consciousness. There is some standard of reason and enjoyment which gives it truth, the standard of the Eternal Man whose experiences are through our experiences.

E. This is a realization of the human [7] entity.

T. Yes, one eternal entity. We have to realize it through our emotions, and activities. We realized the Supreme Man who has no individual limitations through our limitations. Science is concerned with that which is not confined to individuals, it is the impersonal human world of truths. Religion realizes these truths and links them up with our deeper needs; our individual consciousness of truth gains universal significance. Religion applies values to truth, and we know truth as good through our own harmony with it.

E. Truth, then, or Beauty is not independent of Man?

T. No.

E. If there would be no human beings any more, the Apollo of Belvedere would no [8] longer be beautiful.

T. No!

E. I agree with regard to this conception of Beauty, but not with regard to Truth.

T. Why not? Truth is realized through man.

E. I cannot prove that my conception is right, but that is my religion.

T. Beauty is in the ideal of perfect harmony which is in the Universal Being, Truth the perfect comprehension of the Universal mind. We individuals approach it through our own mistakes and blunders, through our accumulated experiences,—through our illumined consciousness—how, otherwise, can we know Truth?

E. I cannot prove that scientific truth must be conceived as a truth that is valid independent of humanity; but I believe it firmly. I believe, for instance, that the Pythagorian theorem in geometry states something that is approximately true, independent of the existence of man. Anyway, if there is a *reality* independent of man, there is also a truth relative to this reality; and in the same way the negation of the first engenders a negation of the existence of the latter. [9]

T. Truth, which is one with the Universal Being, must essentially be human; otherwise whatever we individuals realize as true can never be called truth, at least the truth which is described as scientific and which only can be reached through the process of logic, in other words, by an organ of thoughts which is human. According to Indian Philosophy there is Brahman, the absolute Truth which cannot be conceived by the isolation of the individual mind or described by words but can only be realized by completely merging the individual in its infinity. But such a truth cannot belong to Science. The nature of truth which we are discussing is an appearance, that is to say, what appears to be true to the human mind and therefore is human, and may be called *Maya* or illusion. [10]

E. So according to your conception, which may be the Indian conception, it is not the illusion of the individual but of humanity as a whole.

T. In science we go through the discipline of eliminating the personal limitations of our individual minds and thus reach that comprehension of truth which is in the mind of the Universal Man. [11]

E. The problem begins whether Truth is independent of our consciousness.

T. What we call truth lies in the rational harmony between the subjective and objective aspects of reality, both of which belong to the super-personal man.

E. Even in our everyday life, we feel compelled to ascribe a reality independent of man to the objects we use. We do this to connect the experiences of our senses in a reasonable way. For instance, if nobody is in this house, yet that table remains where it is. [12]

T. Yes, it remains outside the individual mind but not the universal mind. The table which I perceive is perceptible by the same kind of consciousness which I possess. [13]

E. Our natural point of view in regard to the existence of truth apart from humanity cannot be explained or proved, but it is a belief which nobody can lack—no primitive beings even. We attribute to Truth a super-human objectivity, it is indispensable for us, this reality which is independent of our existence and our experience and our mind—though we cannot say what it means.

T. Science has proved that the table as a solid object is an appearance and therefore that which the human mind perceives as a table would not exist if that mind were naught. At the same time it must be admitted that the fact that the ultimate physical reality of the table is nothing but a multitude of separate revolving centres of electric force, also belongs to the human mind.

In the apprehension of truth there is an eternal conflict between the universal human mind and the same mind confined in the individual. The perpetual process of reconciliation is being carried on in our science, philosophy, in our ethics. In any case, if there be any truth absolutely unrelated to humanity, then for us it is absolutely non-existing. [14]

It is not difficult to imagine a mind to which sequence of things happens not in space but only in time like the sequence of notes in music. For such a mind its conception of reality is akin to the musical reality in which Pythagorian geometry can have no meaning. There is the reality of paper, infinitely different from the reality of literature. For the kind of mind possessed by the moth which eats that paper literature is absolutely non-existent, yet for Man's mind literature has a greater value of truth than the paper itself. In a similar manner if there be some truth which has no sensuous or rational relation to human mind, it will ever remain as nothing so long as we remain human beings. [15]

E. Then I am more religious than you are!

T. My religion is in the reconciliation of the Super personal Man, the Universal human spirit, in my own individual being. This has been the subject of my Hibbert Lectures, which I have called "The Religion of Man." [16]

373. "On the Theory of Spaces with Riemannian Metric and Distant Parallelism"

"Zur Theorie der Räume mit Riemannmetrik und Fernparallelismus"
[*Einstein 1930ee*]

PRESENTED 17 July 1930
PUBLISHED 1930

IN: *Preussische Akademie der Wissenschaften* (Berlin). *Physikalisch-mathematische Klasse. Sitzungsberichte* (1930): 401–402.

In the following, a general property of such spaces is demonstrated whereby the question of their physical significance will be left aside for the meantime.[1]

Let $(T^{\mu\nu})$ be a tensor which can have other indices in addition to the contravariant indices μ and ν. Then, the commutation rule for differentiation always holds:

$$T^{\mu\nu}{}_{;\sigma;\tau} - T^{\mu\nu}{}_{;\tau;\sigma} \equiv -T^{\mu\nu}{}_{;\alpha}\Lambda^{\alpha}_{\sigma\tau}\,.^{[2]} \tag{1}$$

By contraction, this becomes

$$T^{\mu\nu}{}_{;\nu;\mu} - T^{\mu\nu}{}_{;\mu;\nu} \equiv T^{\sigma\tau}\Lambda^{\alpha}_{\mu\nu}\,. \tag{1a}$$

With the aid of a simple rearrangement, this results in

$$[(T^{\mu\nu} - T^{\nu\mu})_{;\nu} - T^{\sigma\tau}\Lambda^{\mu}_{\sigma\tau}] + T^{\sigma\tau}\Lambda^{\alpha}_{\sigma\tau;\alpha} \equiv 0\,. \tag{2}$$

Only the antisymmetric part of the tensor T enters into (2). We may thus assume without restrictions that the tensor T is antisymmetric with respect to the indices that we are considering. Then (2) takes on the form

$$\left[T^{\mu\nu}{}_{;\nu} - \frac{1}{2}T^{\sigma\tau}\Lambda^{\mu}_{\sigma\tau}\right]_{;\mu} + \frac{1}{2}T^{\sigma\tau}\Lambda^{\mu}_{\sigma\tau;\mu} \equiv 0 \tag{2a}$$

This relation can be further rearranged by making use of the following identity, which results from the integrability of parallel displacements:

$$\Lambda^{\mu}_{\sigma\tau;\mu} \equiv \phi_{\sigma,\tau} - \phi_{\tau,\sigma} \qquad (\phi_\sigma = \Lambda^{\alpha}_{\sigma\alpha}) \tag{3}$$

or

[1] The content of the article "The Compatibility..." in these *Berichte* 1930: I is presupposed to be known here.[1]

$$\Lambda^{\mu}_{\sigma\tau;\mu} \equiv \phi_{\sigma;\tau} - \phi_{\tau;\sigma} + \phi_{\mu}\Lambda^{\mu}_{\sigma\tau} \tag{3a}$$

Due to (3a), we namely find:

$$\frac{1}{2}T^{\sigma\tau}\Lambda^{\mu}_{\sigma\tau;\mu} \equiv (T^{\sigma\tau}\phi_{\sigma})_{;\tau} - \phi_{\sigma}T^{\sigma\tau} + \frac{1}{2}\phi_{\mu}T^{\sigma\tau}\Lambda^{\mu}_{\sigma\tau}.$$

If we insert the right-hand side of this latter relation into (2a) and, at the same time, introduce the divergence symbol[3]

$$A^{\nu}{}_{/\nu} \equiv A^{\nu}{}_{;\nu} - \phi_{\nu}A^{\nu}, \tag{4}$$

where A^{ν} is a tensor of arbitrary rank with the contravariant index ν, we obtain

$$\left.\begin{aligned} U^{\mu}{}_{/\mu} &\equiv 0, \\ U^{\mu} &= T^{\mu\nu}{}_{/\nu} - \frac{1}{2}T^{\sigma\tau}\Lambda^{\mu}_{\sigma\tau}. \end{aligned}\right\} \tag{5}$$

We can thus derive from any tensor with an antisymmetric pair of indices $\mu\nu$, *by means of a linear differential operation, a tensor* U^{μ} *of 1 lower rank, whose divergence vanishes identically.*

Thus, for example, from the tensor

$$L^{\alpha}_{\underline{\mu\nu}} = \Lambda^{\alpha}_{\underline{\mu\nu}} + a(\phi_{\underline{\mu}}g^{\nu\alpha} - \phi_{\underline{\nu}}g^{\mu\alpha}) + bS^{\alpha}_{\underline{\mu\nu}}, \tag{6}$$

where a and b are arbitrary constants, and the definition

$$S^{\alpha}_{\underline{\mu\nu}} = \Lambda^{\alpha}_{\underline{\mu\nu}} + \Lambda^{\mu}_{\underline{\nu\alpha}} + \Lambda^{\nu}_{\underline{\alpha\mu}} \tag{7}$$

has been used, we can derive the tensor

$$G^{\mu\alpha}{}_{/\mu} = L^{\alpha}_{\underline{\mu\nu}/\nu} - \frac{1}{2}L^{\alpha}_{\underline{\sigma\tau}}\Lambda^{\mu}_{\sigma\tau}, \tag{8}$$

whose divergence / taken with respect to μ vanishes identically:

$$G^{\mu\alpha}{}_{/\mu} \equiv 0. \tag{8a}$$

It follows from this that the system of equations

$$G^{\mu\alpha} = 0 \tag{9}$$

is a compatible system for the $h_{s}{}^{\nu}$, whatever the constants a and b may be.[4]

374. To Friedrich Leppmann[1]

Caputh, 17 July 1930

[Not selected for translation.]

375. To Emil Rupp

Caputh, 17 July 1930

[Not selected for translation.]

376. To Friedrich Schmiedl[1]

Berlin, 17 July 1930

[Not selected for translation.]

377. From Aurel Stodola

Zurich, 18 July 1930

Esteemed Master and kind Friend of Humanity,

Through your genial and prompt reply to my letter,[1] you have acted in a most philanthropic manner, for which I cannot thank you heartily enough. Now, the false complexes in my subconscious have been eliminated, and I believe that I can gradually penetrate deeper into the material. Show me your more than generous favor, and allow me to pose a few additional questions:

(1) How does one determine the change in the clock rate (of the proper time) *when a clock is set in accelerated* motion? There would seem to be two methods

(α) the *equivalence of gravitation and acceleration.* If the latter is given, one need only place a mass so that

$$\frac{\partial\phi}{\partial r} = -b$$

and all the formulas of the gravitational case remain valid. In particular, x_4 would also represent the "time" in the accelerated system, and the deviation of the clock rate would be given by eq. 72 of the "Fundamentals,"[2]

$$x_4 = t = \left(1 + \frac{\kappa}{8\pi}\int\frac{\rho dt}{r}\right)s\,.$$[3]

The deviation $t - s$ would thus be proportional to the elapsed time s.

(β) Or one could, referring to a remark of Prof. Weyl's,[4] make use of the formulas of the gravitational field, but only setting $dx_2 = 0$, $dx_3 = 0$ and, in contrast, $x_1 = \dot{f}(t)$, i.e., $\underline{dx_1 = \dot{f}(t)dt}$, and then calculate the proper time from the general formula

$$ds^2 = g_{11}\dot{f}(t)^2 dt^2 + g_{44}dt^2 .$$

Then, however, the deviation of the proper time as a result of the acceleration $\ddot{f}(t)$ would be proportional in *the first approximation to the integral* $\int \dot{f}(t)^2 dt$. If one wished to reach a certain final velocity $v = \dot{f}(t_e)$, then, *from this, one would have to accelerate as strongly as possible* in order to obtain a small deviation of the clock rate.

Since the results from (α) and (β) do not agree, and I read in the Physikalische Zeitschrift 1921, p. 651[5] that, according to Mie, general relativity *can give no result*(!) on the situation of accelerated motion in itself, I am again in a dire quandary.

(2) If, namely, the role of the acceleration were clearly defined, than one could approach the following important case:

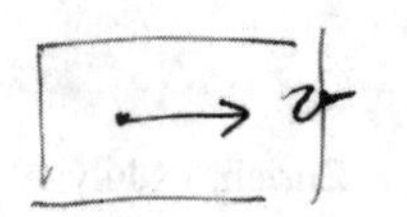

(α) Given: a fixed point in a *coordinate system in uniform motion*, (K_1 = the case of special relativity theory), is known;

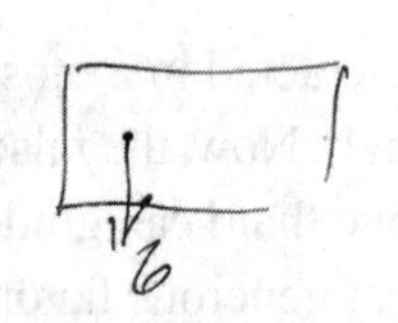

(β) *Given: a coordinate system that is accelerated down ward,* K_2, in which the proper time deviates in a manner that is presumed to be known from the cosmic time;

(γ) *The mass point undergoes the motions* (α) *and* (β) *at same time*: it moves along a segment of a parabola; an infinite number of infinitely short segments of such a parabola can be joined together (as done by Newton) to give a circular motion.

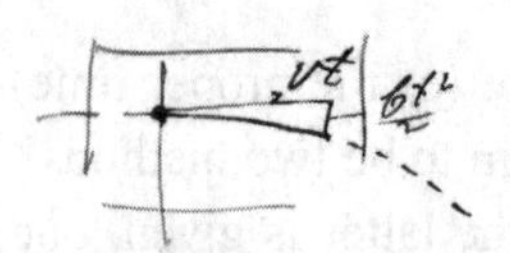

The time correction for this motion *must indeed be composed of the corrections* (α) *and* (β) *together.*

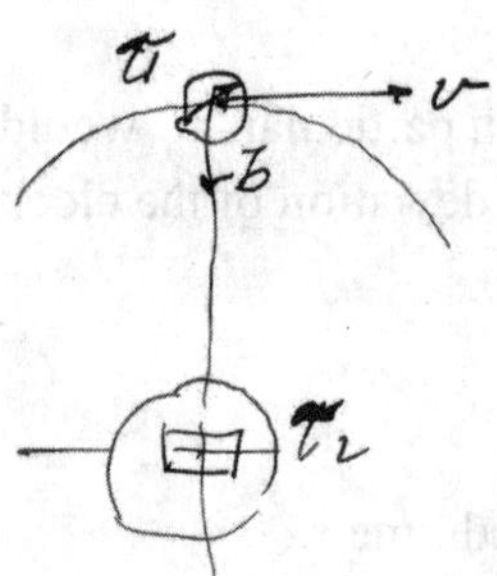

When Pauli, in his demonstration of the time relation in the Enzyklopädie der Mathematik,[6] considers only the velocity *v*, *is he not committing an omission*?

(3) The polemic writings of the two philosophers *Kraus: offene Briefe* and *Lipsius: Wahrheit und Irrtum* i[n] d[er] R[elativitätstheorie][7] have come into my possession, and I can only laugh at their sophism, which, in fact, is, at the same time, dull; *but it still irritates me*. A more precise exposition of the situation, at least of the translational circular motion sketched above, would nevertheless be useful and instructive.

(4) According to the schematic procedure given above, the *energy of a light quantum* of frequency ν is represented by $E = h\nu$. A *light quantum emitted from the Sun* however arrives at the Earth (or somewhere) according to your confirmation with *the same frequency* ν, therefore *its energy* $E' = h\nu$ is also identical to E. But nevertheless, since the quantum represents gravitational mass and has to overcome the gravitational force of the Sun and has thereby performed work, the *equality* $E = E'$ *seems problematic.* I had posed this question to Prof. Weyl, who answered that he had not yet considered that case. I am writing to him today that it appears more correct to me to first present this problem to you and to leave to you the priority of dealing with it. If the result should be that one must consider h to be variable in a gravitational field, then your priority for such a decision is evident. I would have rather turned to you right at first, if I had been *able to overcome the justifiably great respect for you in the first instance.* But one should indeed not overtax your time with minor questions, and only the torments of my own inadequacy finally gave me the courage to do so.

The illness of our mutual friend Großmann, with whom I developed a closer relationship during the war, is among the most horrible that can befall a human being.[8] I advised the professor's wife[9] from the very beginning to request that the Ehrlich Institute in Frankfurt—i.e., to place the means at their disposal which, to be sure, would run to the thousands—scientifically investigate the spirochetes of that treacherous illness (which has nothing to do with syphilis)—which, up to now, has not been undertaken—in order to possibly develop an antiserum. Whether it is now not too late for that, I do not know. So must our poor friend undergo the decline in which, as a result of the decay of his spinal cord, gradually all connections to the external world, sensory and motoric, will be interrupted, and the intact, healthy cerebrum; that is, his inner self, will agonize to death alone in helpless suffering. Gruesome!!

So can those praise God who are still in possession of all of their mental powers. I saw you wrestling with general relativity—that was an heroic undertaking. The new problem which is now fascinating you would be a crowning achievement, but not such a sensation as was relativity. Therefore, permit a pilgrim who is plagued by understandable weariness at the end of his trek to give you a most well-meaning piece of advice: No need for haste—you have already brought forth the unprecedented; one more or less jewel in your many-jeweled crown—there's no hurry.

In lasting gratitude, yours truly,

A. Stodola

378. To Elsa Einstein

Geneva, 22 July [1930][1]

Dear Elsa,

I had a very comfortable journey, as I had the compartment all to myself.[2] A large part of the way to Frankfurt, I spoke with Dr. Riezler, Liebermann's[3] son-in-law, who is the curator of the university in Frankfurt. I enjoyed the food very much and finished it. The suits are already hanging in the wardrobe. Now the dirty business here can get going. I am firmly determined, however, not to get irritated. Riezler told me a nasty story about Krüss.[4] I will swap him with Kessler,[5] regardless. The worst that could happen is that I wouldn't be reelected to the Commission, which would a priori comfort me brilliantly. Now, Luchaire[6] is the first to come to me at the hotel. I am curious to see how he will defend himself.

Until I return, be happy and greeted and kissed by your,

Albert

Best regards to the daughters, Rudi, and Off.[7]

379. To Mileva Einstein-Marić

[Geneva, between 23 and 27 July 1930] [1]

[Not selected for translation.]

380. To Walther Mayer

[Geneva,] 23 July 1930

Dear Mr. Mayer,

I now understand why, in the case we last considered, the series expansion fails only for the positive powers of ε. Then, this is the case when there exists an identity in the equations of first approximation that does not correspond to an exact identity. Our case is in a first approximation like this

$$G^{\mu\alpha} = \Lambda^{\alpha}_{\mu\nu,\,\nu} - (\varphi_\mu \delta_{\nu\alpha} - \varphi_\nu \delta_{\mu\alpha})_{,\nu} = 0 .$$

or $$G^{\mu\alpha} = h_{\alpha\mu,\,\nu\nu} - h_{\alpha\nu,\,\nu\mu} - \varphi_{\mu,\,\alpha} + \varphi_{\nu,\,\nu}\delta_{\mu\alpha} = 0 .$$

In addition to the strict identity

$$G^{\mu\alpha}{}_{/\mu} ,$$

one also has in a first approximation the identity $G^{\mu\alpha}{}_{,\alpha} = 0$.

That's because, in first approximation, $\Lambda^{\alpha}_{\mu\nu,\,\nu,\,\alpha} = \varphi_{\mu,\,\nu\nu} - \varphi_{\nu,\,\nu,\,\mu}$, which is canceled against $(-\varphi_{\mu,\,\alpha} + \varphi_{\nu,\,\nu}\delta_{\mu\alpha})_{,\alpha}$. A corresponding strong identity does not exist, even not in the case when an *S* term with arbitrary coefficients is added.

If one assigns the φ-term another coefficient than –1, then the second identity disappears (in first approximation). Then we again have a series expansion with only positive powers of ε.

In my earlier considerations, where I naturally assumed an expansion possibility according to positive powers of ε, there resulted quadratic equations for the first approximation that did not accord with the linear equations. But with your example, you have made it clear that this is the fault of the impossibility to expand along positive powers. Now, it is interesting that one is thus led to the case where one requires that, in first approximation, the equations

$$R_{ik} = 0$$

have to be valid for the pure gravitational field.

Here, there is a witches' brew of intrigues and trifles, but the thought that it will be over in six days keeps me upright.[1]

With warm greetings, your

A. E.

381. Statement at ICIC Enquiry to Study its Work, Program, and Organization

[*Einstein 1930gg*]

DATED 24 July 1930
PUBLISHED 1930

IN: ICIC Twelfth Session Meeting Minutes

[See documentary edition for English text.]

382. To Walther Mayer

[Geneva,] Monday [28 July 1930][1]

Dear Mr. Mayer,

Tomorrow, the meetings come to an end—praise God.[2] It is unbelievable with which certain instinct these intelligent men discover the minor details that will allow them to kill time in the most effective way.

In the meantime, however, I have pondered a lot without finding anything new. Of the things that occurred to me, the strongest impression was made by the fact that, for that choice,[3] we have

$$L^{\alpha}_{\mu\nu} = \Lambda^{\alpha}_{\mu\nu} - (\varphi_\mu \delta_\nu{}^\alpha - \varphi_\nu \delta_\mu{}^\alpha),$$
$$L^{\alpha}_{\mu\nu/\alpha} \equiv 0.$$

Since[4]

$$\Lambda^{\alpha}_{\mu\nu/\alpha} \equiv \varphi_{\nu,\mu} - \varphi_{\nu,\mu} - \varphi_\alpha \Lambda^{\alpha}_{\mu\nu} \equiv \varphi_{\mu;\nu} - \varphi_{\nu;\mu}$$

and

$$(\varphi_\mu \delta_\nu{}^\alpha - \varphi_\nu \delta_\mu{}^\alpha)_{/\alpha} \equiv \varphi_{\mu;\nu} - \varphi_{\nu;\mu},$$

I see no logical basis for the significance of this identity, but it gives this choice of the constants somehow a privileged position which, at the same time, has the effect that, in first approximation, the equations $R_{ik} = 0$ hold. Therefore, I believe that we must take this case for the centrally symmetric problem most especially into account.

Hearty greetings, and looking forward to seeing you soon, yours truly,

A. E.

383. To Rudolf Kayser and Ilse Kayser-Einstein

Zurich, 31 July 1930

My dears,

I am much more satisfied with Geneva this time than previously.[1] The task is being addressed more substantively than before. I wish you both a happy move, and Margot, as well as Off, much success with the escapade into "real" life.[2] I will probably travel Sunday, and likely without Tetel,[3] who is only coming to Caputh in the autumn. I arrive at midnight in Haberlandstrasse.

Warm regards, also to Else and Maja,[4] from your

Albert

384. To Ove Rode[1]

[Caputh], 5 August 1930

[Not selected for translation].

385. To Joseph Müller[1]

[Caputh,] 5 August 1930

[Not selected for translation.]

386. To Leon Simon[1]

Caputh, 5 August 1[930]

Dear Sir,

I must explain my position somewhat more explicitly. I think that all of my occasional statements [on] Zionism are not worthy of a special publication and therefore believe that I should not contribute, although, of course, I have no right to ban it.[2]

You perhaps are aware that I am not particularly pleased with the attitude of the Zionist leadership in regards to the Arab problem.[3] On the other hand, I do not think it is right for the latter to get involved in the current difficult situation. I also do not consider myself competent to that end either. Therefore, I cannot give the desired statement of opinion at the present time.[4] The idea of a luxury edition should, in any case, be dropped,[5] particularly since such a proceeding would evoke an embarrassingly snobbish impression in the current difficult conditions.

Sincerely yours,

387. To Max Warburg

Caputh, 5 August 1930

Dear Mr. Warburg,

Of the two reports, the remark about the great distance of the markets, or rather the convenient transport lines, made an impression on me.[1] On the other hand, the climate of the area under consideration shouldn't be tropical, by virtue of its altitude. Far be from me to have any opinion regarding the suitability of the envisioned territory. I just think that, in view of the very difficult situation in the East,[2] no possibility coming into consideration should be discarded *unexamined.* Sending a few trusted people there to study is not an undertaking of great risk. Prof. Oppenheimer finds an on-the-spot investigation to be justified, on the basis of basic research about the nature of the region under consideration.[3] Naturally, I am also of the conviction that Palestine has a unique *moral* significance for Jews throughout the world. However, this country is certainly not suited for the absorption of large masses [of people].[4]

This letter should be nothing more than a justification for having permitted myself to trouble you with this matter at all.

Warm regards, Your–

388. To Helmar Lerski[1]

[Caputh, 6 August 1930][2]

The Jews are, today, more of an ethnic community than a religious community.[3] Defining this category—as difficult as it is—thus fulfills a vivid wish: We wish to capture in a vivid way what we mean and feel when we say "we." May the artist succeed in this difficult task.[4]

Albert Einstein to Mr. Lerski

389. Foreword to *Alberini 1930*

"Zum Geleit"
[*Einstein 1930ff*]

PUBLISHED around 8 August 1930[1]

IN: Coriolano Alberini, *Die deutsche Philosophie in Argentinien.* Series: *Argentinische Bücherei.* Berlin-Charlottenburg: H. W. Hendriock, 1930.

[Not selected for translation.]

390. From Jacques Hadamard

[Saint-Cybranet, before 9 August 1930][1]

My dear colleague and friend,

Almost two months ago, I sent you the attached text, asking you if it expresses faithfully the spirit of the correspondence and the conversation we had together last autumn.[2] I have never received your response, and wonder if you even received my letter. I am therefore obliged to ask you to send this response to me again—and, in fact, by return post.[3]

This text must actually appear as a post-script to an article that, as I say, was written one year ago, and therefore, I do not wish to delay the publication interminably.

Excuse me for bothering you—perhaps once again—and, in anticipation of the pleasure of seeing you again soon, I hope, yours sincerely,

J. Hadamard

391. To Jacques Hadamard

Caputh, 9 August 1930

Dear Colleague,

I had, of course, read your answer at the time[1] and was in agreement with the publication. However, I have not changed my view. If indeed, in such cases, a state can be declared to be guilty, it will in any case not be those who will have risked their own necks.[2] Furthermore, I am convinced now as before that only a fundamental rejection of the method of war can help.[3] If those who are respected as leaders for their achievements would publicly profess their unconditional opposition to war, this attitude would rapidly spread among the rest.[4] I believe that, in the current circumstances, taking upon oneself such a goal would *not* represent too great a risk.

Warm regards, your

392. To Auguste Piccard[1]

Caputh, 9 August 1930

Dear Mr. Piccard,

I have just written an urgent letter to the Deutsche Luftfahrt-Verband [German Association for Air Travel].[2]

The force on a rotating mass due to its rotation is of the order of magnitude $\frac{\kappa^2}{c^2} \cdot \left(\frac{v}{c}\right)^2$, which is much less than 10^{-39}, so that its observation is out of the question.[3]

Hearty greetings from yours truly,

A. Einstein.

393. To Paul Winteler

[Caputh,] 10 August 1930

[Not selected for translation.]

394. To Marcel Grossmann

[Berlin,] 14 August 1930

Dear Marcel,

You will receive the offprints that you asked for in the coming days. I trust in the correctness of the path that I have begun to follow, although I do not yet have the field equations, and therefore there can be no serious talk of a comparison with experiments. For completeness, I send you my most recent result, which is currently being printed.[1]

If $L^{\alpha}_{\mu\nu}$ is an arbitrary tensor, antisymmetric in μ and ν, and[2]

$$T{:}^{\nu}{}_{/\nu} = T{:}^{\nu}{}_{;\nu} - \varphi_{\nu} T{:}^{\nu}$$

is the definition of the divergence operator "/", then the tensor

$$G^{\mu\alpha} = L^{\alpha}_{\underline{\mu}\underline{\nu}/\nu} - L^{\alpha}_{\underline{\sigma}\underline{\tau}} \Lambda^{\mu}_{\sigma\tau}$$

fulfills the identity

$$G^{\mu\alpha}{}_{/\mu} \equiv 0.$$

We thus have, expressing L in terms of the Λ and h,

$$G^{\mu\alpha} = 0,$$

which is a compatible system of field equations. This follows from the commutation relation for differentiation. It is evident[3]

$$L^{\alpha}_{\underline{\mu}\underline{\nu}} = \Lambda^{\alpha}_{\underline{\mu}\underline{\nu}} + a(\varphi_{\underline{\mu}}\delta_{\underline{\nu}}{}^{\alpha} - \varphi_{\underline{\nu}}\delta_{\underline{\mu}}{}^{\alpha}) + bS^{\alpha}_{\underline{\mu}\underline{\nu}},$$

where a and b are constants. All this is understandable only after one has read the articles.

With hearty greetings, your

A. E.

395. To Arnold Sommerfeld

[Caputh,] 14 August 1930

Dear Sommerfeld,

I would be very happy to be able to see you, but during the summer, I am [in] Caputh, near Potsdam (twenty minutes by bus from the Potsdam Station).[1] I do not know anyone who is going to Odessa.[2] I cannot give any lecture, and have nothing special to report. I find your book on wave mechanics to be very good.[3] But the whole development does not quite satisfy me despite its great successes.[4]

Hearty greetings from your,

A. Einstein

396. "Tagore Talks with Einstein"

[*Einstein 1931b*]

DATED 19 AUGUST 1930[1]
PUBLISHED March 1931

IN: *Asia* 31, no. 3 (March 1931): 139–143

[See documentary edition for English text.]

397. Address for Opening of the German Radio and Phonographic Exhibition[1]

[*Einstein 1930xx*]

DATED 22 August 1930
PUBLISHED 30 August 1930

IN: *New York Times* and *Daily Mail,* 30 August 1930.

When you are listening to the radio,[2] think also about how humanity came into possession of this wonderful communications tool.

The original source of every technical achievement is the divine curiosity and the playful instincts of the tinkering and pondering researcher, and no less the constructive imagination of the technical inventor.

Think about Oerstedt[3], the first to notice the magnetic effect of electric currents; about Reis[4], who was the first to use that effect to produce sound by electromagnetic means; about Bell[5], who made use of sensitive electrical contacts in his microphone to convert sound waves into variable electrical currents[6]. Think also about Maxwell[7], who demonstrated mathematically the existence of electromagnetic waves; about Hertz[8], who was the first to generate them using a spark[9]. Think[10] especially about Liebens[11], who, with his vacuum electron-valve tubes,[12] conceived an incomparable sensory organ for electrical oscillations which proved also to be an ideal and simple instrument for producing them. Think thankfully of the army of nameless technicians who simplified the instruments of radio communications and made them suitable for mass production, so that they have become accessible to everyone.

Those people should all be ashamed of themselves, who make use of the wonders of science and technology without giving them a further thought, and have understood no more of them than a cow understands of the botany of the plants that she is comfortably eating.

Think also about the fact that it is the technologists who first make true democracy possible. For they not only ease the daily labor of many humans, but also make available the works of the finest thinkers and artists—whose enjoyment was, until only recently, the privilege of the favored classes—to all of humanity, and so have awakened the populace from their drowsy stupor.

What concerns radio in particular: It has a unique function to fulfill, in the sense of international reconciliation. Right up to present times, nations learned about each other almost exclusively through the distorting mirrors of their own daily newspapers. Radio, in contrast, shows them to one another in the most vivid forms and, in the main, from their most agreeable aspects. It will thus contribute to expunging mutual feelings of foreignness, which can so readily turn into mistrust and enmity.

Consider, in this sense, the results of the creative efforts which this exhibition is offering to the astonished senses of its visitors.

398. From Marcel Grossmann

Zurich, 24 August 1930

Dear Albert,

Many thanks for your two letters to me,[1] which made it possible for me to inform myself about your efforts and results. Naturally, I was, in the first instance, interested in the mathematical foundations of the grandiose structure that you want to erect with such tenacious perseverance.

However, I have come to the conviction that this foundation is an illusion, upon which you will certainly not insist.[2] One can understand that the following theorem holds:

When one imposes teleparallelism upon a Riemannian manifold (or something equivalent), then it becomes simply a Euclidean manifold; "pseudo-Euclidean" manifolds which would be different from Euclidean manifolds do not exist.

Proof: You yourself have found, on p. 7 of your paper for the Academy in 1928,[3] that the g_{ik} are constant: therefore, the Riemannian curvature measure vanishes, $R = 0$ (this follows also from p. 5, bottom). The manifold is thus "flat" and, in terms of differential geometry or invariant theory, not distinct from a Euclidean manifold. Nothing of this fact changes if one puts aside R as "too complicated" (*ibid.*, p. 7, top).

I am convinced that you will reach your goal, although not with this manifold. Please answer my letter and accept my heartiest greetings, from yours truly,

G.[4]

399. To Marcel Grossmann

[Caputh?,] 24 August, 1930

Dear Grossmann,

I also have many doubts as to whether the path I have taken is the correct one, from a physical standpoint. But formally, the matter is wonderful and not at all contradictory. That your objection is not correct can be seen as follows:[1] Take some arbitrary surface in Euclidean space. On it, a Riemannian metric holds. At each point, there is a Euclidean local system *K*, in which the Pythagorean theorem is valid. *K* is, however, determined only up to a rotation. But nothing prevents me from assuming that a particular orientation of *K* is preferred, indeed, at each point on the surface. If you then further assume that, in fact, not a complete orientation of *K* is given, but rather that all can be rotated in the same manner, then you have before you the geometric structure which I ascribe to space. Vectors at different points which have the same local components are termed parallel. As you can see, such a surface need by no means be planar. It can, in fact, be quite arbitrarily chosen.

You are thinking of the following theorem: When the Riemann-Christoffel infinitesimal parallel displacement is integrable, then the manifold is Euclidean. That is true, but it is not a question of that here. If one sets

$$\delta A^{\alpha} = -\Delta^{\alpha}_{\mu\nu} A^{\mu} \delta_{\alpha}{}^{\nu} \quad \text{(parallel displacement theorem)},$$

then the $\Lambda^{\alpha}_{\mu\nu}$ [2] become equal to the Christoffel symbols when we require the following:

(1) The magnitude of the vector ($ds^2 = g_{\mu\nu} dx^{\mu} dx^{\nu}$) remains unchanged upon parallel displacement;

(2) the $\Delta^{\alpha}_{\mu\nu}$ are symmetric with respect to μ and ν.

But I do not employ (2); instead, I require integrability (the $R^{i}{}_{k,lm}$ which are derived from the Δ must vanish). This, then, does not yield a Euclidean manifold as a result. That is how the manifolds which I have studied are generated.

Unfortunately, the field equations are not so uniquely formally determined as in the Riemannian case. Mayer[3] and I are sweating, day by day, without having obtained a secure result up to now. I, however, console myself like your Diploma student, with his motto, "Man errs as long as he strives."[4]

Hearty greetings to you and to your wife,[5] from your

A. E.

400. From Arthur Stanley Eddington

Observatory, Cambridge, 26 August 1930

[See documentary edition for English text.]

401. To Deutsches Studentenwerk[1]

Caputh, 28 August 1930

Your article[2] appears, to me, to touch on the right thing, in so far as it connects the overcrowding of the universities with the fact that it can be difficult to provide places for young people in nonacademic professions. After all, the reproach must be made of the school that it has thoroughly proven itself a failure as an instrument of selection. It promotes those privileged by their birth, even if they are not gifted and, automatically closes the intellectual professions to the children not privileged by birth.[3]

But all of this seems to still neglect the most important point: The failure of our economic system in the face of the rationalization of production. I do not believe that this country's production capabilities are too small in proportion to its population. But through the rationalization of production, and through the poor distribution of property and land, a large part of those able to work are excluded from the production process, by which means sales decrease again and, with that, production is again kept down.

Therefore, it seems to me that the problem you have broached cannot be separated from the great problem of economic policy. But this is probably not the place to express oneself on this [topic].[4]

Respectfully yours,

402. To Willi Richter[1]

Caputh, 28 August 1930

Dear Sir,

There can be no democratic research, because "toiling with the mind" will always be an amusement for the few. On the other hand, it could someday happen that each individual will become a self-aware creature with so much practical insight that, gradually, those politics which are directed toward parasitism and dumbing-down will become impossible. This possibility, we genuinely owe to technology. That this state has already been achieved in Russia can truly not be claimed with any justification (Church of Marx, with burnings at the stake).

Friendly greetings from yours truly,

403. A Contemporary Maxim

"Lebensweisheit aus unserer Zeit"
[*Einstein 1930hh*]

PUBLISHED 29 August 1930

IN: *Die Literarische Welt* 6, no. 35 (29 August 1930): 1.

A person's true worth is determined primarily by the degree to which, and in what sense, they have attained emancipation from the self.[1]

A. Einstein

404. From Felix Ehrenhaft[1]

Park-Hotel Vitznau, 31 August 1930

Dear Mr. Einstein,

I am making use of one of my last days on holiday to answer your letter of August 9,[2] which reached me the day before yesterday, in more detail than you probably will have expected.

Your reproaches are quite justified. I, too, have long since known that Dr. W. Mayer,[3] who has been working with you for some time, is a quite excellent man. One of Dr. Mayer's students,[4] who, just like his wife, was given the examination for teachers by me one or two years ago—I unfortunately cannot remember his name, at the moment—is either already with Mayer in Berlin or will soon join him there. Mayer will know to whom I am referring.

That we in Vienna frequently have to lose such talented people,—I have unfortunately become accustomed to that in the past years. But it is often a happy circumstance for those people to escape from the unhealthy Viennese atmosphere.

In Vienna, an intransigently nonobjective, compact majority is at work within the faculty, so that individuals can hardly accomplish anything. Wegscheider[5] is now in his last year [before retirement], and then I will be completely isolated.

When it is a direct matter concerning a physicist, I am tireless in trying to help people obtain what is due to them. In Mayer's case, I could not intervene directly, since it was a matter for the mathematicians. In another case, a young, very competent physicist was also forced, for purely anti-Semitic reasons, to initiate legal proceedings at the administrative court against the decisions of the faculty—in which I stood by the young physicist, energetically and with the mobilization of all my influence.

The court—it indeed had no alternative—ruled *expressis verbis* that the faculty was in the wrong. This case is now going into its fifth year.

That such occurrences have a highly negative influence on our young academics is only too clear.

In October, I plan to travel to Berlin; I am taking advantage of the opportunity to accompany my wife[6] there. She is making an official visit to study certain types of schools in Berlin.

I would be very grateful if you would write to me in Vienna as to whether you will most likely be in Berlin in October.

In the case that you will, I would be happy to visit you.

With hearty greetings, yours truly,

Ehrenhaft

P.S. May I show your letter of August 9 to the Viennese mathematicians Hahn and Wirtinger?[7]

405. Couplet for Hermann Anschütz-Kaempfe[1]

[Caputh, probably on 4 September 1930][2]

[Not selected for translation.]

406. To Abraham Geller[1]

[Caputh,] 4 September 1930

Dear Sir,

Unfortunately, I have not read Maimonides.[2] But I can tell you with certainty that the theory of relativity has nothing to do with the philosophical argument. Every purely scientific theory like the theory of relativity is compatible with any properly philosophical point of view. The term "relative" as applied to space and time is so ambiguous that nothing can be said about it without more detailed explanation. Answering your question would fill books, so I couldn't possibly let myself get into it.[3] I can only briefly say that I am pretty much of Spinoza's[4] opinion and, as a convinced determinist,[5] cannot befriend the monotheistic point of view.

Respectfully yours,

A. Einstein

407. To Longfellow English Debating Club

Caputh, 4 September 1930

Dear Sir,

Below, you will receive the requested statement.

Respectfully yours,

The cultivation of cosmopolitan sentiments, like those practiced by your association, belongs, alongside the struggle against armament for war and military conscription, to the most important duties of our time. I wish your events every success and wide-reaching impact.

408. To Ferdinand Müller

Caputh, 4 September 1930

[Not selected for translation.]

409. From Michele Besso

Le Sepey s. Aigle, 7 September 1930

Dear Albert,

I'm sure you've heard about the Sanatorium universitaire founded by Dr. Vauthier;[1] perhaps you've already been there, in which case the following lines are outdated; you can spare yourself reading them if you wouldn't otherwise enjoy hearing me chatter a little.

I heard from Stodola that you'd been to visit him[2]—this just crossed my mind. It was very nice that you visited him. I'm sure you also enjoyed his idealistic, youthful enthusiasm. It's just a shame that I didn't find out about your presence in Zurich in time; naturally, I would have liked to have seen you again and listen to your conversation with Stodola. Speaking of idealistic, youthful enthusiasm, Dr. Vauthier and his work must also be mentioned. As with Dunant[3] and Miss Nightingale[4] [regarding] the plight of wounded soldiers, with Zangger,[5] who brought those poisoned *in* and *by* industry to full [public] awareness and prompted action, so Vauthier, who helps ill students—and teachers. Vauthier is in a position to mobilize the means of the Swiss universities—and he also entirely put in his own—in order to bring the young people together so that they can experience the joy and stability of intellectual work. His work has even shown that intellectual work, under the guidance, stimulation, and joyfulness that he is in a position to create, is also a factor in healing, just like sunshine and mountain climates. But it also shows the circumstances under which, similar as in the English colleges, the borders of specialization like those between the nations can be negotiated. The half day that I was permitted to spend there was a strange oasis in the silence treatment that I am also undergoing. In particular, I have recently felt the need to socialize, and it has been easier than expected, given that, currently, there are loud English people here, from whose conversations among themselves I far too seldom pick up a snippet to get some motivation and put to the test their stumbling French or my rudimentary English. I thus have barely spoken one hundred words in this time: and that was a mutual letdown, as all possibilities connected to this means—that is to say, with language—of youthful exhilaration, intellectual and emotional stimulation, flared up in an otherwise serious circle. Vauthier fosters the great plan of further expanding his work on an international basis. Just the fact that he could have earned a medical and a peace prize but, as a businessman, has his limits naturally awakens the liveliest sympathy. On this note, I will end my Sunday chat. From Maja[6]—who I telephoned with last evening; she visited her little godchild in Geneva, Vero's youngest (Maja Anna Regina)[7]—I have had good news of you. And of myself, I'm experiencing the limits of written bliss, so I strongly demand to receive a card in response from you (as to) whether I may send you the brochures from the Sanatorium universitaire.

Until then, sincere greetings!

Your

Michele

410. From Jakob Grommer

Minsk, 8 September 1930

Dear Professor,

I would like to inform you of something. I have learned that you are well, thank God,[1] from Mr. Brode and Mr. Simon at the physics congress in Odessa.[2] For my part, there has been a change. I will not be giving regular lecture courses this year. I will probably be inducted into the Academy here.[3]

Our pedagogical faculty is being reduced, and there is more and more technologizing. Large technical and chemical institutes are being set up, here, where real workers are admitted. The demand for teachers is also very great, so that everything is proceeding at great speed. In contrast, I have been squeezed out into the scientific backwater.

I attended the mathematicians' congress in Kharkov.[4] In my talk there, I said:

"Causality is restricted in Newtonian mechanics by assuming a limited system. Mathematically: from $\Delta\varphi = 0$ everywhere and φ regular everywhere, $\varphi =$ const. only follows if one presupposes that φ will vanish at ∞ (in a certain manner).

In the general theory of relativity, however, it would appear that one can find unique solutions even without such conditions at ∞. Nevertheless, here, one has to presume regularity at ∞ in order to avoid the unnoticed solution for the electron

$$(f^2 \text{ is the coefficient of } dt^2,\ f^2 = \frac{a}{r} + br).$$

Such an assumption about infinity does not weaken causality; on the contrary, it strengthens it. Owing to continuity, the equations $R_{ik} - \frac{1}{2}g_{ik}R = -\kappa T_{ik}$ lead inevitably to a singularity from the variation principle. (A variation principle is also not an adequate expression of causality.) In contrast, with the equation $R_{ik} - \frac{1}{4}g_{ik}R = -\kappa T_{ik}$, one can indeed construct a steady electron. The Lorentz four-force can then be derived from a potential, which is sufficient for causality; but it then follows that, along the path of the electron, $\frac{\mathrm{d}}{\mathrm{d}t}(EH) = 0$ (E el. H magnetic field), which is, in fact, not acceptable."

Are you, Mr. Professor, in agreement?

I am now planning to write a textbook on mechanics. I am thinking of starting with continuum mechanics, not point mechanics, and not following the traditional way at all, but rather working with more modern concepts. Dispense with detailed mathematical derivations (perhaps including in particular a mathematical appendix) and treat the ideas, rather, in connection with reality.

Would you, if need arises, write a recommendation for me to the Academy (mainly concerning political matters), or perhaps to Moscow? I would ask you kindly to *reply* immediately. I will hopefully soon be in Berlin.

Yours sincerely, with hearty greetings,

Grommer

411. "Science and War"

[Caputh, 11 September 1930][1]

You ask me about the relationship between science and war.[2] Science is a powerful tool. How it is used, whether for the salvation or as a curse of human beings, depends upon human beings, not the tool. With a knife, one can kill or serve life.

We must not expect salvation from science but only from mankind. As long as it is systematically reared for crimes against people, the mentality thus produced will always lead to further catastrophes. Only the refusal of anyone to act in service of war and its perpetual preparatory actions can help.

412. Statement on Brith Shalom[1]

[Caputh,] 11 September 1930

We Jews constitute a community only through the spiritual and moral tradition that binds us.[2] If we as a community undertake the work of building up Palestine, then we may only do so in the spirit of tolerance and philanthropy if this work is to be a blessing for us. Brith-Shalom is a community that has completely embraced this understanding, and works in this sense. It is therefore my conviction that this association deserves the sympathetic recognition and support of all Jews.

413. To Robert Luther[1]

[Caputh,] 11 September 1930

[Not selected for translation.]

414. On Being an Authority

[Caputh, before 18 September 1930][1]

As punishment for my contempt for authority, fate has made me an authority.

415. To the Jewish Telegraphic Agency

[Caputh,] 18 September 1930

For now, I see in the National Socialist movement only an aftereffect of the current economic crisis and a teething problem of the Republic.[1] I always think Jewish solidarity to be necessary, but a particular reaction to the election result wholly inappropriate.

416. Appeal for the Hebrew Sheltering and Immigrant Aid Society[1]

[*Einstein 1930iia*]

DATED 18 September 1930
PUBLISHED 22 September 1930

In: *Jewish Daily Bulletin*, 22 September 1930, p. 1

Einstein Appeals to American Jewry Not to Abandon Relief Work for European Immigrants

(Jewish Telegraphic Agency)

Berlin, Sept. 20—An appeal to the Jews of America not to abandon in the future the help they have been giving to their fellow Jews of Europe to emigrate was issued today by Professor Albert Einstein. The Jewish leaders of Europe are alarmed at the report that the Hebrew Sheltering and Immigrant Aid Society of New York was about to give up its activities in Europe on behalf of Jewish emigrants because of lack of funds.

Dr. Einstein said, "The situation of East-European Jewry, especially of the immigrant, is in the highest degree critical, and the abandonment of assistance by American Jews would be catastrophic. In the name of social justice for Jews everywhere I join in the call for the continuance of the brotherly help". [2]

A call to American Jews not to let the Hebrew Sheltering and Immigrant Aid Society give up its work on behalf of the emigrants has also been issued jointly by Rabbi Leo Baeck, president of the German B'nai B'rith and Rabbis Ezra Munk and Meyer Hildesheimer,

artists and lecturers into the provincial towns for special performances on Rosh Hashanah. Groups of such performers were sent today to Zhmerinka, Nemirov, Sheptovka, Ovruth and 17 other cities in Ukrainia where the Jews are preparing to meet the High Holidays in the customary religious fashion.

Einstein Appeals to American Jewry Not to Abandon Relief Work for European Immigrants

(Continued from Page 1)

three of the leading rabbis of Germany. The appeal reads:

"We are deeply grieved over the report that the Hebrew Sheltering and Immigrant Aid Society (HIAS), the old emigrant relief organization, is contemplating the abandonment of its great work in Europe. Support for our oppressed co-religionists of Eastern Europe is vital. In the name of human justice and Jewish 'rachmones' (pity) we ask that the relief work on behalf of Jewish emigrants be not abandoned. It is one of the greatest 'mitzvoth' (good deeds) possible. Particularly before Rosh Hashanah we appeal to you American Jews to strain all your efforts to help your brethren".

417. Jews Should Not Despair Over Hitler's Victory

[*Einstein 1930ii*]

DATED 18 September 1930
PUBLISHED 19 September 1930

IN: *Jewish Daily Bulletin*, 19 September 1930, p. 1.

[See documentary edition for English text.]

418. To Armin Weiner[1]

Caputh, 18 September 1930

Dear Sir,

Thank you sincerely for sending the tremendously interesting Mach letter,[2] which I am keeping with your permission. I did not have any meaningful correspondence with Mach.[3] But through his writings, however, Mach had a considerable influence on my development.[4] Whether, or in how far, my life's work has been influenced by this is impossible for me to find out. In his final years, Mach concerned himself with the theory of relativity and, in a foreword to a later edition of one of his works, even spoke out against the theory of relativity, repudiating it quite harshly.[5] But it can hardly be subject to doubt that this was a consequence of diminished capacity due to age because the theory's whole line of thought conforms to Mach's own so that Mach is rightly considered the precursor to the general theory of relativity. I mentioned this in an essay dedicated to Ernst Mach that appeared many years ago in "The Science Of Nature."[6]

Kind regards your,

A. Einstein.

419. From Michele Besso

Bern, 19 September 1930

Dear Albert,

Since my last letter, I have seen Maja.[1] It would seem that the range of possibilities for the identification of the mathematical construct with natural phenomena has already become rather limited. All the more valuable is a result which is finally found to be the only possible one. She planted still another stinger in me: You seem to be once more considering something very fundamental about the relationship between the motions of matter and magnetic fields, which will confirm your lecture in Lucerne,[2] in spite of Piccard's observations.

I am *very* curious!

And also about whether in your conversation with Tagore—"Logic vs. Aesthetics"[3]—there is no room for the thought that: "The continuous sequence: 'Man—filterable Virus—Atom' 'explains' why the table still 'is here,' unmoving—evidently recalling 'Maja's veiled figure'—when no one is touching it."[4]

We will meet in early November in Zurich. Until then, lots of luck on your odyssey!

Nonno Besso
to Nonno Einstein![5]

420. From Arnold Berliner

Berlin, 22 September 1930

[Not selected for translation.]

421. To Annie C. Bill[1]

[Caputh,] 25 September 1930

Journalistic products are dubious documents. I never wanted women excluded from working in intellectual pursuits. But if such statements should be repelled, then it seems to me that the citation of your personality as a counterexample is not exactly effective.[2]

Respectfully yours,

422. Speech to World Congress of Palestine Workers

[*Einstein 1930jj*]

DATED 27 September 1930
PUBLISHED 28 September 1930
IN: *The New York Times,* 28 September 1930, sec. N, p. 5.

[See documentary edition for English text.]

423. From Leo Szilard

Berlin, 27 September 1930

Dear Mr. Professor,

Please excuse my laziness, but this time it was no normal laziness for a change. Since I left you in Caputh, I have been extraordinarily strongly affected by the political situation, stopped all correspondence, and tried to bring myself into equilibrium with the outside world, something, however, that I have not yet succeeded at. For one and a half years (the Schacht speech in Paris),[1] if my nose does not deceive me, new symptoms indicate from week to week that a peaceful development in Europe in the next ten years is not to be expected. But since one cannot build something sensible in a few months, I fear one will not be able to do much here at the moment. I don't even know if we will be able to finish building our refrigeration machine in Europe.

The research institute[2] had its board of trustees meeting yesterday, and, as Prof. Ramsauer[3] told me today, it was decided that 25% of the research institute is to be dismantled. Whether we fall under that or not isn't supposed be reviewed by a committee until the middle of December, which will then more closely examine the prospects of our case. The committee will consist of Prof. Ramsauer, Petersen,[4] and others, whose persons are probably not yet known. Ramsauer is certainly not unfriendly toward us; in other ways as well, it is probably an atmosphere friendly toward us, but I am convinced of the worst in my current mood, and I am worried [about] what will happen to our collaborators[5] who would be immediately terminated, in the event it were decided to dismantle. Of course, we must look to find accommodation somewhere else as rapidly as possible in case the AEG lets us down, and I think that then one should try it with General Electric in America.

In and of itself, one cannot be surprised if AEG. wants to examine the prospects of the case once because it has not already done so yet. Also, yesterday at the inspection, which we were prepared for, the matter was not seriously discussed, but rather the men only looked at the Institute in general, during which they paid us a friendly visit of five minutes. We showed them an experimental machine with potassium-sodium alloy in operation, which delivered cold below zero degrees, and the motor for potassium in broken down state and various details.[6]

Even so, with the overall situation in the background regarding December, I am feeling extremely uncomfortable.

Now for the problem with the switch: Your switch discovery No. 1 seems (I'm writing so "carefully" because I still understand absolutely nothing about switches) to not yet be the final solution. I'm assuming thereby that due to mechanical reasons one would have to use one-tenth of a second's time opening a switch. With ten switches, that would already be a time of one second, and that is in and of itself a lot of time because in the event of a short circuit, which may well be the case, the wiring must switch off immediately. Moreover, if at 100,000 volts one wants to switch off a 10,000 amperes short circuit current and one (with your method) has gotten so far, that the voltage on the resistor has risen to 50,000 volts and the current has fallen to 5,000 amps, so much heat is generated in the resistors during the one-tenth of a second as would suffice to heat roughly 100 kilograms of water to the boiling point. If one now considers that one needs one-tenth of a second for *each* switching process, and that according to your method, several switches will operate consecutively, and consider as well that during thunderstorms flashovers on isolators happen frequently (and for a short time cause a short circuit,) so that under these circumstances, the switches could operate ten times in a short period of time (i.e., before the water would be cool), so one sees that one would need a great deal more water as a thermal storage system. Also the spatial requirement of your multiple switches (so it seems to me at least) would be a multiple of the current simple switch because the individual switches that work together would have to be isolated from each other. In short, I believe that we still have to come up with something else.

Perhaps next week, if I can take time off at AEG, I will ask you again in a postcard when you would have a few hours free for me.

Warm regards from your very devoted,

Leo Szilard

P.S. I forgot to tell you the other day that Mr. von Horvath visited me in the meantime.[7] I spoke with him for a long time, but his company seems to be a typical agency and hardly suitable for us.

424. "How I Look into the World"

[before October 1930][1]

[See Doc. 425 in the documentary edition for the published English version of this text.]

425. "What I Believe"

[*Einstein 1930kk*]

DATED October 1930

IN: *The Forum* 84, no. 4 (October 1930): 103–104.

[See documentary edition for English text.]

426. To Heinrich Zangger

[Caputh, 8 October 1930][1]

Dear Zangger,

Finally, a greeting from someone long silent. Hopefully we will see each other at the Poly-celebration.[2] Life in the countryside blends splendidly with the pursuit of theoretical Don-Quixotery, which still fills up my time. It is so similar to an evening in bed on a cool night with a blanket that is too small, somewhere is always wrong. Good thing that one isn't coming into conflict with windmills in the process.

Tetel[3] has become a dear, wise companion, a real mensch; he departs again the day after tomorrow. Do you understand the present economic difficulties?[4] It seems to me that by reducing working hours in connection with an increase in the compensation for work could help. But I am not certain.

Warm regards, your

A. Einstein

427. From Michele Besso

Bern, 8 October 1930

Dear Albert,

Thank you for your kind letter; I can imagine what a tremendous strain it must have cost you. The communication of even the simplest insights is already unbelievably laborious. I am noticing that just now, as I am trying to contribute some things in preparation for an international toxicological institute with patent-law privileges; although it has been up to now an extended monologue of Zangger,[1] to present the idea to people who are sympathetic from the outset, in order to be animated by them through the most obvious objections: with a completely changed relationship, something of that which my words of twenty-five years ago may have meant to you.[2] I have a very hard time getting beyond the role of a simply reactive, well-meaning person.

We are looking forward to your visit—perhaps the side trip will offer you a few hours of escape from the crowds of people.[3] I would be happy to come to Zurich to pick you up and would also be glad to bring you back, so as to add the travel time to the time that we will have together. Please send a card to Thunstrasse 84 if you find time; otherwise by telegraph or telephone ("Bollwerk 49–61") at the office.

Heartiest greetings, also from Anna,[4] from yours truly,

Michele

428. From Käthe Kollwitz[1]

Berlin, 8 October 1930

Dear Prof. Einstein,

I would be very grateful to you if you could give me some advice.

A few days ago I was asked by Prof. Sering[2] to join a protest against the arrest of Russian scientists.[3] Naturally, you know about the matter and have yourself been asked for a signature and for this exact reason, I am turning to you and asking for your opinion.

To begin with, it goes without saying that one would like to join an act of protest against the barbaric shooting of forty-eight people and further arrests of scientists.

But only a few months ago, when I campaigned for the easing of the fate of Spiridonova,[4] I was most uncomfortably hit over the head by Russia and had intended not to interfere in Russian matters any longer.

How do you stand with respect to this case? Are you signing the protest? I would very much like to know because I know, or believe I know, that you are also not a Communist, but approach Soviet Russia with great sympathy. Please let me know what you decide.

I send warm regards,

Käthe Kollwitz

429. To Käthe Kollwitz

[Caputh?,] 10 October 1930

Dear Mrs. Kollwitz,

I have also signed the protest.[1] Read the enclosed letter from Prof. Frank, whom I know as an absolutely *honest* man, even if he cannot be praised for *objectivity* of judgment. However, he conveys *facts* correctly.[2] Now I consider it as absolutely impossible, that Russian scholars intentionally poison or spoil food items.[3] The frequent accusations of acts of sabotage by Russian intellectuals also cannot correspond to real crimes; I am convinced that this is psychologically unthinkable for various reasons. Explanation: either a desperate act of a regime driven into a corner or mass psychosis or a mixture of both states of affairs. In any case, I consider a protest necessary and unconditionally morally imperative. I am very saddened that this development, which we have watched with hopeful gaze, now leads to such dreadful things.

Warm greetings,

430. To Max Planck

[10 October 1930]

Dear Colleague,

I am just reading a letter from Prof. Frank[1] that describes the goings-on in detail.[2] Although I have previously learned that Prof. Frank is not objective (which one cannot blame him for given his fate), there is, however, no doubt about his honesty and his conscientiousness in the reporting of facts. Now it is completely impossible that Russian scholars poison food intentionally. One can hardly rule out the conviction that here exists [either] the political machinations of a regime that feels itself extremely threatened or a psychosis. The protest is also not only justified but an absolute duty. May Western Europe remain preserved from such terrible destructions. But it is in no way certain. Because people are the same everywhere and the external circumstances here are also developing slowly but steadily toward the threatening side.

Warm regards,

431. To Romain Rolland

[Caputh?,] 10 October 1930

Dear Mr. Romain Rolland,

I am very happy to sign your beautiful text.[1] Also, I will happily write a short article. The oral dialogue with Tagore has been wholly unsuccessful due to communication difficulties and naturally should not have been published.[2] In my article, I want to express the thought that according to my conviction, those personalities recognized for their intellectual achievement have the duty to morally support principled and unconditional conscientious objection. Do you not think that such an action could be launched with success?

With sincere reverence, your

432. From Heinrich Zangger

Zurich, 15 October 1930

[Not selected for translation.]

433. From Jacques Hadamard

[Paris,] 16 October 1930[1]

My dear colleague and friend,

Already some time ago I received from the Dutch mathematician Brouwer[2] an invitation to join the editorial committee of an international mathematical periodical that he wants to found (*Compositio Mathematica*), not only [with] me but some other geometricians from here. It seems to us quite regrettable, you well understand, not to participate in a scientific work of a resolutely international character; and yet the proposition is very troubling to us because of what we know of Mr. Brouwer's character, clouded for a number of our colleagues of different countries, and who, in particular at the congress in Bologna,[3] let out a stream of abuse against Painlevé[4] in a lampoon he distributed.

In another respect, having obtained from him with some difficulty something that I had made a formal condition of my acceptance, the list of people to whom he appealed, I was struck to see neither Hilbert[5] nor Landau[6] included, and I was told precisely that Brouwer, who had fallen out with Hilbert, was only launching this organization in order to thwart the *Mathematische Annalen*.[7]

If it was like that, our feelings of internationalism would no longer compel us to join. It is naturally impossible for me to make inquiries on this subject close to Hilbert or Landau; would it be possible for you to find out what they think about this and generally what the German mathematicians think about this. It would not please us in any case to take a position against Hilbert or against the *Mathematische Annalen*.

I hope that you are well, but I cannot say the same of myself. I have just been tried by an infectious flu from which I actually continue to convalesce.

All apologies for having bothered you, but on this subject that is to be considered seriously, I, we, need advice.

My thanks and best regards,

J. Hadamard

It will be necessary, alas, to speak again of the question of "non-défense" in light of the latest events that have occurred in your country.[8]

434. On the Josiah Macy Jr. Foundation

[*Einstein 1930kka*]

DATED 17 October 1930[1]
PUBLISHED 17 October 1930

IN: *New York Times*, 19 October 1930[2]

Einstein Expresses Gratitude For Aid Given by Macy Fund

Special Cable to THE NEW YORK TIMES.

BERLIN, Oct. 18.—Professor Albert Einstein, when asked how he would use the special Josiah Macy fund offered to provide assistance for him in his work, told your correspondent:

"The Josiah Macy Fund voted me a stipend for an indefinite period, which shall be used to support my esteemed colleague, Dr. Walter Mayer, mathematician of the University of Vienna. The gift has rendered my work an extremely important service. I rejoice at this opportunity to express my gratitude publicly."

435. To Wilhelm Solf[1]

Berlin, 18 October 1930

[Not selected for translation.]

436. To Stefan Zweig[1]

Berlin, 18 October 1930

Dear Mr. Stephan Zweig,

Your intention to devote one of your works to me[2] naturally makes me very happy. The impression made by your opuscule "Die Augen des toten Bruders" was one of the strongest I have ever felt from a modern literary work.[3]

With cordial thanks and regards, your

437. To Elsa Einstein

Brussels, [19 October 1930][1]

Dear Elsa,

The journey went well.[2] Micha[3] was still with me yesterday, stayed the whole time and forced a small packet of expensive fruit on me. She wanted to cry when I resisted. She's an odd duck. I still have not seen anyone. In twenty minutes Mr. Heinemann[4] is coming. His English secretary picked me up from the train. The cracklings were superb but not vegetarian. Because of Micha, I almost exploded as everything had to be dutifully devoured. There will be turbulent days! But in private I am already looking forward to Caputh, where we will hopefully stay the whole winter.

Greetings and kisses to you and Margot from your

Albert.

438. To Walther Mayer

Brussels, 19 October [1930]

Dear Mr. Mayer,

After thinking it over during the trip,[1] it occurred to me that the older interpretation of the electromagnetic field has not yet been tested seriously enough.[2] To be sure, that would mean limiting the considerations to *one* identity. It is really quite remarkable, that from

$$\Lambda^{\alpha}_{\mu\nu,\,\nu} = 0 \qquad \text{(everything in first approximation)},$$

$$\varphi_{\nu,\,\nu} = 0 \qquad \text{(through contraction)},$$

and by taking the divergence with respect to α, it follows that

$$(\varphi_{\mu,\alpha} - \varphi_{\alpha,\mu})_{\alpha} = 0;$$

i.e., in the first approximation, Maxwell.

But this cannot yet be correct, since the antisymmetric equation is

$$S^{\nu}_{\alpha\mu,\nu} - (\varphi_{\alpha,\mu} - \varphi_{\mu,\alpha}) = 0 .$$

This would mean the existence of a second vector potential for the electromagnetic field, which is not permissible. One however also obtains the Maxwell equations by setting, more generally,

$$[\Lambda^{\alpha}_{\mu\nu} + a(\varphi_{\mu}\delta_{\nu}{}^{\alpha} - \varphi_{\nu}\delta_{\mu}{}^{\alpha}) + bS^{\alpha}_{\mu\nu}]_{\nu} = 0 .$$

It must only hold (from the contraction law) that $a \neq -\frac{1}{3}$. If $a = -1$, then φ vanishes from the antisymmetric equation.

Furthermore, the b-term has no influence at all on the equation in first approximation (perhaps however on the exact equations).

It would thus be quite interesting to investigate, more strictly, the ansatz

$$\Lambda^{\alpha}_{\underline{\mu}\underline{\nu};\nu} - (\varphi_{\underline{\mu}}\delta_{\underline{\nu}}{}^{\alpha} - \varphi_{\underline{\nu}}\delta_{\underline{\mu}}{}^{\alpha})_{;\nu} + bS^{\alpha}_{\mu\nu;\nu},$$

with the requirement of only one identity (in μ) and no adjoint equation. The φ_{μ} would be el. potentials.

Probable difficulty: too many centrally symmetric solutions, as long as we do not demand freedom from singularities.

With hearty greetings, yours truly,

A. E.

439. From Emil Rupp

Berlin-Reinickendorf, 20 October 1930

Esteemed Professor,

In addition to the photos of the rotated-mirror experiment that I sent to you some time ago, I want to send you herewith some more photos.[1] Mr. Straub from Munich was here last week.[2] I showed him the interferences and took the enclosed pictures.

Fig. 1 shows interferences of the beam of canal rays (the line at 546 m[m]) at 18 kV with a path difference of $d = 20$ cm. The fluorescence decays, as one can see from the decrease in intensity. The spacing between the lens and the rotated mirror is 39 cm, while the focal length of the lens is $f = 40$ cm.

Fig. 2 shows that a Hg Geissler tube in this interferometer configuration yields no interferences.

Fig. 3 shows the interferences after rotating the mirror for stationary fluorescence.

Fig. 4 refers to the canal rays in fig. 1, only the imaging condition for parallel beams is not fulfilled. Therefore, no interferences.

Mr. Straub made no factual objections to these photos; but he refused to withdraw anything before he had spoken with Prof. Gerlach.[3]

The behavior of the gentlemen from Munich thus far has enriched my knowledge of human nature considerably, but I have gained no scientific benefits from it.

With best greetings, yours very sincerely,

E. Rupp

440. From Chaim Weizmann

Oakwood, 21 October 1930

Dear Professor,

I much regret that you cannot accept my invitation. I regret it all the more as it is impossible for me for political reasons to attend the dinner of the OZE.[1] I hope nevertheless to have an opportunity of talking to you. You are coming to London at a time which is very grave for us and—I believe—for Jewry as a whole. It is not *we* who do not want peace with the Arabs, but the Arabs (at least their leaders) and, I am afraid, the English (at least their people in Palestine), and the government's latest white paper makes this very clear.[2] I was all the more painfully struck by your interview and the statement that soon "decent" people will turn away from Zionism, etc., etc.[3] You will certainly be besieged by journalists here, and I beg you in this grave hour to spare me a few minutes so that we can talk things over. Any time on [October] 28 or 29 would suit me.

Kindest regards to you, yours,

Ch. Weizmann

441. To Elsa Einstein

Brussels, [22 October 1930][1]

Dear Else,

I was very happy to receive your dear letter. Here the congress[2] is very exhausting. One notices that one has many years under one's belt and is no longer as elastic as in youth. If the more distant people knew this, they would no longer associate with me, which would be really fine with me. I like the prospect of staying in Caputh for the time being. It may turn out that winter there isn't as bad as one imagines.

Now it's three more days to endure and then comes the abominable London[3] despite the tiredness. But I'll rest as much as possible at Mrs. Karr's[4] and look to remain completely reclusive. How I look forward to it! I like Dima a lot, too. Margot[5] has good instincts despite her apparent remoteness from life. If you spoil Rudi[6] a lot, he will certainly react in the opposite way, and redirect his unpleasant feelings at you. That's just how he is made, even if he is good-natured.

Greetings from your

Albert.

442. To Walther Mayer

Brussels, Wednesday [22 October 1930][1]

Dear Mr. Mayer,

My treatment[2] was falsified by a computational error. Precisely when the φ_μ do not enter into the antisymmetric equation $(\Lambda^{\alpha}_{\mu\nu} - (\varphi_\mu \delta_\nu{}^\alpha - \varphi_\nu \delta_\mu{}^\alpha))$, one does *not* obtain Maxwell's equations $(\varphi_{\mu,\alpha} - \varphi_{\alpha,\mu})_{,\alpha} = 0$ in first approximation from the field equations by differentiation with respect to α. Therefore, the φ_μ cannot be interpreted as electric potentials. Instead, one is forced to interpret the electrical quantities differently (e.g., through the antisymmetric $h_{s\nu} - h_{\nu s}$). Then, however, one must definitely prefer the case with two divergences.

What remains is thus only the case of $R_{ik} = 0$ and the Parisian case. Thus, from the formal point of view, the Parisian case is without qualification preferable for the often-quoted reasons. The objections to the Parisian case taken from the static special case are only then pertinent if one assumes that the macroscopic static case has no dynamic structure. One is thus again confronted with the question of the existence of nonstatic singularity-free solutions, which leave us for the moment helpless.

The congress[3] is very tiring, but it is becoming apparent that the quantum-mechanical theory reproduces such fine features of physical reality that it must contain a considerable quantity of truth. It is therefore quite justified for us to study Dirac.[4] Perhaps that study will lead us to a workable ansatz.

Hearty greetings from yours truly,

A. E.

443. From Thomas Mann[1]

Munich, 22 October 1930

[Not selected for translation.]

444. To Walther Mayer

[Brussels, 24 October 1930][1]

Dear Mr. Mayer,

Because the magnetic force on a moving charge is expressed in three dimensions as a vector product, i.e., as a tensor, the objection which I had expressed against the new interpretation of the field

$$\underbrace{\begin{matrix}\varphi_{23} & \varphi_{31} & \varphi_{12}\\ \mathfrak{e}_x & \mathfrak{e}_y & \mathfrak{e}_z\end{matrix}}_{\text{electric}} \qquad \underbrace{\begin{matrix}\varphi_{14} & \varphi_{24} & \varphi_{34}\\ j\mathfrak{h}_x & j\mathfrak{h}_y & j\mathfrak{h}_z\end{matrix}}_{\text{magnetic}} \quad j = \sqrt{-1},$$

is not valid.

One can namely interpret the electrical density not as a scalar, but just as well as an antisymmetric tensor of fourth rank, ρ_{iklm}; and the current density as an antisymmetric tensor of third rank:

$$\rho_{ikl} = \rho_{iklm}\frac{dx_m}{ds}.$$

The Maxwell equations with current density are then given by

$$\varphi_{\mu\nu,\,\nu} = 0 \qquad \text{...(I)}$$

$$\varphi_{\mu\nu,\,\sigma} + \varphi_{\nu\sigma,\,\mu} + \varphi_{\sigma\mu,\,\nu} = \rho_{\mu\nu\sigma} \qquad \text{...(II)}$$

The derivation of the ponderomotive forces k_μ is just as simple, and that of the Maxwellian voltages even simpler than in the earlier version.

One needs to set

$$k_\mu = \rho_{\mu\nu\sigma}\varphi_{\nu\sigma}.$$

From II, taking account of (I), we then obtain

$$k_\mu = \varphi_{\mu\nu\sigma}\varphi_{\nu\sigma} = \left(\frac{1}{2}\varphi_{\sigma\tau}^2\delta_{\mu\nu} - 2\varphi_{\mu\sigma}\varphi_{\nu\sigma}\right)_\nu$$

Maxwellian voltages.

I am now firmly convinced that this interpretation is the most natural, already in the old theory. For the consideration regarding time reversal is equally well justified there as in the new theory. But we of course still have to take into account the nonequivalence of the electron and the proton.

With hearty greetings, yours truly,

A. E.

445. To Elsa Einstein

Brussels, Sunday [25 October 1930][1]

Dear Else,

Thanks for all the news. We are finished here, thank God. Today there is still the reception at the Königs'[2] and tomorrow it continues. I really don't like meetings of this kind, especially since in company one can't think. Also, I don't feel at ease in large gatherings. It wasn't right for the OSE[3] to announce that I would give a speech, and in that way place me at the center of things, as you can read in the JTA.[4] The next time I won't allow myself to be persuaded.

I find the British government's actions comprehensible, and I even find the grounds for them to be comprehensible. That is why I do not endorse the steps taken by Weizmann and Mond.[5]

Nothing is achieved in that way. But N.[6] does not want to abandon this matter! He might do that only if he no longer had the Zionists' trust.

London and Zurich still sit heavily on my stomach.[7] But everything eventually passes, and I am after all only minimally concerned with my innards.

I hope all is well in your household. Best regards to you and the children,[8] including those offstage.

Albert

446. Speech for ORT-OZE Dinner[1]

[London, 28 October 1930][2]

[Not selected for translation.]

447. Speech for ORT-OZE Dinner[1]

[*Einstein 1930ll*]

DATED 28 October 1930[2]
PUBLISHED 29 October 1930

IN: *The New York Times*, Wednesday, 29 October 1930, sec. 12, p. 2.

[See documentary edition for English text.]

448. From Paul Ehrenfest

Pasadena, Caltech. Norman Bridge Laboratory, 28 October 1930

Dear Einstein,

This afternoon at two P.M. we were able to listen to your London talk[1] here (and all over America). Whereas only a few fragments of Shaw's speech were comprehensible,[2] I was able to understand practically every word of yours, and now feel a need to tell you how extraordinarily valuable I found each of your remarks. Just in these last weeks I have repeatedly thought about what would have to happen in order to shape us Jews *correctly*. This question forces itself on me so vigorously because in the course of this year I have met again and again closed groups of Jews (Holland, Copenhagen, Berlin, Russia, Canada, and now at last the whole trip across the continent and, finally, the whole Pacific coast).[3] Again and again I met "us Jews," and in the most various social positions. And over and over I felt that "the wood we are made of" is of the best kind (the closed *purely* Scandinavian groups in the agricultural areas of the northern "Midwest" also made an excellent impression on me, and so did a large part of the NOT *FATTEND* half of the Chinese population of San Francisco.[4]

At the same time, however, I felt that this splendid Jewish human material (1) is abominably badly shaped, (2) that I absolutely do not see how it really *ought* to be shaped (I mean: on the one hand, corresponding to our striking idiosyncrasies, but on the other hand harmoniously adapted to the modern tasks of humanity as a whole. HOW TERRIBLY HAPPY I WOULD BE TO HEAR WHAT YOU THINK AT THIS POINT! And particularly which concrete Jewish figures you would identify as more or less exemplary in this respect—today you mentioned only two names: Spinoza and Marx[5]—I would be so happy to hear how you would assess in greater detail whole ranges of splendid Jewish men. I *don't* mean *exclusively* in the traits that you would regard positively.

(3) That closed Jewish population groups can so terribly quickly and greatly degenerate as soon as they (to cite the expression you used today) "are lying on a bed of roses."[6]

(4) That I can't see how a recently emerged Jewish "prophet" could rally behind him a sufficiently great part of the Jewish youth, UNLESS A SUBLIME CATASTROPHE CAME TO HIS AID! (You've seen enough of the youth in Palestine,[7] and you may have seen things there that I don't know about?)

Perhaps it can indeed be said that in all these respects Jews are no worse than other peoples. But even if I could learn to believe that—one *should* not seek to calm oneself in this way.

I believe that you could do Jewish youth of the coming generations a [very] great service if you outlined YOUR ideal of the Jew (e.g., by citing examples of outstanding Jewish men and women whom you have met). Of course, I understand very well that in the end only a little that is SPECIFICALLY Jewish can remain, but the way to the goal will still be strongly determined by the uniqueness of the Jew.

It is such a pity that modern Jews (in contrast to the old prophets) find it so difficult to castigate us Jews LOVINGLY and SEVERELY, worrying that they MIGHT thereby be putting into their persecutors' hands materials with which to attack them. But why, among all the other communities, do such "self-accusers" stand up, even though they have similar concerns and difficulties? (It is interesting that you have never hesitated, under the most adverse circumstances, to hold up a mirror in front of EVERY community to which you have belonged, as soon as it became necessary and without worrying about your "popularity" or the danger of giving "external enemies" new ways of attacking that community. With regard to only one community that you belong to have you never—SO FAR AS I KNOW—acted in a similarly PUBLIC way with regard to the Jews—if I am not mistaken, why? Have you here finally allowed yourself to be restrained by concern about helping external enemies (especially of the Jews of Eastern Europe)? Dear, dear Einstein! Be healthy and happy! Hearty greetings to you and your family! I hear that Margotl was in Russia with Tagore.[8] Very curious to learn what impressions she came back with.

Your comments on Shaw were really incredibly subtle and affectionate. "One almost forgets that these persons were created not by Nature, but by B. Shaw"[9] Was that JUST praise?

Warm regards from your

449. To Elsa Einstein

Brussels, [30 October 1930]

Dear Else,

The thing in London was worth the trouble.[1] It was not just exhausting, but also amusing. The voyage by way of Ostende was splendid. Today I'm going to see the king and queen.[2] I will arrive in Zurich tomorrow at 8:30 P.M.[3] There's a light rain, but it's constant, unfortunately. Thank Herbert Samuel,[4] on my behalf as well, for all the kindness he has shown me. I also liked his son, Hadasserich, very much.[5] Here I am staying in a funny little room that the porter arranged for me. You would all turn up your uncomprehending noses at it.[6] Albert sent me a telegram to tell me that I should visit him. But I can't tootle around so much.

I thank you for your lovely report, and also Margot,[7] who floats on the wind.

I am happy to be a relatively free man again. The OSE people were very considerate and cordial, especially Brutzkus,[8] of whom I became quite fond.

Regards,

Albert

450. To Hans Albert Einstein

Brussels, [30 October 1930]

Dear Albert,

Your telegram gave me much pleasure.[1] However, after days in Brussels and London, I'm so dead-tired that I wouldn't be good company anyway.[2] I will come the next time my travels bring me near you. It's too bad that I have to do so many things that may be, in and of themselves, very useful, but are empty for me, and yet cost me so much effort.

With best regards to you three,[3] your

451. From Eduard Rüchardt[1]

Munich, 31 October 1930

Dear Professor,

Mr. Gerlach told me that you have doubts as to whether in the "rotated-mirror" experiment with canal rays, the condition that you mentioned in the Berlin Academy: *focal length of the lens = distance to the rotated-mirror lens* must necessarily be fulfilled.[2] Mr. Straub made a point of adhering most strictly to that condition in his experiments.[3] For our discussions with Mr. Rupp,[4] it is important to understand that question as precisely as possible. For Rupp, on the one hand, never fulfills the condition strictly (only to within a few cm); but on the other hand, he now maintains that when he does not meet that condition, his interferences vanish. In my opinion, the condition must either hold *strictly*, or not at all. Following the arguments that you gave in the reports of the Berlin Academy, we believed that this condition must indeed necessarily be fulfilled, since otherwise the image of the "infinitely distant canal rays," as a result of the rotation of the mirror, in addition to the rotation, it also pivots, and the equality of the optical paths cannot be established for the coherent image points of different colors. Mr. Gerlach however said that you hold this sort of thinking to be overly specialized, since one need only consider the general case of a small rotation of *both* mirrors around a very distant rotational axis.

Mr. Straub has, by the way, tried to carry out your first experiment: canal-ray interferometer without lenses. In that case, according to your theory, the interference pattern from moving fluorescence should be shifted by an angle v/c relative to the interference pattern from fluorescence at rest.[5] This experiment is particularly simple, because the result does not depend upon the path difference. But Mr. Straub could observe interferences here also with a decaying beam only up to a path difference of 0.7 mm. The interference rings are then rather broad and an angular shift is hard to detect; but Mr. Straub wants to repeat the experiment more precisely. That the coherence lengths are so short with moving fluorescence is certainly due to the inhomogeneity of the velocities of the canal-ray particles. Homogeneous beams can in fact be produced, but unfortunately up to now not with the necessary intensity for these experiments.

If you are still interested in the experiments, and want to express any wishes with respect to them, Mr. Straub would certainly be willing to try out some other variations. He is thoroughly experienced in both the rather difficult canal-ray techniques and in the observation of interference, both necessary for these experiments.

With most respectful greetings, yours truly,

E. Rüchardt

452. To Eduard Rüchardt

[after 31 October 1930]

[Not selected for translation.]

453. To Elsa Einstein

Zurich, Saturday [1 November 1930][1]

Dear Else,

Now the hardest part is behind me. I'm talking about my experience in England. The trip with Brutzkus[2] was very interesting. He is a man who has experienced a lot and has arrived at a great inner peace. The OSE people and the locals picked me up in Dover.[3] In London, at the train, a half-circle of photographers and a written interview.[4] Herbert Samuel was wonderfully kind to me; and so was his whole family.[5] In the evening I sat up past one A.M. writing my speech.[6] I finished it the following morning. Then Bruzkus came for the translation, then Weizmann[7] and took me off to the Zionist office, where I had a reception with the staff. At noon I went with H. Samuel and his son to see Bernard Shaw, who despite his age,[8] is uncommonly limber and witty. He was fascinated by my article in Forum.[9] In the afternoon I attended the opening session[10] in the chamber. A small hall with wood decorations, strict ceremonial, an enormous impression. Then I went home and lay down for an hour. Then we drove to the banquet.[11] First I had to stand with the fat Rothschild[12] and shake hands with about 400 little men and women. This was psychologically interesting (faces, bearing, who first?) I was much amused by this, although most of them were commonplace money-people, good only to be milked. After dinner, H. Samuel spoke very vigorously against the government, on account of Palestine.[13] Shaw delivered, without notes, a masterly speech whose manuscript he had given me beforehand;[14] he is admirable, quite apart from his great age (over seventy). On that evening more than 100,000 M was pledged, and I am glad to have done it. Wednesday morning I drove around London again with Cohn and a very interesting Jewish-English military man.[15] I visited Weizmann and Sokolow[16] in their apartment. Then we drove to Regent Street, walked down it, and then through the jurists' quarter (all Gothic).[17] We also visited a law court, where a divorce case was going on. Then H. Samuel took me the train station. L. Cohn also came with his young wife[18] and the officials who organized the meeting. Brutzkus and the chairman of the OSE[19] accompanied us as far as the (Belgian) ship, which took me to Ostende. I found the voyage so refreshing, finally being alone after so many people.

At 10:30 in the evening I arrived in Brussels in good spirits, but didn't know where to go. I found a porter and told him that he should take my luggage to a decent hotel. He did so, and in that way I came to know the life of ordinary people: the place was called the Siegers Hotel, and I shall certainly go there again when I come to Brussels and want some peace and quiet. I went to bed early and did not get up until around nine A.M. Then I went to the train station across the street to take care of business (sending tickets for Thursday to Lux, calling the king and queen).[20] The latter took a long time because the line was constantly busy. At noon I ate a vegetarian meal in my little hotel, and around three P.M. went to the king and queen, where I was welcomed with moving cordiality. Both these people are of a purity and goodness that is seldom found. First, we conversed for about an hour. Then a British musician came, and than the four of us or the three of us (a musical gentlewoman[21] was also there) played music together for a few hours and enjoyed ourselves greatly. Then they all left and I remained alone at the king and queen for dinner: without servants, vegetarian, spinach with fried eggs and potatoes, period. (It was not arranged in advance that I should stay there.) I liked being there enormously, and I'm sure the feeling is mutual. My journey here yesterday was pleasant. Albert[22] had sent me a telegram in London to tell me that I should come to Dortmund. But I couldn't make up my mind to do that, because I wanted to finally have a little time to myself, and I was afraid of the daughter-in-law[23] and her effect on me. Lux was at the station and is very cordial. I am staying until the 4 [November] in the evening and will then go to see Mileva,[24] so that she would not be short-changed. Unfortunately, the ceremony[25] is only on 7 [November]. It's too bad that I have to sit around so long without little Meyer.[26] For that reason I am including here a letter for him,[27] since unfortunately I don't have his address. Please send it to me immediately.

Warm regards to all from your

Albert

454. To Walther Mayer

[Zurich,] 1 November [1930]

Dear Mr. Mayer,

I really long to see you. But I will have to be here on the 7th.[1] I will at least be happy if I can depart on the 8th.

I can tell you something interesting. I still believe in the tele||, but I have found a reason to retreat from the requirement of *two* identities.[2]

The essence of quantum mechanics lies in the existence of the De Broglie waves, if a field-like interpretation is possible at all. To a point at rest corresponds a temporally oscillating function, not dependent on the spatial coordinates:

$$\frac{\partial\psi}{\partial x_4} \neq 0 \qquad \frac{\partial\psi}{\partial x_1} = \frac{\partial\psi}{\partial x_2} = \frac{\partial\psi}{\partial x_3} = 0.$$

For the electron, it must be an object with several components.

I now require that the field equations be such that the equations in first approximation have solutions of this type. This is however not the case either for the Parisian solution or for the Riemannian solution $\left(R_{ik} = 0 \quad \frac{\partial S^{\alpha}_{\underline{\mu\nu}} h\psi}{\partial x_\nu} = 0 \quad F^{\mu\nu} = 0\right)$. It indeed does exist, when only *one* identity is required. Then, the system

$$L^{\alpha}_{\mu\nu/\nu} - \frac{1}{2} L^{\alpha}_{\underline{\sigma\tau}} \Lambda^{\mu}_{\sigma\tau} = 0 \qquad \ldots(a)$$

is the most natural.

The existence of such wave fields requires that

$$L^{\alpha}_{\mu 4} = 0 \qquad \ldots(1)$$

The coordinate system can in this case be chosen so that

$$0 = h_{14} = h_{24} = h_{34} \qquad h_{44} = 1\,.$$

If one takes $L^{\alpha}_{\mu\nu} = \Lambda^{\alpha}_{\mu\nu} + A(\varphi_\mu \delta_{\nu\alpha} + \varphi_\nu \delta_{\mu\alpha}) + BS^{\alpha}_{\mu\nu}$, then the calculation shows that (1) demands, when $A = -\frac{1}{3}$ is excluded, that $\boxed{A = -1}$, which in first approximation is equivalent to $R_{ik} = 0$. If this is the case, then such waves will be possible for which h_{41}, h_{42}, h_{43} are non-vanishing. B can initially be chosen arbitrarily.

A further special case occurs when $B = -\frac{1}{2}$ is chosen. Then, namely, also *spatially* wavelike components of the type $h_{sm} - h_{ms}$ are possible. This case, $B = -\frac{1}{2}$, however is eliminated for physical reasons. For the antisymmetric equation would become $F^{\mu\nu} = 0$ in first approximation, which, due to $F_{\mu\nu,\sigma} + F_{\nu\sigma,\mu} + F_{\sigma\mu\nu} \equiv 0$,[3] would imply the existence of a quadratic equation for the first approximation. We now have the possibility of studying these kinds of waves when they are superposed onto a static field. One should then find something like the Schrödinger equation.

The principal weakness of the method lies in the large number of field equations with only one identity. I however believe that those which are based on the commutation rule are the most plausible. The constant B could be related to the ratio of the masses of the electron and the proton. h_{41} h_{42} h_{43} would indeed be associated with the magnetic field, and $h_{sa} - h_{as}$ (spatially) with the electric field.

It is certainly very tempting to try to attribute some physical reality to the de Broglie waves, in particular since those waves make themselves so immediately apparent in the observations of crystal interference.

Think about this matter likewise. I believe that the reality of the de Broglie waves in fact leads to the conclusion that I must give up my hobbyhorse of the overdetermination.

Hearty greetings from yours truly,

A. Einstein

455. To Hans Albert Einstein and Frieda Einstein-Knecht

[Zurich, 5 November 1930]

My beloved children,

I've just pitched my nomad's tent in the old homeland. The best part was a musical afternoon with Tetel and the temperamental baroness.[1] Stodola is trying very hard to understand relativity[2] and gives me the nuts to crack that he can't open. Friday is the Zurich Polytechnic's anniversary,[3] in which I am involved. Then I can go back to my work at home.

Warm greetings from your

Papa

456. To Walther Mayer

[Zurich, 6 November 1930][1]

My dear Mr. M.,

I already miss you and our function. You too are now being persecuted by the slaves of hack journalism;[2] soon you will have enough of that. What I recently wrote[3] was madness. The special cases I mentioned are precisely the ones in which the first approximation becomes indeterminate. I hope I can leave on Saturday, so that we can see one another, hopefully already on Sunday afternoon or Monday morning. Too bad that it is no longer in Caputh.

Warmest greetings also to your wife,[4] from your

A. E.

457. To Eugen Meyer-Peter[1]

[Zurich,] 6 November 1930

Dear Colleague,

I heard by chance that a civil engineer is to be hired by the Federal Hydrological Institute in the foreseeable future. That gave me the idea that this could possibly be a field of employment for my son Albert, who was your student at the Poly.[2]

The young fellow has been working for Klönne in Dortmund for about three years as a civil engineer (high-rise construction, bridge construction, hydraulic engineering), and his contract there will continue until next summer. His career there so far shows that he has proven himself very well in his work. I also know that he has had ideas there which have led to patents for the firm.[3]

I would like to put in a good word for him with this letter, for the case that he might be considered for your institute. He is a competent lad (twenty-six years old) and a Swiss citizen.

With collegial greetings, yours truly,

A. Einstein

458. Poem for Frieda Huber[1]

[Zurich, 7 November 1930]

[Not selected for translation.]

459. From Myron Mathisson

Warsaw, 7 November 1930

[Not selected for translation.]

460. Poem for Luise Karr-Krüsi

[Zurich, 8 November 1930] [1]

[Not selected for translation.]

461. To Carl Friedrich Geiser[1]

Zurich, 8 October [8 November] 1930[2]

Dear Prof. Geiser,

I must especially thank you once again for giving me the opportunity yesterday to see you and shake your hand.[3] Your lectures (and even your exams)[4] compose the most beautiful memories from my student days. It was only later that what you had planted properly came alive while at the time, the productive power of mathematical methods had not yet dawned on me. I believed that the understanding of natural processes grew automatically out of observation, so to speak.

Warm regards and thanks, your

A. Einstein.

462. "Religion and Science"

"Religion and Science"
[*Einstein 1930nn*]

DATED on or before 9 November 1930[1]
PUBLISHED 9 November 1930

IN: *New York Times Magazine*, 9 November 1930, p. 1.

EVERYTHING that men do or think concerns the satisfaction of the needs they feel or the escape from pain. This must be kept in mind when we seek to understand spiritual or intellectual movements and the way in which they develop. For feeling and longing are the motive forces of all human striving and productivity—however nobly these latter may display themselves to us.

What, then, are the feelings and the needs which have brought mankind to religious thought and to faith in the widest [2] sense? A moment's consideration shows that the most varied emotions stand at the cradle of religious thought and experience.

In primitive peoples it is, first of all, fear that awakens religious ideas—fear of hunger, of wild animals, of illness and [3] of death. Since the understanding of causal connections is usually limited on this level of existence, the human soul forges a being, more or less like itself, on whose will and activities depend the experiences which it fears. One hopes to win the favor of this being by deeds and sacrifices, which, according to the tradition of the race, are supposed to appease the being or to make him well disposed to [4] man. I call this the religion of fear.

This religion is considerably stabilized—though not caused—by the formation of a priestly caste which claims to mediate between the people and the being they fear and so attains a position of power. Often a leader or despot, or a privileged class whose power is maintained in other ways, will combine the function of the priesthood with its own temporal rule for the sake of greater security; or an alliance may exist between the interests of the political power and the priestly caste.

• • •

A SECOND source of religious development is found in the social feelings. Fathers and mothers, as well as leaders of great human communities, are fallible and mortal. The longing for guidance, for love and succor, provides the stimulus for the growth of a social or moral conception of God. This is the God of Providence, who protects, decides, rewards and punishes. This is the God who, according to man's widening horizon, loves and provides for the life of the race, or of mankind, or who even loves life itself. He is the comforter in unhappiness and in unsatisfied longing, the protector of the souls of the dead. This is the social or moral idea of God.

It is easy to follow in the sacred writings of the Jewish people the development of the religion of fear into the moral religion, which is carried further in the New Testament. The religions of all the civilized peoples, especially those of the Orient, are principally moral religions. An important advance in the life of a people is the transformation of the religion of fear into the moral religion. But one must avoid the prejudice that regards the religions of primitive peoples as pure fear religions and those of the civilized races as pure moral religions. All are mixed forms, though the moral element predominates in the higher levels of social life. Common to all these types is the anthropomorphic character of the idea of God. [5]

Only exceptionally gifted individuals or especially noble communities rise *essentially* above this level; in these there is [6] found a third level of religious experience, even if it is seldom found in a pure form. I will call it the cosmic religious sense. This is hard to make clear to those who do not experience it, since it does not involve an anthropomorphic idea of God; the individual feels the vanity of human desires and aims, and the nobility and marvelous order which are revealed in nature and in the world of thought. He feels the individual destiny as an imprisonment and seeks to experience the totality of existence as a unity full of significance. Indications of this cosmic religious sense can be found even on earlier levels of development—for example, in the Psalms of David and in the Prophets. The cosmic element is much stronger in Buddhism, as, in particular, Schopenhauer's magnificent essays have [7] shown us.

The religious geniuses of all times have been distinguished by this cosmic religious sense, which recognizes neither dogmas nor God made in man's image. Consequently there cannot be a church whose chief doctrines are based on the cosmic religious experience. It comes about, therefore, that we find precisely among the heretics of all ages men who were inspired by this highest religious experience; often they appeared to their contemporaries as atheists, but sometimes also as saints. Viewed from this angle, [8] men like Democritus, Francis of Assisi and Spinoza are near to one another.

How can this cosmic religious experience be communicated from man to man, if it cannot lead to a definite conception of God or to a theology? It seems to me that the most important function of art and of science is to arouse and keep alive [9] this feeling in those who are receptive.

Thus we reach an interpretation of the relation of science to religion which is very different from the customary view. From the study of history, one is inclined to regard religion and science as irreconcilable antagonists, and this for a reason that is very easily seen. For any one who is pervaded with the sense of causal law in all that happens, who accepts in real earnest the assumption of causality, [10] the idea of a Being who interferes with the sequence of events in the world is absolutely impossible. Neither the religion of fear nor the social-moral religion can have any hold on him. A God who rewards and punishes is for him unthinkable, because man acts in accordance with an inner and outer necessity, and would, in the eyes of God, be as little responsible as an inanimate object is for the movements which it makes.

* * *

[11] SCIENCE, in consequence, has been accused of undermining morals—but wrongly. The ethical behavior of man is better based on sympathy, education and social relationships, and requires no support from religion. Man's plight would, indeed, be sad if he had to be kept in order through fear of punishment and hope of rewards after death.

It is, therefore, quite natural that the churches have always fought against science and have persecuted its supporters. But, on the other hand, I assert that the cosmic religious experience is the strongest and the noblest driving force behind scientific research. No one who does not appreciate the terrific exertions, and, above all, the devotion without which pioneer creations in scientific thought cannot come into being, can judge the strength of the feeling out of which alone such work, turned away as it is from immediate practical life, can grow. What a deep faith in the rationality of the structure of the world and what a longing to understand even a small glimpse of the reason revealed in the world there must have been in Kepler and Newton to [12] enable them to unravel the mechanism of the heavens in long years of lonely work!

Any one who only knows scientific research in its practical applications may easily come to a wrong interpretation of the state of mind of the men who, surrounded by skeptical contemporaries, have shown the way to kindred spirits scattered over all countries in all centuries. Only those who have dedicated their lives to similar ends can have a living conception of the inspiration which gave these men the power to remain loyal to their purpose in spite of countless failures. It is the cosmic religious sense which grants this power.

[13] A contemporary has rightly said that the only deeply religious people of our largely materialistic age are the earnest men of research.

463. On Kepler[1]

[Berlin?, before 9 November 1930]

In anxious and uncertain times like ours,[2] when it is difficult to find pleasure in humanity and the course of human affairs, it is particularly consoling to think of the serene greatness of Kepler. Kepler lived in an age in which the reign of law in nature was by no means an accepted certainty. How great must his faith in a uniform law have been, to have given him the strength to devote ten years of hard and patient work[3] to the empirical investigation of the movement of the planets and the mathematical laws of that movement, entirely on his own, supported by no one and understood by very few![4] If we would honor his memory worthily, we must get as clear a picture as we can of his problem and the stages of its solution.[5]

Copernicus had opened the eyes of the most intelligent to the fact[6] that the best way to get a clear grasp of the apparent movements of the planets in the heavens was by regarding them as movements round the Sun conceived as stationary. If the planets moved uniformly in a circle round the Sun, it would have been comparatively easy to discover how these movements must look from the Earth. Since, however, the phenomena to be dealt with were much more complicated than that, the task was a far harder one. The first thing to be done was to determine these movements empirically from the observations of Tycho Brahe.[7] Only then did it become possible to think about discovering the general laws which these movements satisfy.

To grasp how difficult a business it was even to find out about the actual rotating movements, one has to realize the following. One can never see where a planet really is at any given moment, but only in what direction it can be seen just then from the Earth, which is itself moving in an unknown manner round the Sun. The difficulties thus seemed practically unsurmountable.

Kepler had to discover a way of bringing order into this chaos. To start with, he saw that it was necessary first to try and find out about the motion of the Earth itself.[8] This would simply have been impossible if there had existed only the Sun, the Earth and the fixed stars, but no other planets. For in that case one could ascertain nothing empirically except how the direction of the straight Sun–Earth line changes in the course of the year (apparent movement of the Sun with reference to the fixed stars). In this way it was possible to discover that these Sun–Earth directions all lay in a plane[9] stationary with reference to the fixed stars, at least according to the accuracy of observation achieved in those days, when there were no telescopes. By this means it could also be ascertained in what manner the line Sun–Earth revolves round the Sun. It turned out that the angular velocity of this motion went through a regular change in the course of the year.[10] But this was not of much use, as it was still not known how the distance from the Earth to the Sun alters in the course of the year. It was only when they found out about these changes that the real shape of the Earth's orbit and the manner in which it is described were discovered.

Kepler found a marvelous way out of this dilemma. To begin with it was apparent from observations of the Sun that the apparent path of the Sun against the background of the fixed stars differed in speed at different times of the year, but that the angular velocity of this movement was always the same at the same point in the astronomical year, and therefore that the speed of rotation of the straight line Earth–Sun was always the same when it pointed to the same region of the fix stars. It was thus legitimate to suppose that the Earth's orbit was *a self-enclosed one,* described by the Earth in the same way every year—which was by no means obvious *a priori*. For the adherent of the Copernican system it was thus as good as certain that this must also apply to the orbits of the rest of the planets.[11]

This certainty made things easier. But how to ascertain the real shape of the Earth's orbit? Imagine a brightly shining lantern *M* somewhere in the plane of the orbit. We know that this lantern remains permanently in its place and thus forms a kind of fixed triangulation point for determining the Earth's orbit, a point which the inhabitants of the Earth can take a sight on at any time of year. Let this lantern *M* be further away from the Sun than the Earth. With the help of such a lantern it was possible to determine the Earth's orbit, in the following way:

First of all, in every year there comes a moment when the Earth *E* lies exactly on the line joining the Sun *S* and the lantern *M*. If at this moment we look from the Earth *E* at the lantern *M*, our line of sight will coincide with the line *SM* (Sun–lantern). Suppose the latter to be marked in the heavens. Now imagine the Earth in a different position and at a different time. Since the Sun *S* and the lantern *M* can both be seen from the Earth, the angle at *E* in the triangle *SEM* is known. But we also know the direction of *SE* in relation to the fixed stars through direct solar observations, while the direction of the line *SM* in relation to the fixed stars was finally ascertained previously. But in the triangle *SEM* we also know the angle at *S*. Therefore, with the base *SM* arbitrarily laid down on a sheet of paper, we can, in virtue of our knowledge of the angles at *E* and *S*, construct the triangle *SEM*. We might do this at frequent intervals during the year; each time we should get on our piece of paper a position of the Earth *E* with a date attached to it and a certain position in relation to the permanently fixed base *SM*. The Earth's orbit would thereby be empirically determined, apart from its absolute size, of course.

But, you will say, where did Kepler get his lantern *M*? His genius and Nature, benevolent in this case, gave it to him.[12] There was, for example, the planet Mars;

and the length of the Martian year—i.e., one rotation of Mars round the Sun—was known. It might happen one fine day that the Sun, the Earth and Mars lie absolutely in the same straight line. This position of Mars regularly recurs after one, two, etc., Martian years, as Mars has a self-enclosed orbit. At these known moments, therefore, *SM* always presents the same base, while the Earth is always at a different point in its orbit. The observations of the Sun and Mars at these moments thus constitute a means of determining the true orbit of the Earth, as Mars then plays the part of our imaginary lantern. Thus it was that Kepler discovered the true shape of the Earth's orbit and the way in which the Earth describes it, and we who come after—Europeans, Germans, or even Swabians,[13] may well admire and honor him for it.

Now that the Earth's orbit had been empirically determined, the true position and length of the line *SE* at any moment was known, and it was not so terribly difficult for Kepler to calculate the orbits and motions of the rest of the planets too from observations—at least in principle. It was nevertheless an immense work, especially considering the state of mathematics at the time.

Now came the second and no less arduous part of Kepler's life work.[14] The orbits were empirically known, but their laws had to be deduced from the empirical data. First he had to make a guess at the mathematical nature of the curve described by the orbit, and then try it out on a vast assemblage of figures. If it did not fit, another hypothesis had to be devised and again tested. After tremendous search, the conjecture that the orbit was an ellipse with the Sun at one of its foci was found to fit the facts. Kepler also discovered the law governing the variation in speed during rotation, which is that the line Sun–planet sweeps out equal areas in equal periods of time.[15] Finally he also discovered that the square of the period of circulation round the Sun varies as the cube of the major axes of the ellipse.[16]

Our admiration for this splendid man is accompanied by another feeling of admiration and reverence, the object of which is no man but the mysterious harmony of nature into which we are born. As far back as ancient times people devised the lines exhibiting the simplest conceivable form of regularity. Among these, next to the straight line and the circle, the most important were the ellipse and the hyperbola.[17] We see the last two embodied—at least very nearly so—in the orbits of heavenly bodies.

It seems that the human mind has first to construct forms independently before we can find them in things.[18] Kepler's marvelous achievement is a particularly fine example of the truth that knowledge cannot spring from experience alone but only from the comparison of the inventions of the intellect with observed fact.

(As published in Einstein, *The World As I See It*, pp. 40–46)

464. To Heinrich Dehmel[1]

[Berlin,] 10 November 1930

Dear Dr. Dehmel,

I am in complete agreement with the drift of your pamphlets,[2] both from the human-moral standpoint and from the practical standpoint. Over-population is a heavy danger for any country.

It has become a disaster for China and India,[3] and threatens to become a disaster for Europe as well.

Your plan for self-denunciation seems to me promising if you are able to realize it as a mass movement.

Wishing for your sake and for us all that your meritorious campaign[4] succeeds, I am your

465. To Abraham Adolf Fraenkel[1]

[Berlin,] 10 November 1930

Dear Colleague,

I hasten to answer your questions, especially since I may hope thereby to contribute directly to the success of the project that is dear to both of us.

(1) It seems to me absolutely more correct at this point to appoint a physicist who is ⟨trained⟩ oriented primarily *theoretically*.[2] My reasons: for now, an experimental physicist will hardly be given adequate means to do valuable research, whereas the standard working conditions for successful theoretical work are fulfilled completely. Also, it is easy to find and hire competent theoretical physicists who are Jewish, whereas in the experimental area things are not favorable in that respect.

(2) Mr. Epstein at Pasadena is a theoretical physicist of international renown who has already done capable work in theory.[3] He is also sufficiently familiar with experimental physics to guide the gradual incorporation of that branch of science into the university in appropriate ways, especially with regard to teaching. Personally, he is very likable and irreproachable; in this respect as well, he would certainly be a valuable acquisition for the university.

Cordial greetings, your

466. To Charles Hayden Church[1]

[Berlin, 12 November 1930][2]

Neither on my deathbed nor before will I ask myself such a question.[3] Nature is not an engineer or an entrepreneur, and I am myself a small part of nature.

A. Einstein

467. To Victor Fraenkl[1]

[Berlin,] 12 November 1930

Dear Sir,

I am very sorry to say that I am convinced that animal experiments are an indispensable research tool for biology. It must however be insisted upon that cruelty be avoided, which is indeed possible given the methods available today.[2]

With greatest respect,

468. To Semen L. Frank

Berlin, 12 November 1930

Dear Prof. Frank,

In the light of recent events I, too, consider it proven that the Russian authorities are having people killed on political grounds, and on the basis of false allegations,[1] in an attempt to escape their responsibility for unsuccessful activities. I am now convinced that you were right the first time.[2] I hardly need tell you that I condemn these methods without qualification. I am also convinced that no good goal can be achieved in such ways.

Cordial greetings from your

A. Einstein

469. To Herbert Wunsch[1]

[Berlin,] 12 November 1930

Dear Mr. Wunsch,

I have long striven not to believe that the Soviet regime sacrifices guiltless people for capricious political reasons.[2] I have resisted this belief, even though such accusations have long been made by very trustworthy sources and although the lack of an orderly judicial procedure necessarily arouses distrust.[3] But the allegations underlying the recent executions are so fatuous that, for me, there is no doubt. No matter how lofty, no goal can justify such crimes; it can only be damaged and discredited by the latter.

With the greatest respect

470. To Hans Albert Einstein

[Berlin,] 13 November 1930

Dear Albert,

Enclosed is a copy of a letter from your former teacher.[1] Mama[2] recommended that I write to him, which I did;[3] please inform her immediately, since she is naturally very interested and I am too busy to write. I believe that it would be a good route to follow, but the decision is yours. If I were you, I would prefer to carry on the major part of the negotiations with him in person, since written documents are rigid and easily lead to breaking off the discussion. The position is for an engineer at the Federal Hydrological Institute in Zurich.

Hearty greetings from your

Papa

P.S. I am traveling for two months to Pasadena (near Los Angeles).[4]

471. To Marcus Friedmann[1]

[Berlin,] 13 November 1930

Dear Rabbi,

My standing with Russian officials has been tarnished by my endorsement of the protest against the execution of people who were doubtless innocent.[2] Consequently, it is unlikely that my plea would still count for much there. In an article that appeared in America,[3] I have also spoken out against the regime based on force.

However, I would be glad to add an endorsement to your petition, if you think it would be helpful. From the materials you sent, it isn't clear whether the petition concerns all the persons named, or only the two women; in the latter case, it would probably have a better chance of success. It would probably be advantageous if I could add my writing to the petition myself, or—if it has already been submitted—if you sent me a copy of the petition, or its reference number. Otherwise, my recommendation will fall into the abyss of the doubtless huge multitude of such emigration requests.

With special regards

472. To Marcel Grossmann

Berlin, 13 November 1930

Dear Grossmann,

This paper is not stolen; indeed, I am really writing from there.[1] You are a very stubborn man, but so am I; so let's go![2]

Think of a Riemannian manifold with the metric

$$ds^2 = g_{\mu\nu} dx^\mu dx^\nu \tag{1}$$

This manifold we associate with ($s = 1$ to 4) vector fields $h_{s\mu}$, in such a manner that

$$g_{\mu\nu} = h_{s\mu} h_{s\nu} \left(\sum_s h_{s\mu} h_{s\nu} \right) \tag{2}$$

holds. This is possible for any choice of the $g_{\mu\nu}$. The $h_{s\nu}$ are not yet uniquely determined by the $g_{\mu\nu}$, *but for the choice* of the $g_{\mu\nu}$ *they imply no limitations*. The h field is thus richer in structure than the g field.

Now, according to (l) and (2), we have

$$ds^2 = \sum_s \left(\sum_\nu h_{s\nu} dx^\nu \right)^2 \tag{la}$$

The quantities $h_{s\nu} dx^\nu = dx_s$ thus fulfill the Pythagorean theorem. We call them the local components of the vector (dx^ν). We can call vectors with the same local components "parallel." This is, however, a different parallel relation from that defined by the Christoffel symbols for infinitesimals. It is, according to its definition, integrable.

One can easily prove that this new "parallel relation" is defined in the infinitesimal by

$$\left.\begin{aligned} \delta A^{\nu} &= -\Delta^{\nu}_{\alpha\beta} A^{\alpha} \delta x^{\beta} \\ \Delta^{\nu}_{\alpha\beta} &= h_s{}^{\nu} \frac{\partial h_{s\alpha}}{\partial x^{\beta}} \end{aligned}\right\} \quad \begin{array}{l} h_s{}^{\nu} \text{ are the normalized} \\ \text{subdeterminants of the } h_{s\nu}. \end{array}$$

Its "integrability" is expressed by the identity

$$0 \equiv R^{\iota}{}_{\kappa;\lambda\mu} \equiv -\Delta^{\iota}_{\kappa\lambda,\,\mu} + \Delta^{\iota}_{\kappa\mu,\,\lambda} + \cdot - \cdot,$$

which, however, has nothing to do with the Riemannian curvature formed from the $\left\{ \begin{matrix} \alpha\beta \\ \nu \end{matrix} \right\}$ and $\Gamma^{\nu}_{\alpha\beta}$, resp.

The field equations are expected to be the "simplest" differential equations for the $h_{s\nu}$. The condition for the Euclidean case with Euclidean parallelism are the equations

$$\Lambda^{\nu}_{\alpha\beta} = \Delta^{\nu}_{\alpha\beta} - \Delta^{\nu}_{\beta\alpha} = 0.$$

Hearty greetings from your

A. E.

Friendly greetings to your wife, too.[3]

473. To Myron Mathisson

[Berlin,] 13 November 1930

Dear Mr. Mathisson,

I am very sorry that our application to the Rockefeller Foundation[1] was for the time being unsuccessful. But you should not let yourself be drawn by that into hastily writing up some publications.[2] I am in fact firmly convinced that the method for solving the quantum problem which you have suggested is not practicable.[3] It is in this respect equivalent to the earlier quantum rule $\int pdq = nh$.[4] This is an integral over time, which is to be taken over a *period* of the motion. That means that the starting and end points of the integration are characterized by the fact that the point is at the same position in *three-dimensional space*. This conception is however meaningless in terms of relativity.[5]

It would be more reasonable at this point to introduce a (*superluminary*) wave in the sense of De Broglie-Schrödinger, which, just because of this property, would single out discrete solutions of the oscillation.[6] This theory is indeed very useful. But the rigid manner in which the potential energy is taken as a predetermined given shows the theory's preliminary character. It leads on the one hand to the necessity of introducing multidimensional spaces for systems of several particles, and on the other hand, it is in principle non-relativistic. *Finally, so to speak, the real objects vanish from the theory*. It has been attempted, erroneously, to make this defect into a strength of the theory!

In short, I believe that we are still quite far from a true knowledge of the real natural laws. Only a construction of the greatest mathematical naturalness and simplicity can succeed, which is not produced so to speak by squinting after reality and applying mathematical patchwork. As an example, I am thinking of the general theory of relativity as a theory of gravitation, without patching on electromagnetism $(R_{ik} = 0)$, which corresponds to the mathematical requirements; physically, it is of course insufficient, because it does not relate to reality *as a whole*.

Do not think that I am only eager to tout my new theory based on teleparallelism.[7] That is a lovely mathematical construction, which however probably has nothing to do with reality. I mention it only because it fulfills the justified architectural and/or logical requirements. We will have to search for a space structure which involves the Riemannian metric but also contains another, distinct structural element that is intimately linked to it.

Hearty greetings from your

474. On Autocracy

[*Einstein 1930pp*]

DATED 15 November 1930
PUBLISHED 20 December 1930

IN: *Die Wahrheit*, 20 December 1930, p. 3[1]

I am an absolute opponent of every oppressive regime because sooner or later they always lead to the abuse of power. The shortcomings of democracy arise wherever a politically underdeveloped people are suddenly expected to make decisions for which they are not sufficiently mature. Mistakes, which materialize from such conditions, prove nothing against the justification of the goal. Examples like Switzerland or Holland prove convincingly that politically healthy democracies are very much possible under current conditions, while the counterexamples of similar, healthier, and more stable autocratically governed states are currently completely lacking.

475. To Jacques Hadamard

[Berlin,] 15 November 1930

Dear Mr. Hadamard,

You are completely right to be concerned.[1] It was an ugly dispute between Brouwer and Hilbert,[2] for which I think Hilbert was mainly to blame.[3] However, in this instance Brouwer behaved in such an immoderate and insistent manner that he seemed to me someone who was pathologically irascible. Since it was no longer possible for Brouwer and Hilbert to work together, all other editors left the editorial board to satisfy Brouwer.

I am convinced that Brouwer's intended enterprise sprang less from an objective need than from a resentment that plagued him from then on. I am also firmly convinced that such an enterprise, under the leadership of the unpredictable and hypersensitive Brouwer would be in constant danger. I would therefore absolutely avoid getting involved in this, despite great respect for the brilliance and honorable character of Brouwer,[4] who is himself probably not aware of the depths of his temperament.

Regarding the "nondefense,"[5] I have not changed my conviction because of a couple of noisy opponents who are presently making people talk about them. What is the most prudent for the moment is not always the most reasonable in the long run.

Greetings from your

P.S. This letter is delayed due to my longer stay abroad.[6]

476. To Antonina Luchaire-Vallentin[1]

[around 15 November 1930][2]

[Not selected for translation.]

477. To Max Planck

[Berlin,] 15 November 1930

Dear Colleague,

I am sending you a questionnaire that I cannot answer,[1] though I'm more than willing. Who is now to be the official director of the Kaiser Wilhelm-Institute of Physics seems to me to be an unsettled question. In the event that no special opposing considerations exist, it would be all right with me if Laue[2] were also appointed as director—which he has already been, for a long time, de facto.

I would be very grateful if someone other than I could substitute for Haber in giving the public lecture in the Academy,[3] because my heart is being a little rebellious again and I am not up to the physical strain involved.

Finally, it moves me to tell you once again how wonderful I found your presentations on positivism with respect to the modern phase of theoretical physics.[4]

Warm regards, your

478. To Selig Eugen Soskin[1]

[Berlin,] 15 November 1930

[Not selected for translation.]

479. Diplomatic Passport

Berlin, 18 November 1930

[Not selected for translation.]

480. From Myron Mathisson

Warsaw, 18 November [1930]

Esteemed Professor,[1]

The newspapers have brought the news that you will leave for California at the end of November.[2] I therefore permit myself to ask you for a tip as to whom I could address myself for scientific questions during your stay in the U.S.A. I indeed want to publish my works on relativistic mechanics.[3] (They are not related to the notes on quantum theoretical problems which I have sent to you.[4] They date from the past three years). In fact, I succeeded in carrying on the solution to the mechanical problem much further than in my first work on this subject, with which you are familiar (and which was done four years ago).[5] The goal, above all, was to settle the uniqueness of the singularity solution, to prove why overdetermined equations do not occur, and to connect the constancy of mass and electric charge with mechanics. The method of successive approximations is thereby justified, and I discuss the extension of the solution to areas in which the original development fails.

However, it has happened that the integrals of the gravitational equations based on a Euclidean substrate are all divergent when the region of integration includes the entire world. Taking into account the damping effect of a cosmology becomes unavoidable. If I now take a De Sitter universe as the substrate, I am by no means making a special assumption—for the gravitational equations may change their form in some manner in the future, but this most simple, completely homogeneous cosmological substrate will remain unchanged by the possible transformations. Once one has accepted the necessity of taking the cosmology into account, there can be no more talk of an approximate mapping of the curved substrate onto a Euclidean one: this causes great mathematical difficulties, which I have completely solved. This solution appears to be of first-rate mathematical interest.[6] It shines a new light on the earlier methods of solution of the differential equations of wave type, and is based on a new integration relation, different from the Gauss-Stokes relation. It permits us to relate wave equations in Riemannian spaces to Fredholm integral equations. The purely mathematical progress can, I presume, be published in a mathematical journal.

The purely physical advantages of the cosmological version are the following: Firstly, one is freed of the incongruous presupposition that the elementary particles all had parallel worldlines in the distant past, and produced static fields. (This presupposition leads essentially to the cosmology of Anaxagoras: parallel particle rain as the original state of the universe,[7] [κάτω→ἄνω]).[8] Second, it is found that our solutions exist only within a bounded region of the universe hyperboloid. At the boundaries of this region, they decay (along with all their derivatives at the same time), permitting a completely continuous transition to the homogeneous background. The limitation of the region of existence *is, with respect to the direction of time, nonsymmetric*. This means that the solutions are represented by light cones which are opened only toward one time boundary. The cones that open toward the other boundary lead to no solutions. This result has inestimable consequences and seems to be the first concrete indication that the causal structure of the world (past⟶future *and not the reverse*) is not to be found in the differential equations and not in the world-globe of the substrate, but rather in the *solutions* (of the differential equations on the world globe). One has to penetrate into the solutions and not flutter around the differential equations in order to arrive at a true foundation of mechanics.

Third, the cosmology permits us to pose the question of the numerical relations of the constants describing the particles in a concrete manner, and thus to arrive at the necessary completions of the gravitational equations. One has thus far attempted to settle this problem only dialectically, by speaking of a correlative connection between the constant curvature of space and the mass of the electron.

The problems listed above are treated in my works. I have already essentially completed their formulations in German, only copying remains to be done. It cannot be finished very quickly, since I am still an "evening physicist."[9] But I can deliver a finished article once a week. May I send those articles directly to the Zeitschrift für Physisk? I have made great efforts to keep the description simple and to allow the leading ideas to come clearly into the foreground. I have no fears as to the supposed decrease in the actuality of relativistic questions.

The relation of my works to quantum mechanics would require a longer discussion. I will not attempt that in this letter, since it would otherwise grow toward the infinite. And I do not want to try your patience all too much.

With still more right to admire you than the rest of the contemporary world has, is the best consolation,

Your most respectful

Myron Mathisson

481. To the Academy of Sciences Allahabad

[Berlin, 22 November 1930][1]

Mr. M. N. Saha,[2] to whom physics, in particular, astrophysics, owes so many valuable contributions, has informed me about the ⟨founding and the⟩ high goals set by the ⟨newly founded⟩ Academy of Sciences which has been established in Allahabad. I greet this founding with the warmest sympathy and with great expectations. In the brief time since individual Indians have applied themselves to the modern exact sciences, these sciences have already received valuable stimuli from these men. I recall here only the physicist and physiologist Bose.[3] But it has by far not yet been accomplished that the great treasures ⟨of human⟩ of scientific talent have been recovered or attained full development, which are slumbering in your ancient cultural nation. May your efforts ⟨contribute to⟩ effectively advance not only ⟨your country⟩ the intellectual and economic development of your country ⟨promoted⟩ but also ⟨the international development understanding the penetration into the riddles of nature be promoted the understanding⟩ the clarification of the deep questions which eternally enigmatic nature has addressed to us.[4]

A. Einstein

482. To Arnold Berliner

[Berlin,] 22 November 1930

Dear Mr. Berliner,

I have read both of the papers from Mr. Skreb.[1] The article "A Relationship between Arithmetic, Geometry, and Physics"[2] is reasonable, even though I do not believe that much can be accomplished with the methods that he suggests. The article on the foundations of mechanics seems to be to be directly unreasonable. An atomization [quantization] of momentum would in my opinion face insuperable difficulties. When a material point is accelerated, the *kormons* which form it must be continually replaced through the changing velocities, or else they must be accelerated (curvature of the worldline). How one is to manage this without something like the concept of force is not understandable. I would not accept the article, and also would not recommend it to anyone else.

I read the lecture by von Mises with great interest.[3] The presentation is excellent. The content corresponds to the generally accepted opinions at present, which I however do not share.

With hearty greetings, yours truly,

483. To Wolfgang Elert[1]

[Berlin,] 22 November 1930

Dear Colleague,

I completely understand why you find your situation painful.[2] And yet I can't really say that you are right. Because it is really correct, that up to this point school has made our youth unfit for life by cramming too much stuff into their young heads.[3] As for yourself, in the present situation of things[4] it seems impossible to give everyone interested in science a position in science. For example, I myself worked for the Swiss patent office for seven long years.[5] Maybe you will still succeed in keeping the pursuit of a teaching career and scientific ambition cleanly separate from one another, and so achieve something in science. You will be aided in doing so by the relatively long school vacations. It must not be easy to find another practical line of work that will give you more inner freedom. We have to accept the fact that most of our contemporaries are obtuse.

If you are publishing, or have already published, something that you consider particularly original, please send it to me.[6]

Perhaps then I could make it possible for you to do a Habilitation at the university or otherwise help you find a vocation that corresponds better to your particular nature.

Friendly regards,

484. To Jüdisches Gemeindeblatt[1]

[Berlin,] 22 November 1930

[Not selected for translation.]

485. To Conrad Matschoss[1]

[Berlin,] 22 November 1930

Dear Colleague,

I am, of course, in full agreement with the content of your pamphlet.[2] The research grant was made personally for the mathematician Dr. Walter Mayer,[3] on my recommendation and for an unlimited period, by the Josiah Macy Jr. Foundation in New York. I owe this foundation extraordinary gratitude, because without it long-term collaboration of this colleague of mine, for whom I have the greatest esteem, would not have been possible.

I would like to take this opportunity to make another suggestion. A large number of people with theoretical-scientific talents are lost because after completing their university studies they are forced to work in practical occupations that leave them no time or energy for thinking. These days, there are certain official positions that require the incumbent to be present but leave a great deal of free time for thought. For example: proctors in telephone amplification centers, lighthouse operators, etc. Such positions should be made available to young theoreticians, so that they can pursue intellectual work in peace. I am sure that many more occupations in this category could be found without any sacrifice. Such a position is more valuable than a grant, because it is not burdened with the moral pressure that comes with grants and jobs in research institutes, and can easily tempt employees to engage in large amounts of shallow scribbling.[4]

With great respect,

486. To Edward Adams Richardson[1]

[Berlin], 22 November 1930

Dear Sir,

I consider it to be absolutely necessary to leave the difficulties of the quantum problem aside for the moment, when one is dealing with problems of relativity theory.[2] Mr. de Sitter[3] is thus quite right[,] when in his considerations he does not worry about the quantum structure of light. The theory of relativity can deal with all the facts of physics in the macroscopic world. We must require of quantum theory that it should someday find a formulation which is compatible with the principle of relativity. This is also generally accepted by all theoreticians.

Yours very respectfully,

487. To Hermann Simons[1]

[Berlin,] 22 November 1930

Dear Sir,

Bread cannot, and should not, be replaced by a religion.[2] The community has a duty to protect individuals from perishing. The extreme parties can therefore make only harmful use of this justified feeling, because those who have held power up to this point have been lacking in social conscience.

With great respect,

488. To Heinrich York-Steiner

[Berlin,] 22 November 1930

Dear Mr. York-Steiner,

Your article[1] is, once again, outstanding; I thank you for it. In the meantime, the British government has recognized its mistake[2] and partly rectified it. Work, patience, and justice will achieve more than political trickery.

Warm regards, your

489. From Marcel Grossmann

Zurich, 23 November 1930

Dear Einstein,

I ruminated for a long time over how to find stationery which would be even halfway equivalent to what you are using.[1] To no avail.

I did not need to think so long about the factual part of your letter. The ansatz is simply false, for one cannot construct a Riemannian manifold from 4 vector fields, since, then, $g_{ik} \neq g_{ki}$. This is, however, an essential assumption of Riemann's, that the tensor g should be symmetric.

I will soon write down what I have against Levi-Cività, Cartan, you, Weitzenböck, etc.[2] Only, I do not want to cause you any difficulties, since I am aware that there are all too many out-and-out dumb people around.

So, what suits you better: That I send you my manuscript[3] (about ten printed pages) at the beginning of December to be submitted to the Berlin Academy and published in its meeting reports, or that I publish it here in Switzerland?? Please answer soon.

Hearty greetings, yours,

G.

P.S. In any case, I am not in need of special protection and can still defend myself against reasonable arguments.[4]

Prof. Dr. M. Grossmann[5]

490. “Space, Ether, and Field in Physics”

“Raum, Äther und Feld in der Physik”
[*Einstein 1930oo*]

Dated before 24 November 1930
Published between 24 November and 31 December 1930[1]

In: *Forum Philosophicum* 1 (1930): 173–180.

Prescientific concepts have always been the object of philosophical dispute. This is in particular true of the notion of *space*. From whence does this notion come? To what extent is it based on experience, and to which experiences is it related? To these questions, we can answer: Considered logically, concepts never originate in experience; i.e., they are not derivable from experience alone. And yet, their gestation in our minds occurs only with reference to what we have experienced with our senses; and the elucidation of such fundamental concepts is to be sought in the identification of the character of our sensory experiences which led us to the formation of those concepts.

In the case of space, this connection is easy to comprehend. It is preceded by the emergence of the notion of an objective material world. I can recognize material objects through sensory impressions, without having grasped them spatially. Once the notion of a material object has been conceived, our sensory experience forces us to ascertain the locational relations among various objects, i.e., relations of mutual contact. What we interpret as spatial relations between objects are precisely these. Therefore: without a concept of objects, we have no concept of the spatial relations between objects, and without the concept of spatial relations, we have no concept of space.[2]

But how does the concept of space itself come about? If I imagine that all the material objects have been removed, the empty space still remains, does it not? Should this concept indeed be dependent on the concept of material objects? In my opinion, quite certainly! In considering the locations of objects relative to one another, the human mind in fact finds it simpler to relate the locations of all the objects to that of one single object, rather than to imagine the confusing multiplicity of the relation of each object relative to all the others. This *one* object, which must exist everywhere and be permeable for all the other objects, in order to be in simultaneous contact with all of them, is to be sure not detectable to our senses; but we feign it for convenience in our thinking. In our practical everyday experience, the surface of the Earth plays such a role in our perception of the spatial relations between objects, so that its existence may well have facilitated the formation of a concept of space as described above.

This interpretation of the concept of space as having emerged from our recognition of the totality of the positional relations between objects is furthermore confirmed by a consideration of the development of the scientific study of space, i.e., geometry. For in the earliest geometry, which was bequeathed to us by the ancient Greeks, the subject of investigation is limited to the positional relations between idealized objects, which are termed "point," "line," and "plane." In the concepts of "congruence" and "measurement," the reference to the positional relations of objects is clearly apparent. A spatial continuum, in short "space," does not occur at all in Euclidean geometry, although that concept was already quite familiar, even in prescientific thinking.

The extraordinary significance of the geometry of the Greeks lies in the fact that—so far as we know—it represents the first successful attempt to capture a whole complex of sensory experiences conceptually within a logical deductive system.

The spatial continuum as such was introduced into geometry only in modern times, by Descartes, the founder of analytic geometry.

The value of the service which Descartes performed by introducing the concept of a spatial continuum into geometry cannot be overestimated.[3] First of all, that step made it possible to describe geometric figures using the resources of the calculus. And second, it deepened geometry as a science in a decisive way. For from then on, straight lines and planes were no longer given precedence in principle with respect to other lines and planes, but rather all lines and all surfaces could be treated in a similar manner. In place of the complicated axiomatic system of Euclidean geometry, a single axiom emerged, which in modern terminology can be expressed as follows: There are coordinate systems in which the distance *ds* between neighboring points P and Q can be expressed in terms of the differentials of the coordinates dx_1, dx_2, and dx_3 by means of Pythagoras' law, i.e., by applying the formula

$$ds^2 = dx_1^2 + dx_2^2 + dx_3^2 \,.$$

From this fact, that is from the Euclidean metric, all the concepts and theorems of Euclidean geometry can be deduced.[4]

However, the most important fact is perhaps that without the introduction of a spatial continuum in the sense of Descartes, a formulation of *Newton*'s mechanics would not have been possible at all. The concept of acceleration, which is fundamental to that theory, must be supported by the concept of Cartesian coordinates;

for the concept of acceleration cannot be derived from the *relative* positions of objects or material points and their temporal evolution alone.[5] One can therefore quite rightly say that in Newton's theory, space plays the role of a physically real object; and Newton was well aware of this, although most who came after him overlooked it.

The Cartesian-coordinate space thus had, from the point of view of physics, two independent functions. It determined the possible positions of practically rigid solid bodies and the inertial motions of mass points (through the Pythagorean theorem or the Euclidean metric). It seemed to be absolute, in the sense that it affected objects, but conversely, nothing could affect it in a way that would modify it, that boundless, eternally unchanging vessel of all being and happening.[6]

Newtonian physics was founded solidly upon the concepts of space, time, and mass. It presumed that all natural phenomena could be described in terms of those concepts. Later, they were joined by the special concepts of electric and magnetic mass, which were indeed analogous to the original concept of mass in several respects, but did not appear to share the principal characteristic of mass, i.e., its inertia. Apart from that, however, this interpretation of physical reality appeared to be the only one possible.

However, in the nineteenth century, there was a gradual turnaround, which slowly but drastically changed our view of physical reality. After the discovery of the wave nature of light by *Young* and *Fresnel*, based on the phenomena of diffraction and interference, it seemed that a medium for light-wave propagation which would permeate all of space was required; it was termed the "ether." It was initially thought to be of the same character as the bodies treated in Newtonian physics. But its ubiquity, its imperceptibility, and the absence of friction with respect to ponderable objects gave it a strange aspect, I am tempted to say a certain ghostliness; it seemed to be indubitable, uncanny, and unfathomable, all at the same time. And—what was still worse—the mechanical properties which were ascribed to it were contradictory.

The theoretical framework of Newtonian physics appeared to be almost blown away by the Faraday-Maxwell field theory of electromagnetic phenomena. For gradually, the realization grew that the localized electromagnetic fields in empty space could not be interpreted consistently, or satisfactorily, as mechanical states of the ether. One became accustomed to considering electromagnetic fields as fundamental objects of a nonmechanical nature. Nevertheless, they were still taken to be states of the ether, which however could no longer be considered to be analogous to ponderable matter; all the less so when around 1900 the molecular structure of the latter became more and more accepted, while the ether seemed to fill all space as a continuous medium.

While the fields themselves were by then accepted as fundamental entities, not describable in terms of mechanical aspects, the question of the mechanical properties of their presumed medium, the ether, remained unanswered. It was clarified by H. A. Lorentz, who stated: All the electromagnetic phenomena force us to the assumption that the ether is at rest relative to Cartesian or Newtonian space.[7] How readily might one then say that the fields are states of space itself; space and the ether are one and the same. That this was indeed *not* said was due to the fact that space, as the seat of the Euclidean metric and Galilean-Newtonian inertia, was held to be absolute, i.e., unchangeable, a rigid skeleton for the world, which so to speak was there before all of physics and could not be the carrier of variable states. With this surrender of any mechanical interpretation of electromagnetic fields, they became independent physical entities, alongside but distinct from the material particles.

The next step in the evolution of the concept of space was the development of the special theory of relativity. The law of light propagation in empty space, together with the principle of relativity with respect to uniform motion, necessarily led to the conclusion that space and time must be fused into a unified four-dimensional continuum. Thus, one recognized that the entirety of simultaneous events did not correspond to anything in reality. This four-dimensional spacetime must possess a Euclidean metric, as Minkowski was the first to realize;[8] it is completely analogous to the metric of three-dimensional Euclidean space if an imaginary time coordinate is added to the latter. The later development of what has become known as the "general theory of relativity" was based upon the existence of a spacetime structure characterized by a Euclidean metric.

After it had become clear that not only velocity but also acceleration cannot be ascribed with any absolute character, it was evident that there is nothing real in nature corresponding to the notion of an inertial system. It became clear that the natural laws must be formulated in such a way that those laws would retain their validity with respect to any Gaussian coordinate system of four-dimensional spacetime (this requires a general covariance of the equations which express the natural laws).[9] This is the formal content of the relativity principle. Its heuristic power lies in the question: What are the simplest generally covariant, reasonable systems of equations, i.e., systems which are independent of the choice of coordinates?

This question, in its most general form, is not yet fruitful. Some specification about the character of the spatiotemporal structure must be made. They were provided by the special theory of relativity, whose validity locally (for infinitesimal neighborhoods) must be retained. This means that there must be a spatial structure which can be expressed mathematically in terms of a Euclidean metric in the infinitesimal neighborhood of each point. Or: spacetime possesses a Riemannian metric. For physical reasons it was clear that this Riemannian metric would at the same time be the mathematical expression of the gravitational field.

The mathematical question corresponding to the problem of gravitation was thus the following: What are the simplest mathematical conditions which can be applied to a Riemannian metric in a four-dimensional spacetime? This allowed the field equations for gravitation to be found, and they have passed all the empirical tests, as is well known.[10]

The significance of this theory for our knowledge of the essence of space can be characterized as follows: Space loses its absolute character within the general theory of relativity. Up to this stage of development, space was held to be an entity whose inner structure could not be influenced by anything; it was intrinsically unchanging. Therefore, a special ether had to be assumed to be the carrier of localized field states in empty space. Now, however, the most intrinsic property of space—its metric structure—was found to be variable and subject to external influences. The states of space took on the character of a field; geometric space had become analogous to the electromagnetic field in this respect. The separation of the concepts of space and the ether was so to speak annulled automatically, after special relativity had already cleansed the ether of any last traces of materiality.

The principal accomplishment of general relativity lies, in my opinion, by no means in the fact that it has predicted several very small observable effects but rather in the simplicity of its fundamentals and in its consistency. It does away with the absolute character of the spacetime continuum and ascribes the metric, i.e., geometry, inertia, and gravitation, to a single property of four-dimensional spacetime, namely to its Riemannian metric.

But even this theoretical construction leads necessarily above and beyond itself. It indeed interprets the gravitational field as the metric structure of space. But it requires the introduction of particular conceptual elements in order to comprehend the electromagnetic field within itself. It thus fails to account in a logically satisfying way for precisely that phenomenon which lies at the origin of the whole of relativity theory.

It is now the most eagerly pursued goal of theoreticians to relate the essence of the gravitational and the electromagnetic fields to a unified structure of spacetime. What might this structure be? How can we find the mathematical field laws which govern that structure? Can matter particles (electrons and protons) be understood as regular solutions of those field laws? These are the questions with which the theory is currently grappling.

As long as these questions have not been satisfactorily answered, there will be legitimate grounds for doubts as to whether such extensive deductive physical methods are to be credited at all. Only what follows can decide about this with certainty.

Here, it should finally also be mentioned that there is a spatiotemporal structure which offers itself as a natural extension of the Riemannian metric, and which I have called the "unified field theory," hoping that its investigation will lead to the sought-after success. We should at least make the acquaintance of that structure.[11]

Let P and P' be two arbitrary points on the continuum, while $\overline{PQ}$ and $\overline{P'Q'}$ represent two line elements which extend away from those two points. The premise of the metric structure states that the two line elements can be meaningfully considered to be equal, or, more generally speaking, that line elements are comparable in terms of their lengths. The Riemannian character of the metric is expressed through the requirement that the square of the length of a line element can be expressed in terms of a homogeneous function of second degree of the differentials of the coordinates. In contrast, within the framework of Riemannian geometry, a statement about the *directions*, e.g., about the parallelism of two line elements $\overline{PQ}$ and $\overline{P'Q'}$, makes no sense. If we now add the condition that it is meaningful to consider a parallel relation between the line elements, then we arrive at the formal basis of the unified field theory. For completeness, we need only add the condition that the angle between two line elements which originate from the same point will not be changed by their parallel displacement.

The mathematical expression of the field laws should be the simplest mathematical conditions which can be governed by such a spatiotemporal structure.[12] Finding such laws would appear to have been accomplished, and they indeed agree with the empirically known laws of gravitation and electricity, at least in the first approximation.[13] Whether or not those field laws also yield a workable theory of material particles and their motions will have to be shown by further mathematical investigations.

According to the viewpoints presented here, the axiomatic basis of physics is as follows: Reality is conceived as a four-dimensional continuum with a unified structure of a particular kind (metric and direction). The laws are differential equations which obey that structure, i.e., the fields which appear in nature as gravitation and electromagnetism. The material particles are locations of high field density without singularity.

In summary, we can say symbolically: Space, which we perceive through material objects, and which was elevated to the status of physical reality by *Newton*, has consumed the ether and time in recent years, and appears to be in the process of consuming also the fields and particles, so that it will in the end remain as the sole bearer of reality.

491. To Ismar Freund[1]

[Berlin,] 24 November 1930

Dear Dr. Freund,

I have had the enclosed rectification published in the Jewish press.[2] You have placed my name under the invitation to your rally, even though you knew from our telephone call that I did not wish to take part in the political battle within the community.[3]

I ask you immediately to strike my name as a member of the executive board and not to make any further use of my name in the future.[4]

With great respect,

492. From David Reichinstein

Berlin, 24 November 1930

Dear Mr. Einstein,

Before Planck's lecture,[1] you were so kind as to express your wish by telephone that I should send you a written summary of the problem of reversibility in adsorption. While this previously did not appear to me to be easy, I believe that now, after Mr. Schrödinger[2] in a discussion has accepted half of my contentions (called point 1 below), I will have more success in explaining the problem.

I thus give you the facts herewith:

The envisaged Experiment 1 (in the sense of my dynamic displacement theory)[3]

The solid material A with its [cross-sectional] area $Q2 - Q_2$ is in contact with the liquid solution B, which consists of the solvent a and the solute b, whose concentration has a high value c_2. In a finite but relatively long time interval, the solution B permeates the solid material up to the depth 3–4, without a separation of the solute; thereafter, it further diffuses only slowly (practically not at all = mock equilibrium state).

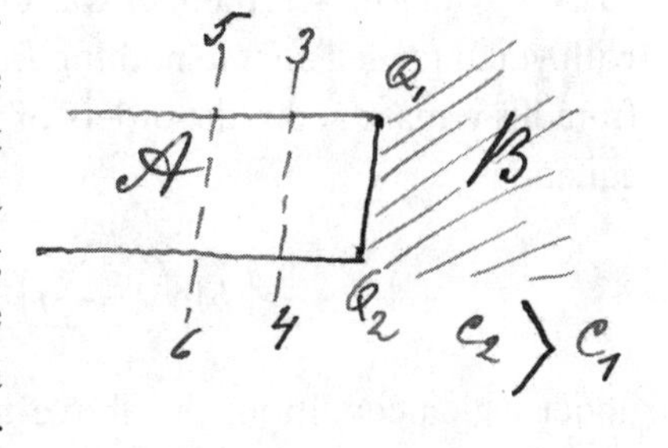

Thought experiment 2. A fresh piece of the solid A comes into contact with the solution B, which however now has a smaller concentration c_1; then in a finite time, the solution permeates the solid material A up to a depth 5–6.

Thought experiment 3. The property of reversibility now requires that when the concentration in the liquid phase is increased from c_1 to c_2, the depth of the permeated liquid within the solid should decrease from 5–6 to 3–4. The question is, does that indeed happen? In the following, I will quote the discussion with Mr. Schrödinger as accurately as I can.

Mr. Schrödinger: "I would rather think that on increasing the concentration (from c_1 to c_2) additional amounts of the solvent would permeate the solid, but the liquid phase would not retreat from it."

I remarked to that: first, that the stopping point at 3–4 in experiment 1 corresponds to a greater absolute quantity of the solvent, only the depth is less; and second, in the case of an osmotic process, the solvent moves toward the more concentrated solution and not *vice versa,* so that we may assume that this occurs because the velocity with which the solute passes through the semipermeable membrane is less than the velocity with which the solvent passes through it *(point 1)*.

Mr. Schrödinger: "I can understand that *the solvent* would retreat back out of the solid material into the liquid phase but not *the solute.*"

I: The displacement theory was derived assuming conditions such that the "Goppelsroeder effect" (= segregation of the diffusing solution *B* during its motion into or out of the solid) does not occur.

This condition thus implies that the solvent, during its diffusion toward the now more concentrated solution, *will carry the solute along with it* (*point 2*), and after the concentration has equalized everywhere, the now more concentrated solution c_2 will diffuse back into the solid material, but reaches only a smaller depth 3–4 in it.

Mr. Schrödinger was not in agreement with this point 2, so that the actual problem which is occupying my attention in continuing the last chapter of my book never came to be discussed.

It is superfluous to mention that my enjoyment of the one and a half hours of discussion with Mr. Schrödinger was only very limited in comparison to a conversation with you. I gained next to nothing from the conversation with Mr. Schrödinger, apart from his remark about the orders of magnitude of the coefficients in my logarithmic equation

$$x_t = b\ln\left(1+\frac{\lambda_3}{R}t\right) \quad \text{… book, page 412, eq. 171}^{[4]}$$

under which conditions the above-mentioned experiments 1 and 2 are carried out. I would have arrived at the same results myself on considering the equation more closely.

I am now turning to you with the request that you inform me on the enclosed card of what you think of points 1 and 2, especially whether you consider point 2 to be valid.

I thank you in advance for your trouble and remain your very respectful

Reichinstein

493. To Herbert Runham Brown[1]

[Berlin,] 28 November 1930

Dear Mr. Runham Brown,

I have signed the letter to the Yugoslavian court, but not the other one,[2] for the following reason: I think it's not right to say that this kind of petition proceeds from a private individual, because that seems presumptuous. Instead, the letter should be sent in the name of the association and signed by several persons. If you can adopt that way of going about this, I would be very willing to sign such a petition. You should also ask Romain Rolland and H. G. Wells whether they wish to join in supporting your praiseworthy effort in the same way. My participation does not, of course, depend on that.

With the greatest respect,

494. To Abraham Adolf Fraenkel

[Berlin,] 28 November 1930

[Not selected for translation.]

495. To the Jewish Colonization Association[1]

[Berlin,] 28 November 1930

I recently wrote to Mr. S. Reinach to tell him that, for the time being, it is not appropriate to interview Mr. von Maltzahn.[2] However, I must frankly admit that I was disconcerted by your reply.[3] The reason you gave for your rejection is absolutely inadequate factually. One might almost conclude that the JCA was there to find grounds for regarding the establishment of Jewish colonies as not advisable. But in my opinion, Jewish colonization work in Palestine has proved that Jews can colonize successfully, if they are not scared off by the unfortunate use of methods as were employed by the JCA in Argentina,[4] according to the judgment of experts. The Jewish people should be able to expect that the administration of thc JCA will not refuse or shy away from a meeting with men who have been selflessly and conscientiously engaged in Jewish colonization plans. I believe this much can be required of you.

With great respect

496. To Gustav Radbruch[1]

Berlin, 28 November 1930

[Not selected for translation.]

497. To Moritz Schlick[1]

[Berlin,] 28 November 1930

Dear Mr. Schlick,[2]

I immediately read your work and found it in large part correct.[3] Roughly I can say the following:

Science seeks general relation statements which link possible sensory experiences in such a way that those statements can prove to be apt or not apt empirically, or rather suitable for correct predictions. No more than this is claimed by the demand of lawfulness[4] in the most general sense.

I disagree with you factually on the following points:

(1) Quantum theory too knows those same relation statements, which are not of a statistical nature, but which state something very specific when applied to a particular occurrence (e.g., law of conservation of energy and momentum applied to an elementary process).[5]

(2) I don't believe that the concept of a "statistical law" is contradictory. It is just a limiting statement which refers to the frequent repetition of an arrangement that is defined in a certain way. Whether one wishes to describe such laws as deterministic or not is a question of nomenclature. Commonly they are described as nondeterministic.

(3) Temporal statements, insofar as they can be conceived as relation statements which refer directly to sense experiences, should not play a *special role* compared to other relation statements. (I approve of the criticism leveled at Reichenbach concerning the irreversibility of the temporal.)[6]

Generally speaking your conception does not correspond to my way of seeing things in that I find your whole point of view too positivist,[7] so to speak. Physics does indeed *supply* relations between sense experiences, but only indirectly. *Its essence* is certainly not exhaustively characterized for me in this statement. I'll tell you point blank: physics is an attempt[8] at a conceptual construction of a model of the *real world* as well as its law-like structure. However, it must exactly represent the empirical relations between the sensory experiences that are accessible to us, but only in *this* way is it chained to the latter.[9]

I, too, admire the accomplishments of quantum theory in the Schrödinger-Heisenberg-Dirac mold, but I am certain that in the long run we cannot and will not get by with this way of thinking. This theory does not offer a model of the real world at all. (The elements functionally connected in the theory do not represent the real world, but only probabilities which refer to our experiences.)[10] In short I suffer from the impure distinction between the reality of experience and the reality of being. And I am also firmly convinced that the "statistical law" as the *basis* of the expression of physical laws will be overcome one fine day.[11] As I've said, I do not share your view that a "statistical law" is no "law" at all.

You will wonder about the "metaphysician" Einstein. But every four- and two-legged animal is de facto a metaphysician in this sense.

With warmest greetings (and in a hurry) yours,

P.S. Address in January and February: Institute of Technology, Pasadena Cal./U.S.A.[12]

498. To Alfred Weber[1]

Berlin, 28 October [November, 1930][2]

[Not selected for translation.]

499. To Benno Hallauer[1]

[Berlin,] 29 November 1930

[Not selected for translation.]

INDEX

References are collected under the appropriate English heading. Certain institutions, organizations, and concepts that have no standard English translation are listed under their foreign designation (with cross-references from an English translation). For the meaning of abbreviations, see the List of Abbreviations in the documentary edition.